THE LOST ART
OF FINDING
OUR WAY

THE LOST ART

OF FINDING

OUR WAY

JOHN EDWARD HUTH

The Belknap Press of Harvard University Press
Cambridge, Massachusetts
London, England
2013

Copyright © 2013 by the President and Fellows of Harvard College
All rights reserved
Printed in the United States of America

Library of Congress Cataloging-in-Publication Data
Huth, John Edward.
 The lost art of finding our way / John Edward Huth.
    pages cm
 Includes bibliographical references and index.
 ISBN 978-0-674-07282-4 (alk. paper)
 1. Navigation—History. 2. Naval art and science—History. I. Title.
 VK15.H87 2013
 629.04'509—dc23
                        2012044083

Dedicated to the memories of
Sarah Aronoff and Mary Jagoda
No one is lost . . . to God

Contents

THE LOST ART

OF FINDING

OUR WAY

1. Before the Bubble

. . . .

ONLY A FEW YEARS AGO you would have sworn the commuter next to you was crazy. He talks into the air, as if to an imaginary friend, while playing with a tiny box he holds in front of his face. His whole world is a bubble two feet around his head. If, by a miracle, you get his attention and ask him questions, he can only answer by manipulating his tiny box. He checks the box and can tell you tomorrow's weather, which way is north, and the name of that bright star in the sky. You hide the box, and he becomes helpless. You ask him, "Why is it warm in the summer and cold in the winter?" By way of explanation, the commuter gestures with his hands in front of his face. His left hand becomes the Sun and his right index finger becomes the Earth as it orbits his fist. He mumbles something about the Earth's being closer to the Sun in the summer and farther away in the winter. His voice has a rising inflection that begs your affirmation.

Now let's suppose it is three thousand years ago and you have transported yourself into a fisherman's boat, far from land. He sings to himself as he hauls in a net from the sea. You ask him questions about the weather and the sky; he thinks you are crazy but humors you anyway. He confidently rattles off tomorrow's weather, points north, and tells you a story about the bright star in the sky. He has no magic box to hide. You ask him, "Why is it warm in the summer and cold in the winter?" His outstretched arm traces an arc toward the sky, and he says that the Sun takes a high path in the summer, and the days are long; but in the winter the Sun has a low path, and the days are short. His voice has no hint of uncertainty.

Whom would you consider the more primitive of the two: the commuter or the fisherman? As information technology has grown, our ability to perceive and think independently without help from devices may be compromised to the point where we, not our forebears, are the primitive ones. Long before the advent of the Global Positioning System (GPS), our distant ancestors traveled vast distances using what seem to us to be primitive technologies. In many cases explanations for the effectiveness of their systems only arose in the twentieth century; some remain a mystery. Systems of navigation developed by different cultures may not have been based on an understanding of science as we conceive it today, but they made sense, and they worked. While the device-addicted commuter may seem oblivious to and clueless about his surroundings, his ancestors proved that humans *can* absorb exceedingly subtle environmental clues to find their way.

To many of us, the word "navigator" conjures up the image of a naval officer on board a ship with a sextant pointed at the stars. But the term means far more than that. Each of us is a navigator; we are constantly finding our way in our environment. The skill set needed to journey long distances is influenced by a number of factors — mode of transport, destination, the environment in which the journey takes place — and depending on the elements involved, different skills come into play, creating what I call "cultures of navigation" that are specific to the environment and mode of travel. In this book I examine these cultures and the various ways humans are able to navigate, using simple instruments and environmental clues.

Dissecting older cultures of navigation is a challenging task. Navigators, as a group, were not inclined to document their travels. Much of what we can gather about their techniques is based on fragmentary information: archaeological remains and traces of earlier languages, sagas, and chants passed down for generations. Ancient historians have left us narratives, but we ought to view them with appropriate skepticism. Cultures that developed in relative isolation before and even during Western European contact retain many of their traditional navigational schemes. By piecing together our knowledge of cultures of navigation, we can begin to appreciate a kinship we all share in the very human art of way-finding.

My work with students has convinced me that people *can* develop their own intuitive navigational skills. Many of these techniques, often presented in mysterious nuggets in diverse "how-to" books used by outdoor enthusiasts and survivalists, are frequently workable schemes, but whose subtleties can only be acquired through experience. What is needed is the time outdoors to allow a learning process. I will try to convey these techniques to you in this book, but there is no substitute for leaving the bubble and experiencing the Sun, the stars, the wind, and the waves. As you become more familiar with techniques of navigation that employ natural phenomena, you will gain an appreciation of the sophisticated skills developed by many cultures in the past.

Three of the most remarkable early cultures of navigation are those of the Norse, of medieval Arab traders, and of Pacific Islanders. Although I will not focus exclusively on these three cultures in this book, I will return to them many times to illustrate the schemes and adaptations I will discuss here.

The Norse began to colonize Iceland in the ninth century AD. According to Norse sagas, when they arrived they found Irish monks already living in established settlements. The monks may have found their way there by following the paths of migrating birds in crude boats called *currachs*.[1] The *Landnámabók* (*Book of Settlement*) describes the first Norse contact with Iceland as occurring because of the accidental drift off course by a sailor named Naddoddr who was sailing between Norway and the Faroe Islands. Subsequent deliberate voyages established permanent Norse settlements in Iceland.

The violent storms and large waves that sweep the North Atlantic make navigating in the waters off Iceland and Greenland a challenge. Also, fog banks obscure the Sun for days at a time. Certainly, life on the open sailing vessels was miserable and exhausting. To sail deliberately to a destination in those waters, a navigator would have to be able to read the weather to gain a favorable wind and construct a vessel capable of sailing into the wind if necessary. Voyages were possible only in the summer months, when, unfortunately, few stars were visible at the high latitudes of Iceland and Greenland. Navigators had to rely mainly on the Sun and other natural signs to gain their bearings.

The first sighting of Greenland, somewhat before AD 1000, also appears to have been due to an accident, when Gunnbjörn Ulfsson, sailing from Norway to Iceland, was blown off course. By legend, Erik Thorvaldsson (Erik the Red) established the first colony in Greenland soon thereafter. A number of settlements along the south coast of Greenland were established but died out around the fourteenth century.

The first documented European sighting of North America also apparently came about by accident, just like the discoveries of Iceland and Greenland. *The Saga of the Greenlanders* describes a voyage from Iceland to Greenland by a Bjarni Herjolfsson, who was lost, along with his crew, for several days in a fog bank. When the Sun enabled him to regain his bearings, he encountered a land covered in forests that did not resemble Greenland. So he returned. Word reached Leif Eriksson, who, according to the *Saga*, embarked on a voyage to find the land that Bjarni encountered. *The Saga of the Greenlanders* describes a rich land, teaming with forests and salmon, along with hostile natives who attacked the Norse. Many of the descriptions of parts of the coast that Leif sailed along correspond well to parts of the North American coast.

A significant archaeological finding provides evidence of Norse attempts to colonize North America. In 1960 Helge Ingstad discovered the remains of a Norse village on the northernmost edge of Newfoundland, called L'Anse aux Meadows. Whether this was a primary Norse settlement or a staging area for forays farther to the south is uncertain, but the find does establish a Norse presence in North America in the eleventh century. This may have functioned

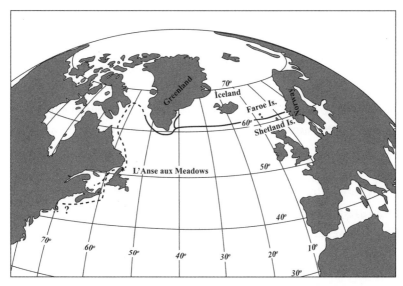

Figure 1 Routes of Norse from Norway to settlements in Greenland (solid) and Leif Eriksson's possible route to North America (dashed) around the eleventh century AD.

as a wintering and repair stop for more explorations to the north and south.[2]

Figure 1 shows the regions of the North Atlantic traversed by the Norse. These include a direct route from Norway to settlements in Greenland (solid line) along with a plausible route followed by Leif Eriksson (dashed line). The location of L'Anse aux Meadows is indicated on the northeast tip of Newfoundland. Judging from the Norse sagas, their stay in North America was short-lived. Many writers have speculated about the southerly extent of their voyaging, with Cape Breton, the coast of New England, and even Long Island suggested as destinations.

During the Middle Ages and into the Renaissance, Arab voyagers sailed the Indian Ocean conducting a lucrative trade involving silk, slaves, and spices. The alternative to the Indian Ocean was an arduous overland route, where goods changed hands many times, with a markup at each of the many cities along the Silk Road. Figure 2 shows the geography and approximate routes of Arab traders in the Indian Ocean from China, the

Spice Islands, and Zanzibar to such trading cities as Alexandria and Baghdad.

Earlier forays by the ancient Persians, Greeks, and Romans often hugged the coast in vessels powered by galley slaves using oars and supplemented with crude sails. Voyages such as these were costly and time consuming. With the development of sail configurations making it possible to sail *into* the wind, and by using stars to steer at night, traders sailing out of sight of land for days at a time with small crews could substantially reduce the overhead needed to bring goods to Europe. In many cases they used the height of stars above the horizon to sail along a constant latitude toward a destination.

Traders crossing the Arabian Sea confronted seasonal wind patterns that blow from southwest to northeast during the monsoon, from June through September, and reverse direction the rest of the year. Voyages to and from India could exploit these wind patterns.

Figure 3 shows the central Pacific Ocean, a lonely expanse scattered with islands and atolls. Archaeological evidence suggests that a people called the Lapita who originated in the Bismarck Archipelago, northeast of New Guinea, began the colonization of islands to the east approximately thirty-six hundred years ago. Some of the voyages involved bold excursions of up to three hundred miles across gaps between major island groups.

The voyages of the Lapita are even more remarkable when we consider the pattern of currents and winds in the central Pacific. To travel eastward, these voyagers would have had to sail *against* both the prevailing wind and the current, meaning they built sailing vessels with this capability and could take advantage of seasonal shifts in weather. The pattern of colonization of the Pacific Islands from west to east against wind and current seemed counterintuitive at first and prompted scholars such as Thor Heyerdahl, author of *Kon-Tiki,* to speculate that the Pacific Islands were largely inhabited through accidental

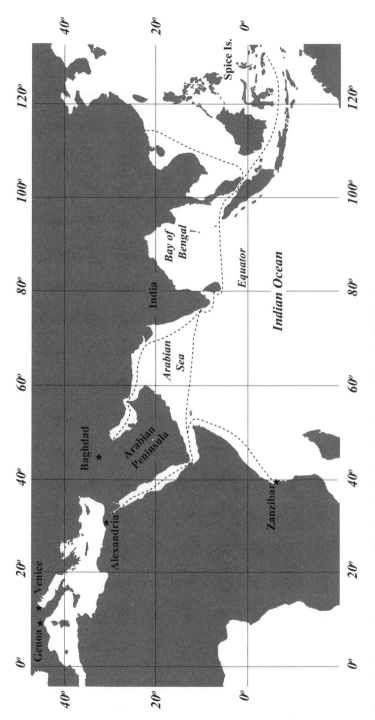

Figure 2 Approximate routes of Arab trade in the Indian Ocean during the Middle Ages.

drifts originating in South America and proceeding westward, as opposed to deliberate eastward voyages.

On the other hand, anthropologists and archaeologists such as Geoffrey Irwin point out that an initial voyage into the wind is far safer than an initial voyage downwind.[3] By sailing into the wind on an outward voyage, a safe homeward passage is assured, while an outward voyage downwind is fraught with the uncertainties of whether a return is possible.

Irwin and others base their argument on archaeological findings that show an initial eastward expansion in the direction of the Marquesas Islands, followed by longer voyages to Hawai'i and New Zealand. The ancestors of today's Pacific Islanders left no record of their navigational techniques, but many navigational practices in use from the time of first European contact are documented, and I will describe these in the chapters that follow.

As I said earlier, Norse, Arab, and Pacific Island navigational strategies are by no means the only ones I will describe. In fact, one could argue th   mastery of navigation and sailing technologies was a m   n Western Europeans established global hegemony   h century onward. I will describe these naviga   ell.

                          to deal with similar challenges: spa   timate distances and find position fr   . These are common to navigation on land   ariners, in addition there is the need to predict th   weather, tides, and ocean currents. These challenges are t   topics I cover in this book. I use the experiences of navigational cultures to illustrate the ways these challenges have been met. By focusing on the underlying science and the struggle to understand the forces of nature that confront a navigator, this book offers a tribute to the human desire to understand the world we inhabit.

There is no one "proper" way to navigate. Many individuals and cultures emphasized different skills. Often these skills were

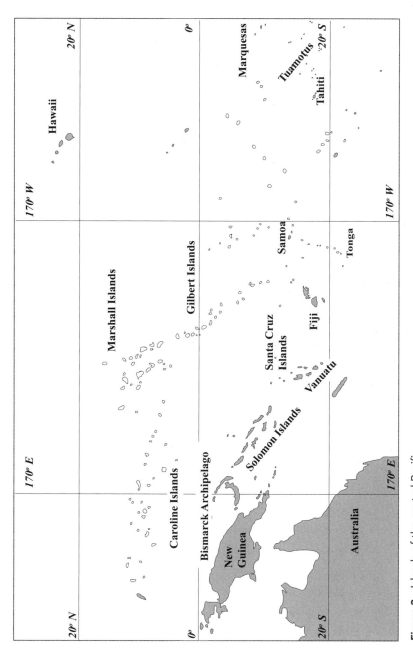

Figure 3 Islands of the central Pacific.

tuned to the local environment and handed down from generation to generation. A good navigator will call on multiple sources of information to complete a successful journey. No one sign or skill can be 100 percent reliable, but an array of skills, when taken together, creates a coherent picture that reinforces success.

You can approach this book in one of two ways: On the one hand, you can read it for an understanding of the challenges faced by voyagers trying to find their way across long distances and the ways they solved complex problems using their own cunning and available tools. On the other hand, you can use this book as an opportunity to test your own observational skills, increase your appreciation of your environment, and perhaps develop a lifetime practice. If you are interested in the latter, be forewarned: it takes practice and patience. I have found that students can become adept at reading star patterns, following the arc of the Sun across the sky, and predicting the weather. But to acquire these skills you absolutely must leave the bubble and look at the stars, the clouds, and the Sun.

The book is divided roughly into two parts. The first addresses what is commonly evoked by the term "navigation": using a log of travels, compasses, stars, and the Sun to find one's way. The second part focuses on factors that might seem at first blush to be more subtle but that end up being critical to navigation in the era before GPS: predicting the weather, reading ocean waves, compensating for tides and currents, and the construction of sailing vessels. Throughout, I examine the historical evolution of way finding. As with many human enterprises, navigation is filled with its tales of woe, hubris, and random walks toward enlightenment.

Although the chapters that follow will not provide the qualifications necessary for the rating of a merchant marine navigator, I hope they will give you some appreciation for the art of navigation as a human skill and the human ability to create and adapt schemes for successful way-finding.

2. Maps in the Mind

· ·

I WAS ALONE, crossing a large open stretch of water off the Cranberry Islands in Maine when the fog closed in. I had no map and the small recreational kayak I'd rented didn't come with a compass. With panic rising in my throat, I told myself to calm down and pay attention. Which way was the wind blowing? The wind was from the southwest. Good. Which way was the swell running? The swell was from the south. Again, good. Could I hear anything noteworthy? I could hear waves breaking on the steep rocky beach to the northwest. Although I didn't have a map with me, I did have a mental image of my surroundings from other paddles in the area and maps I'd seen. I steered southeast toward a narrow passage between two islands and, between the wind, the swell, and the sounds, maintained a steady course.

When the fog descended, I could see nothing except a small patch of blue sky directly overhead. I had to work to calm myself. Paddling, I kept track of time and judged my progress against what my mind's eye saw as my position, as if I were hovering above the seascape. After fifteen minutes, I could hear waves breaking as I approached the narrow passage. Rocks appeared, signaling the gap. After crossing the passage, I noticed that the lobster buoys had little wakes from the incoming tide. The "Vs" of these wakes guided me around the island and back into a safe harbor.

Two months later I was out kayaking in the fog again. This time it was Nantucket Sound off the coast of Cape Cod. Before launching I noted the wind direction, swell direction, and prominent sounds, including the low growl of a buoy two miles offshore. My prior paddles in the area gave me a mental map and a sense of direction that guided my progress through the seascape. If the fog obscured the sight of land, I could steer to the coast using the wind and waves to orient me. Even without visual clues, I had an internal image of my position as I made my way along the coast.

Unknown to me at that moment, two young women were lost in the same fogbank less than half a mile away, disoriented and struggling for their lives. Before they set out on what was supposed to be a quick paddle in Nantucket Sound, Sarah Aronoff, 19, and Mary Jagoda, 20, told their boyfriends that they would be back in ten minutes. When they didn't return forty-five minutes later, the boys contacted the authorities, triggering a massive search effort. The next day Coast Guard helicopters flew back and forth across the Sound, eventually finding their two empty kayaks. The day after, Sarah's body was found. Mary's body was never recovered.

Weeks later I was crushed when I saw a memorial to Mary on the beach, reading "No one is lost . . . to God." What happened? No one really knows, but they probably got disoriented in the fog and mistakenly paddled out to sea rather than back to shore.

While the search was in progress, the *Cape Cod Times* reported an interview with some of the parents of the lost girls. Mary's father told the reporter that she had had kayak training.[1] I later spoke with one of the searchers, wondering aloud why I was out, enjoying myself, and they died instead. His response was laconic, "You're experienced, and they weren't." At the time of the incident I had no kayak training, so clearly that didn't make a difference. What did make the difference?

In that area the coastline runs along an east-to-west line, with land lying on its north side. When I launched my kayak, I noted that the wind direction was from the southeast. Whenever the fog closed in and landmarks vanished, I used the wind as a natural compass and steered back into sight of land. Since I was familiar with landmarks along the coast, a quick sighting of a house or a moored boat was enough for me to figure my location, so if the fog closed in again, I had a rough way of orienting myself. The coastline effectively formed a one-dimensional path for me and made it easy to assess my progress as I paddled.

My guess is that the girls did not know much about the coast; they lacked an internal map. They probably also didn't note the wind or wave direction when they launched, so once the fog closed in on them, they had no way of orienting themselves and paddled out to sea, perhaps thinking that the growl of the buoy two miles out represented the entrance to a harbor.

Many peoples in challenging environments have evolved their own ways of establishing their orientation and mentally measuring progress during journeys. Neuroscientists and psychologists are beginning to unravel the mechanisms of how people track their location — which I will get to — but it's worth taking a short look at how people in different environments organize their space. Below I briefly describe three groups: Inuit hunters, Norse, and navigators in the Caroline Islands. While the days of Norse voyaging are long gone, some Inuit hunters and Caroline Island navigators persist in their traditional ways, although modern modes of transportation and navigation have penetrated much of their cultures.

THE NETSILIK

The Netsilik Inuit (or Netsilingmiut) live in Canada's Nunavut Territory (Figure 4). At the time of their first contact with western Europeans, they were nomadic hunters. Their hunting range (Figures 4 and 5) then included part of the Canadian archipelago to the north: the Boothia Peninsula, King William Island, Somerset Island, and the Adelaide Peninsula. The land is largely flat tundra, spotted by lakes, rivers, and inlets. In the winter the traditional Netsilik hunters travel mostly by dogsled. In the summer they travel by foot on land and by kayak over water. Their life is based on subsistence: fishing for salmon or hunting seals and caribou in migrating herds.

During the European exploration of North America, the Netsilik remained one of the more isolated of the Inuit. Navigation

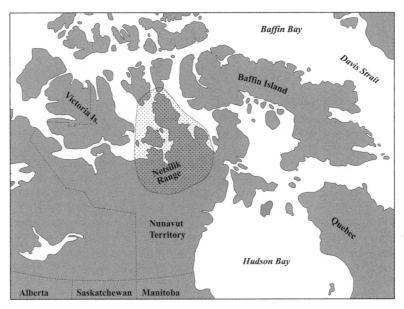

Figure 4   The hunting grounds of the Netsilik Inuit in Canada's Nunavut Territory.

across this barren landscape was, and still is, a challenge, yet they evolved practices that allowed them to negotiate a vast terrain that seems featureless to most Western eyes.

Danish ethnographer Knud Rasmussen, who led the Fifth Thule Expedition across the roof of North America from 1921 to 1924, documented the Netsilik culture. He recorded the ways of the Inuit at a time when their culture remained relatively undiluted by European influence. According to Rasmussen, the Netsilik hunters had in memory a detailed spatial knowledge of their vast flatlands, streams, and inlets that allowed them to find their way.

As reported by Rasmussen, "It is astonishing how much the Netsilingmiut know about the land they live in, be it natural conditions and fauna or its early history. Though they had no previous knowledge of paper and pencil, they were remarkably quick in outlining the shape of their great country, and having done so, could put in all the details with remarkable certainty. Obviously the distances in these hand drawings cannot always be correct; but all islands, peninsulas, bays and lakes are reproduced so accu-

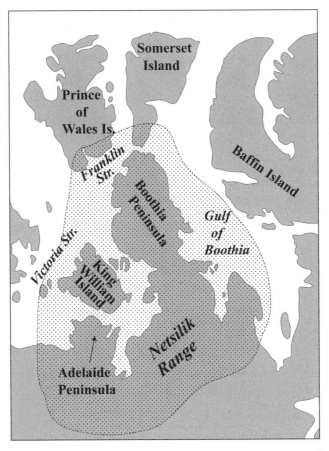

Figure 5 A closer view of the Netsilik hunting territory and geographic features.

rately, that finding one's way from place to place is an easy matter even to an utter stranger to the country."[2]

Figure 6 illustrates a side-by-side comparison of the shape of King William Island as hand-drawn by a Netsilik hunter for Rasmussen to an accurate outline. As Rasmussen noted, the similarities are remarkable.

A catalog of place names Rasmussen recorded with their associated meanings reveals the minute detail of the hunters' memories that thoroughly covered the landscape. Characteristically, the names assigned to different geographies paint a visual description of the landmark, describe game that can be found, or describe an

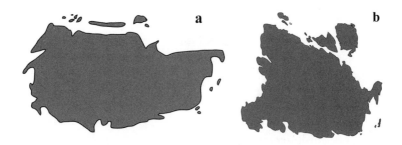

Figures 6a and 6b Outline of a map of King William Island drawn by a Netsilik hunter for Knud Rasmussen (a) compared with the modern outline (b). The left-hand image is based on a drawing by a Netsilik hunter in Knud Rasmussen's *Report of the Fifth Thule Expedition 1921–24* (Copenhagen: Gyldendalske Boghandel, Nordisk Forlag 1930) 8:1–2, 91.

activity at that place. For example, *tuk'Artorwik*, which translates as "the place where one tramps" is next to a lake in the Boothia Peninsula where the Netsilik fish in the winter. Here, Rasmussen guessed that the fishermen have to constantly stomp their feet to keep warm. Another location is *nacilik*, a place stocked with fiord seals. There's *qineq*, "the nose," used to indicate the innermost cove in a fiord. Rasmussen recorded over 413 place names covering eight distinct geographic areas.

The Barrens of Nunavut are swept by whiteout blizzards from October to May and often fog shrouded in the summer. Two recent authors, Colin Irwin and David Pelly, reported on the traditional navigational strategies of the Netsilik. A number of themes become apparent from their reports.[3,4] In traveling the Barrens, the Netsilik often use wind direction for orientation on long journeys. Wind isn't the only means of orientation. They often use the Sun, Moon, and stars as references to update their bearings, but these are rarely visible during extended storms or periods of fog. The prevailing winds are reasonably consistent and form a natural compass to such an extent that wind from different directions has different characters, and each is designated by its own name. The wind creates ripples in the snow perpendicular to the wind direction and allows a hunter to follow the ripples even as the wind drops off. Observant Netsilik hunters keep track of the weather patterns over the course of several days. By combining this knowledge with the local vagaries of the wind produced by valleys and ridges, they can hold a course that is accurate to within 10 degrees.[5] The length of journeys is measured in how many sleeps it takes to go from one place to another. With the scarcity of game across the Barrens, typical journeys in search of food can be as long as two hundred miles, approximately seven sleeps.

Irwin quotes Kako, an Inuit, who describes the use of the wind: "When it's foggy and I can't see anything it's very hard to tell direction. But if I can get a bearing on the wind when it is clear I can then use the wind to judge direction when the fog descends. This would be in the summer when there are no snow ridges, or in a boat."[6]

Often, holding a course to an accuracy of 10 degrees isn't sufficient to reach a camp that is a small speck in an undistinguished landscape. To refine their paths the Netsilik employ tracking. Humans, dogs, and sleds all create markings in the snow that last for days and leave important clues that an experienced hunter can read. Rather than trying to find a single speck on a vast two-dimensional surface, they travel in the approximate direction of an encampment. Eventually, the hunter will encounter tracks leading to or from his destination. When a track is crossed, knowing the recent weather history, he can determine the direction to travel to find camp. This reduces the problem of finding a point in a two-dimensional plane to the problem of finding a line, then following the line to a destination. While simple in concept, the practice of tracking through an ever-changing landscape requires an intuition developed through experience and training.

The nature of the tracks tells a story about the passersby: when they traveled, their size and number, whether they ran or walked next to a sled. From the sled marks you can tell the weight of its load. The dogs leave scat and scent at regular intervals, forming a trail other sled dogs can pick out. Upon finding a track the driver can modify his wind-directed course and follow it in the most likely direction to a camp.

Reducing the difficult problem of finding a point in a plane to the task of finding a point on a line is a common strategy among navigators. According to Robert Rundstrom, a geographer who has studied Inuit navigation, "Given the nature of the barren ground terrain itself, linear conceptualization of the territory may be the easiest way to bring a sense of order to an otherwise chaotic landscape, an order which allows human beings to think and act as a successful part of that landscape."[7]

Caroline Island Navigators

Pacific island navigators have many strategies for orientation and land finding, including the use of wind direction, stars, wave patterns, and even birds. Although they employ all of these, there is typically one or two primary means of orientation. The use of stars for orientation is fairly common, and practitioners of traditional navigation in the Caroline Islands use a curious scheme called the *etak* system to keep track of their progress on long voyages out of sight of land.

The Caroline Islands consist of a widely separated group of islands and atolls, appearing in the northwest sector of Figure 3 in the previous chapter. In this sparse collection of atolls, stars serve as the main reference for the orientation of navigators in sailing canoes. Much of what we know of the etak system comes from reports of authors like David Lewis and Thomas Gladwin.[8,9]

To understand the etak system, the reader needs to know first about something called a "star compass." Near the equator, stars always rise at the same position with respect to north and always set at the same position with respect to north, regardless of the location of the observer. The rising and setting stars form a natural compass. For example, the bright star Sirius rises in the

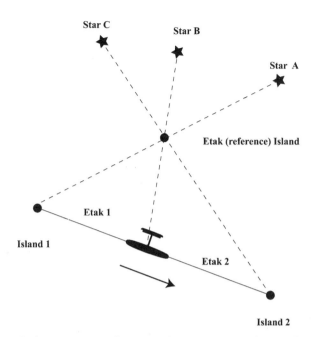

Figure 7 Etak organization of voyages for navigators in the Caroline Islands. The navigator keeps track of position by the location of a reference island against a fixed background of rising and setting stars.

southeast for a Caroline Islander and sets in the southwest, while the star Capella, in the constellation Auriga, rises in the northeast and sets in the northwest. Polaris always hovers approximately above due north.[10] Traditional Carolinian navigators divide the horizon into regions called etaks, based on the angular swath associated with a particular rising or setting star.

In the etak system (Figure 7), the navigators picture in their minds a reference island located between their origin and their destination. Although the reference or etak island could be real or imaginary, it isn't visible during the voyage. The navigator keeps its location relative to the background of rising and setting stars in his or her mind. If you drive down a highway, you might see that a line of telephone poles along the road will move relative to a distant landmark such as a faraway mountain. The telephone poles appear to move against the fixed background. Likewise, the Carolinian navigators visualize the reference island moving against the fixed background of stars. Their position at any time can be related to the particular etak star that was behind the reference island.

Typical voyages could be up to several hundred miles. The length of a voyage is often described in terms of the number of etaks crossed. In addition to stars traditional Caroline Island navigators also use secondary references for orientation, such as wind and swells. As with the Netsilik the etak system effectively reduces the visualization of the journey from a complex two-dimensional problem to a simpler system, in which progress can be measured against a linear reference kept in the navigator's mind.

If the etak system seems arcane to the reader, it apparently was also arcane to many of the Caroline Islanders themselves in the past. To become a successful navigator, an apprentice would have to spend many years learning how the system worked, and lessons were often conducted in secret. Recently, with a rebirth of interest in traditional navigational strategies, the techniques are regaining interest among the inhabitants.

THE NORSE

One of the commonly cited strands of Viking navigational lore is a set of sailing directions from Norway to Greenland from the fourteenth-century *Hauksbók*: "From Hernar in Norway one should keep sailing west to reach Hvarf in Greenland and then you are sailing north of Shetland, so that it can only to be seen if visibility is very good; but south of the Faeroes, so that the sea appears halfway up their mountain slopes; but so far south of Iceland that one only becomes aware of birds and whales from it."[11]

The passage describes the use of a landmark (Island of Hernar) on the Norwegian coastline as a jumping-off point for a voyage due west. The sailor finds the coastal point, then sails west, most likely guided by the wind or views of the Sun. The sighting of the

Faroe and Shetland Islands provides a useful waypoint to verify that the navigator is on course. The distance from the Faroes to the voyager is given by the appearance of the mountains: the lower half is covered by the horizon because of the curvature of the Earth. The navigator targets a piece of the southern coast of Greenland that is well south of Iceland. Although Iceland won't be visible, the nearby ocean is relatively shallow, and upwelling currents bring nutrients close to the surface, attracting fish, birds, and whales. The ocean life provides another useful waypoint indicating that the sailor is on the proper course.

At the time this passage was written, the length of voyages was reckoned in days, and the passage of waypoints was used to measure progress. Note that, again, a two-dimensional search problem is reduced to identifying a series of waypoints along an approximate line. The solid line in Figure 1 in the previous chapter shows the path described in the passage from the *Hauksbók*. As the reader can see, it proceeds along a constant latitude. Once Greenland comes into sight, the sailor can decide where to proceed to reach a particular settlement on the coast.

Norse sagas have references to the use of the Sun to orient the sailor. One appears in the account of the accidental discovery of North America by Bjarni Herjólfsson. Although he had never been to Greenland, he outfitted a ship and crew to sail there from Iceland. This is described in the *Saga of the Greenlanders*:

> Bjarni then spoke: "Our journey will be thought an ill-considered one, since none of us has sailed the Greenland Sea."
>
> Despite this they set sail once they had made ready and sailed for three days, until the land had disappeared below the horizon. Then the wind dropped and they were beset by winds from the north and fog; for many days they did not know where they were sailing.

After that they saw the sun and could take their bearings. Hoisting the sail, they sailed for the rest of the day before sighting land. They speculated amongst themselves as to what land this would be, as Bjarni said he suspected this was not Greenland.[12]

The voyaging season of the Norse was May through September to avoid the severe storms that sweep the North Atlantic in the winter. During these late spring and summer months, stars were rarely visible because of the glow of the Sun in those high latitudes. The Sun was the best celestial body to take a bearing on and was low enough in the sky that it could be relied upon to provide directional information. The Pacific Islanders, on the other hand, couldn't rely on the Sun because it was usually too high in the sky in the equatorial waters, except at sunrise and sunset, when it would be used to take a reliable bearing.

A few features stand out about all three navigational cultures. First, there were few direct and accurate measures of distance in the modern sense; only the number of days or etaks a journey would take. A second prominent characteristic was the memorizing of waypoints that pointed to mental maps that allowed the traveler to measure progress. Third, a primary means of orientation was employed, whether it was the wind, the Sun or the rising and setting stars. All three groups apparently attack the challenges of navigation by reducing the difficult two-dimensional search problem into the identification of a series of waypoints along a route.

THE ORIGIN OF MENTAL MAPS

When we walk around our neighborhood, or drive to a friend's house, we usually don't think twice about the route. A stranger may ask for directions, and we'll tell him or her to drive another mile and turn left at the convenience store. Sometimes we might be more challenged. Say you travel to a city and wake up in a dark hotel room. At first you may be disoriented, but then you remember the events leading up to that moment and begin to visualize your surroundings.

Above we saw that the Netsilik could draw a detailed map of their surroundings from memory. The Caroline Islanders developed an elaborate system of navigation based on the relative positions of islands they couldn't see but visualized. All of this indicates that some part of the mind is able to assimilate, store, and recall what are effectively mental maps. How does this work?

Researchers have been studying mammals for decades, slowly piecing together the elements that make up the internal mapping mechanism. There are many important clues, but we have yet to see a full story.

Roughly speaking, there are three elements in human (and mammal) navigation:

1. Dead reckoning
2. Perceptions
3. A mental map

The term "dead reckoning" comes to us from the British Royal Navy and dates from the seventeenth century. The term "dead" is related to such terms as "dead aim" and "dead right," denoting precision. The concept is simple: you deduce your location through a history of your travels from a starting point to your current location. This requires some sense of orientation, speed, and time elapsed in the journey.

Our perceptions function in two roles: First, the sight of familiar landmarks helps us update our location in the internal map. Second, the sight of objects drawing nearer as we approach them and receding into the distance behind us gives us a sense of speed and motion. We usually take all of this for granted.

One important sense that is often discussed in the context of animal navigation is *proprioception*. This is a sense of the body itself. The weight and relative orientation of your limbs may signal that you're walking or running. If you are in a car at a stoplight and it suddenly accelerates, you sense a kind of force pushing you back into your seat. Inside your ear are three semicircular canals that broadcast information to your brain about the relative orientation of your head and whether you're changing speed or direction.

All of these senses are integrated into the sense of motion and direction that's critical to dead reckoning. When you accelerate from a stoplight, you will see that objects in front of you seem to get larger as you approach. You hear the engine revving. You feel the acceleration. You don't process each bit of information, but you have an overall sense of motion that comes from the integration of these sensations happening simultaneously. As you travel, landmarks that you remember pass by, and you mentally update your position with respect to your expectations. Progress is measured against an internal vision of the environment.

Psychologists often talk about a distinction between *route knowledge* and *survey knowledge*. These two concepts are fairly easy to understand. If we have route knowledge, we know routes and landmarks along these routes. We know of a network where different routes join each other but not what lies between the routes themselves. The quip attributed to the crusty Maine lobsterman, "You can't get there from here," would seem to indicate route knowledge, where no combination of routes connect from "here" to "there."

Survey knowledge, on the other hand, is a complete familiarity with an environment. In your mind you see the region as if you are hovering over the landscape and seeing everything below in miniature. How can we decide whether mammals possess survey or route knowledge of their environments? Shortcuts between heavily traveled paths indicate survey knowledge, as opposed to route knowledge.

In 1948 psychologist Edward Tolman reported an experiment using rats in a maze to test this. He set up a maze that involved a long, convoluted path from the rats' starting position to a source of food. Tolman gave them enough time in this maze to develop a good sense of the location of the food source relative to the starting position. He then replaced that maze with a different layout but retained the original convoluted path. In the new layout there were shortcuts introduced that the rat could take to the food. In Tolman's analysis if rats possessed survey knowledge, they would find shortcuts, while if they had only route knowledge, they would continue to try the longer path. As it turned out the rats did find shortcuts, indicating that perhaps they did have survey knowledge.[13]

Since Tolman's work many studies have been carried out on different species to see if there was evidence of shortcutting. Wolves, for example, hunt over large territories and have evolved strategies for marking paths in and around their territory. When necessary, adults take shortcuts, but pups will either follow established trails or stay near an adult.[14] The habit of dogs urinating on trees is a distant evolutionary habit of trail marking. They can find their back-trail from the periodic scent mark they leave. This comes at the price of attracting predators, so canines clean their feet by rubbing them on the ground to remove the scent from their paws. Frequently, dog owners confuse this behavior with "burying" the droppings.

Recently, the home of the mental map has been discovered. Deep inside the brain is a region called the limbic system. It is more primitive in an evolutionary sense than the more recent additions such as the neocortex. In addition to being the source of the internal map, the limbic system provides a seat for long-term memory and emotions elicited by both thought and external perceptions.

Neuroscientists have found ways to probe individual nerve cells with electrodes that can display whether a cell is firing (active) or not (inactive). By equipping rats with a flexible tether connecting the electrodes to external instruments, they can monitor what's going on inside the rat's brain while it navigates an environment.

Two kinds of cells appear to create the mental map: *place cells* and *grid cells*. Place cells were first reported in 1970 by neuroscientist John O'Keefe.[15] They're a dense array of cells in the limbic system that have a large number of connections with nearby parts of the brain. Each place cell seems to be programmed to fire at a high rate when the rat is at a specific location. If the rat changes location, another place cell starts to fire.

The arrangement of place cells and how they represent the environment to the rat isn't entirely known and involves other parts of the brain. If the rat is moved to a new setting, the firing pattern changes, but specific locations are still represented by specific cells. It is as if a new map has been downloaded into the array of place cells.

The representation of space in the array of place cells also seems to be related to how fast we travel through the environment. For human navigation the environment and means of transportation are crucial, as we've seen with the Inuit, the Caroline Islanders, and the Norse. We walk, we run long distances, we've learned to sail around the world, we drive cars, and we fly in airplanes. How does the mode of transportation affect our perception of space? If we walk around where we live, we're probably familiar with the details of our house, the sidewalk outside, and the trees down the block. If we walk farther away, we may recognize most houses, but we've lost some of the details; we might not realize if someone's changed his mailbox.

If we drive around town, we may know the major intersections and even shortcuts to avoid traffic, but we probably won't know the names of all the side streets. If we drive (without a GPS device) to visit an aunt who lives five hundred miles away, we'll remember some of the major waypoints on the interstate, but if we haven't visited her in a while, there's a chance we might get lost after we've gotten off the freeway.

A Caroline Islands navigator wandering around his home island likely has a detailed knowledge of his neighbors' homes, their gardens, and the shed where the canoes are built. This may form one internal map, but when the navigator voyages to a distant island, the etak system becomes the new mental map, and the scale of distances changes in the mind.

Aviator and author Antoine de Saint-Exupéry captured this sense of distance shifting that is due to the invention of the airplane: "Our very psychology has been shaken to its foundations, to its most secret recesses. Our notions of separation, distance, return, are reflections of a new set of realities, though the words themselves remain unchanged. To grasp the meaning of the world of today we use a language created to express the world of yesterday. The life of the base seems to us nearer our true natures, but only for the reason that it is nearer our language."[16]

This wasn't the first time that the psychology of distance was shaken. The invention of the magnetic compass and subsequent development of celestial navigation vastly expanded the horizons of western Europeans from the thirteenth through the eighteenth centuries, creating new meanings for the sense of distance and separation.

The concept of scale related to mode of transportation was tested with place cells by a research group at the University of Arizona.[17] Rats were trained to drive a mobile platform around a circular track that they could also walk on (Figure 8). The firing pattern of place cells changed, depending on whether the rat walked or drove. When the rat drove, the representation of places became more spread out, just as flying in a plane distorts our internal representation of the distances around the world.

The widespread use of GPS devices creates a related issue: What becomes of cognitive maps in a GPS era? Will the ability to visualize large distances atrophy or even fail to develop for generations raised with these devices?

Recently, Edvard Moser and colleagues discovered another clue

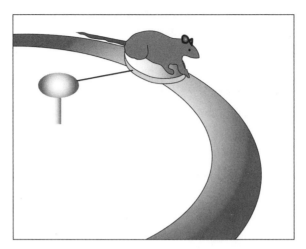

Figure 8 Rat navigating a region on a mobile cart. Figure adapted and reprinted by permission from Macmillan Publishers Ltd: *Nature Reviews Neuroscience*, McNaughton et al., Path integration and the neural basis of the "cognitive map" © (2006), 670.

to the brain's mapping mechanism called grid cells.[18] As O'Keefe had done, Moser allowed rats to wander in an environment, but this time his group monitored the nerve cells in a region adjoining the location of place cells in the brain. This region, called the entorhinal cortex, is home to the grid cells. Here, Moser and colleagues found that a single grid cell would fire when the rat was located at multiple, but quite specific, locations in its environment.

The locations where the grid cell fired created a pattern that looks much like the tiling pattern on a bathroom floor. In mathematics there are only three regular geometric shapes that can tile a plane: equilateral triangles, squares, and hexagons. Moser's group found that grid cells represent space as an array of equilateral triangles. One grid cell will fire at a high rate when the rat is physically located at one of the points in this regular pattern. If the rat moves to another point in this regular pattern, the same grid cell will also fire at a high rate. If the rat moves to yet another spot that is slightly displaced and not on its grid, the cell's firing rate will fall off, but another grid cell will start to fire at a high rate, corresponding to a slightly displaced grid of triangles. It is curious that the brain somehow represents space as an array of triangles. In the western European tradition of mapping and mathematics, we usually create maps and graphs using a regular grid of squares.

Perceptions and dead reckoning affect the way the grid and place cells respond to the environment. When dead reckoning changes or visual clues are altered, the mapping of the grid and place cells changes. Evidently, the mental map is continually updated by these inputs. These include visual clues and the shape of the region the rat explores. The grid and place cells have a dense network of interconnections, and their interactions appear to create the mental map.

Lest you think that this is restricted only to rats, the analogous regions of the human brain light up when engaged in navigation. In a recent study humans were asked to navigate a virtual landscape while the blood flow in their brain was observed. As they mentally "navigated," the regions containing the place and grid cells showed increased activity, pointing to a mechanism similar in humans to that in rats that creates the mental map.[19] The parts of the brain that hold the place cells are also responsible for a form of long-term memory, sometimes called *declarative memory* because a person can recount a past episode verbally.

While languages and navigational strategies may vary from one culture to another, the wiring of the brain to create language, mental maps, and other skills appears to be an intrinsic feature that we're born with. Navigational skill depends on the interplay between memory, perceptions, and the mental map.

3. On Being Lost

. . . .

PEOPLE GET LOST in many different ways, but their reactions can be remarkably similar. A storm descends in the mountains, obliterating landmarks to a climber. A hunter obsessively concentrates on following the track of an elk and loses his way in deep woods. A kayaker paddles into a fog bank. A pilot loses his sense of the horizon. A sailor can't find the way back to land. They can all experience a sense of panic brought on by the ensuing disorientation.

It's one thing to get lost in a parking lot where help is nearby, or a GPS device is handy, but it's quite another to get lost where there is no way to reestablish your orientation. Professor Ken Hill from St. Mary's University in Nova Scotia, an expert on lost persons, defines them thus: *"The lost person is unable to identify or orient his present location with respect to known locations, and has no effective means or method for reorienting himself."*[1]

The first half of the definition is probably familiar: you don't know your location. Most of us have been in a situation where we're temporarily turned around. The second part of the definition is more serious: if there's no way to reorient yourself, you may be in big trouble.

Most of the time we move through familiar territory. Recognizable landmarks passed on a journey give us details that keep us oriented. We're more likely to get lost in unfamiliar terrain. Perhaps we'll start out in a town or city without a map on a familiar road, but we take a wrong turn into a strange and deserted warehouse district, and soon everything looks strange and oddly the same. In a more extended trip, in a wilderness setting perhaps, we might have a physical map where the correspondence between details on the map and our experience on the ground are usually checked, but we get lost in a conversation and suddenly nothing visible corresponds to any features on the map.

The correspondence between a mental or physical map and our perceptions helps us stay oriented, but one of the first stages of getting lost involves a process called "bending the map." The phrase comes from the sport of orienteering, in which competitors find their way around a series of waypoints that are revealed to them on a map at the start of the race. Competitors can become lost and believe they are in one place indicated on the map and mentally try to force features they see to line up with ones indicated on the map even when the correspondence is poor.

Denial is an effective psychological defense mechanism, and map bending is one form that lost persons often engage in. A lost person might first believe he is located at a certain point on the map, but things around him do not seem quite right. He pays attention to details that confirm what he already believes to be true, ignoring all evidence to the contrary. A lost person may be looking for a creek that flows south on the map. In his mind he's sure that he has arrived at the creek. It flows east, yet he conveniently ignores this fact and follows it anyway. It can take some time, but there comes a moment when an eerie realization hits him that something is wrong and he doesn't know why.

Suddenly, the lost hiker recognizes that his map and perceptions don't line up. Panic sets in. The emotional centers of the brain send out warning signals, and perceptions get distorted with a fight-or-flight reaction. Massive amounts of adrenaline flood the mind and body. Breathing and heart rate increase. The person refuses to believe that he's lost and runs frantically in a direction that he's sure will lead back to the trail, only to get deeper into trouble. First one possibility, then another races through his overtaxed mind, and yet he cannot gain any certainty.

"Woods shock" is the term for this kind of anxiety attack brought on by the realization that the subject is lost. Because woods shock can lead to counterproductive behavior, it's best for the lost person to stay where he is and engage in a quiet activity, such as making a small campfire, which can alleviate anxiety. A lost person may abandon a backpack so he can travel quickly, only to become more lost, and then is without means to stay warm and dry.

The experience of anxiety over being lost happens in settings other than the woods. Aircraft pilots can be particularly susceptible to the effects of spatial disorientation. While a hiker or sailor deals with two-dimensional environments, a pilot has to retain his orientation in three dimensions. Oftentimes, the integration of perceptual clues, as described in the previous chapter, does not work normally, and pilots can lose their sense of horizontal and vertical orientation. The spatial disorientation in pilots can create a "break-off" phenomenon, in which the pilot feels detached from his body, the aircraft, or the Earth. This often happens to pilots at high altitudes on long flights, particularly when the horizon is poorly defined.

In the Marshall Islands there is a word, *wiwijet*, which means both "being lost" and "panic" in Marshallese. Anthropologist Joseph Genz, in his PhD thesis, captured the meaning of wiwijet from one of his informants, named Isao.[2] According to Genz, "During subsequent interviews, Isao elaborated the meaning of *wiwijet*. A navigator who becomes lost and does not receive help [from a benevolent spirit] to discern the direction toward land will lose his mind and continue to sail disoriented until his death." Genz also related the story of a sailor who, lost at sea, experienced wiwijet, yet successfully returned to land. Even after returning to land, he still experienced some degree of wiwijet.[3]

Behavior of Lost Persons

We often can learn a lot about systems when they fail. The state of being lost is a failure, a breakdown, in the mechanisms of orientation. While we cannot peer inside the brains of lost persons, we can gain some insight from a wealth of data accumulated over the years about their behavior. It may at first seem surprising that data about lost people would be well documented, but when the report of a lost person reaches authorities, it invariably triggers a large search-and-rescue (SAR) operation. The searchers themselves can be put in harm's way, and the deployment of resources is costly. If an SAR team understands the likely behavior of a lost person, it can concentrate the search efficiently.

One of the largest databases of information on lost people is found in the International Search and Rescue Incident Database (ISRID) project,[4] which was begun in 2002 to help create statistical models for the behavior of lost people. Important factors documented include age, background, terrain, weather, and the activity a person was engaged in just before becoming lost. One finding is that, on average, lost people are found relatively close to their last known position, even though they may have wandered a considerably longer distance in a convoluted path to arrive there.

The most useful information for an SAR team is the probability of finding a person in a particular region. In looking for a lost person, a search-and-rescue team will first construct a probability map where the center is generally taken to be the last known position of the lost person. The map describes the likelihood of finding the person relative to this starting point.

Over the years databases such as the ISRID have become increasingly sophisticated. Although this reduces thousands of life-and-death struggles into the seemingly banal terms of mathematics, it allows for a clear-headed assessment of the likely conduct of those who become lost and the probable locations where they will be found.

One of the simplest statistical models for behavior of a lost person is called a *random walk*. In a random walk a person might walk some distance in a straight line, then suddenly change direction. He'll walk some more, hit another turning point, and randomly change direction again. As the random walk progresses, the most likely location to find the person is still at his starting point, but the region where he might be found spreads out over time.

Possible locations after 1 day

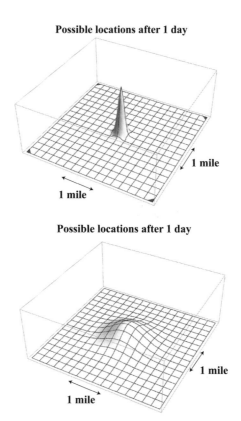

Possible locations after 1 day

Figures 9a and 9b In a simple random walk, the probable distance of a lost hiker from his starting point grows slowly with time.

In reality lost persons don't execute true random walks as described above, but the random walk is useful as a simple model, illustrating how statistics can aid a search. The *total* distance traveled by a lost person along a winding path can be quite long, but the average *direct* distance from the starting point to where he can be found is surprisingly small. In a study of one hundred lost hunters, although they walked a total distance of eight miles on average, they ended up only one and a quarter miles as the crow flies from the point where they were last seen.[5] Figure 9 shows the probability of finding a person making a random walk after one hour and after twenty-four hours.

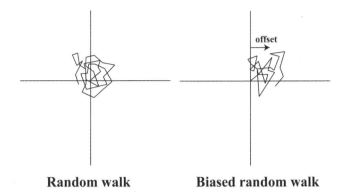

Random walk   **Biased random walk**

Figure 10   Left: In a random walk a person travels in a straight line for some distance, then changes direction, moves in a straight line, and changes direction again, randomly. Right: In a biased random walk, factors such as a downhill slope can systematically push a person in one direction.

One variation on the random walk is called a *biased random walk* (Figure 10). In this case, although the changes in motion are mostly random, some factors can push the individual slightly in one direction over another. One element is a downhill slope. Even if a person is intent on holding a specific course, the constant tug of gravity can systematically pull a person downhill, whether or not she realizes it. Some lost wanderers, such as hunters or backpackers, are more likely to be found downhill rather than uphill from

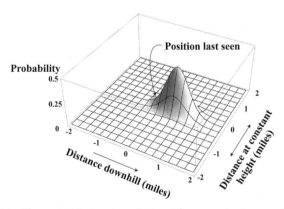

Figure 11   The probable location of a lost hiker on a sloping terrain for a biased random walk. In this case the hiker may have been on a ridge, while roads and a trailhead are located at a lower elevation.

the point where they were last seen. There can be other factors in the downhill bias: men and women frequently recall folkloric advice that they can walk to civilization by following water flowing downhill. In a biased random walk, the region of possible locations also spreads out, but an offset from the starting point creeps in. This knowledge can help a search team concentrate on a downhill area or a region where they believe the person could be seeking shelter or help.

In reality the paths traveled by the lost are rarely random walks as described above. The closest case is when someone is trying to find his way out in a featureless forest with no sunlight or means of maintaining a constant direction. Lost persons will typically create a loopy path that can often double back on itself. This is quite common, but it can be exceedingly unnerving to think you are moving in a straight line but end up in the same place after several hours on the move.

In October 2001 Jason Rasmussen set out alone to hike portions of the Pow Wow Trail in upstate Minnesota. What started as a weekend jaunt became a struggle for survival. After becoming lost, he camped at a lake overnight and decided to walk to the trailhead the next morning. After tramping through the woods for two hours, he came upon a second lake. Author Cary Griffith relates this in *Lost in the Wild*:[6]

> And then he sees it. There is a clear sparkle of water through the dense branch weave. At first he thinks it could be Pose Lake. But by now, hiking this long he should be much farther east than Pose Lake. And there is something about this water, the way it shimmers through the branches, that is eerily familiar.
>
> He cannot put his finger on it, but he suspects the airy tumult in his stomach is because this scene — the sparkle through the branches — looks so similar to the water he saw earlier this morning.

He takes a few more steps toward the lake. He looks at the water, peering through the branches. And then it hits him, not in a pleasant way, or with mild surprise. It hits him like someone placing a well-positioned fist in the center of his abdomen.

"Christ!" he thinks. "That's the same lake. I'm in the exact same spot I was in two hours ago!"

Something like a slow burning panic starts to unravel him. Now he is genuinely scared. When he thinks about how far he's hiked, pushing through the damp woods for the last two hours, hiking in some kind of big circle that has brought him back to his exact point of departure, he's dumbfounded. It is like some kind of science fiction movie in which the universe is confined to the narrow reaches of a wild wood, and there is no way out. He is condemned to walk in huge circles through the woods.

This description also hints at the onset of woods shock. Eventually, Jason was found after a massive search operation.

Outdoor literature is full of reports of lost persons who try to walk out of their predicaments, only to find themselves recrossing their paths at some point. Recently, the anecdotal idea of walking in circles was put to a more stringent test by investigators. Jan Souman of the Max Planck Institute in Tubingen, Germany, and

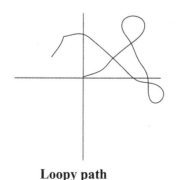

Figure 12 A smooth loopy path is more characteristic of a person who is lost in a featureless terrain with no direction indicators.

Loopy path

his coworkers tested a number of subjects for this in three different settings.[7] Some of the subjects were placed in a large, flat, wooded area on an overcast day and asked to walk in a specific direction. They had difficulty walking in a straight line. Their paths were smooth, yet looped back on themselves a number of times, like the one shown in Figure 12. Older studies attributed the tendency to walk in circles to a difference in leg lengths that caused the lost people to move systematically either right or left.[8] Souman et al. believe that it's more likely due to a buildup of noise in the parts of the brain that help with dead reckoning, a topic discussed in the next chapter.

The simple random-walk model works only up to a point. ISRID is a much more sophisticated instrument to predict the behavior of lost people. While the most likely place to find a lost person is close to where she was last seen, as in the simple random-walk model, the probability map doesn't grow with time as that simple model predicts. According to Robert Koester, a main organizer of ISRID, this arises because lost persons tend to loop back to a familiar point as time goes on, making the probability map relatively static.[9] The ISRID models are more predictive than the simple random-walk models because they're based on a compilation of real-world data.

Personalities and Behavior

Search and rescue operations are aided by knowing the type of missing persons because different people become lost for different reasons and behave differently. Researchers Syrotuck, Hill, and Koester have detailed many of the common types of lost people that search-and-rescue missions have to organize to find.[10] Here are some of the common personalities of lost persons that emerge in their work:

Very young children (ages two to six): Children aged two to six are likely to be distracted by something and become separated from friends or family. They can't wander too far from the point they were last seen. Oftentimes, their instinct for comfort will save them, as they try to find a safe location and stay put.

Older children (ages six to twelve): The ages of six to twelve can be dangerous ages because these children have begun to develop cognitive maps but have not yet acquired the skills to maintain direction or reorient themselves. They realize they're lost, and they panic, traveling a circuitous path in an attempt to find safety, thereby making them difficult to find.

Hikers and backpackers: Sometimes hikers will accidentally follow game trails or lose direction when a known trail covers a rocky area without clear marking. In a party of backpackers, often only one person carries a map and compass or knows the country. Another member of the party lags behind, becoming separated from the group. When this happens, he may not have any means of orientation and might wander from the main trail.[11]

Hunters: Hunters will often become engrossed in following game, focusing on animal signs. When they're wrapped up in the chase, they often forget to check their compasses and maps, ending up lost. Often, the best chance of finding game will be in remote locations. After being on the move, hunters often underestimate how far they've traveled.

Dementia sufferers: Those with dementia, mainly Alzheimer's victims, wander through the woods without much of a strategy or even awareness that they're lost. Something may trigger a long-lost memory, which leads to a behavior similar to a random walk.

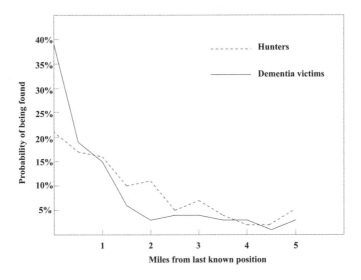

Figure 13 Average distances of lost hunters and dementia victims from the start of the search to where they're found, based on data from Koester, *Lost Person Behavior*, 53. Dementia victims tend to be found much closer to the start of the search than hunters. Lost hikers have a similar distribution to that of hunters, while small children have a similar distribution to that of dementia victims.

There are many other possibilities: cross-country skiers, snowboarders, mushroom pickers, even fugitives. The list above represents just a few of the documented categories.

As a group hikers and hunters are found much farther away (approximately three miles in 75 percent of cases) from their last known location than are dementia victims or young children (who are found approximately one mile away in 75 percent of cases).[12] The distributions, taken from the ISRID data, are shown in Figure 13. This result is not surprising, given the difference in the mobility and experience of the two groups.

Getting Unlost

With the exception of dementia victims and very young children, most lost persons use a number of strategies to find safety. Many

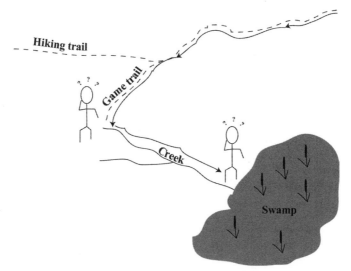

Figure 14 Goal-directed behavior of a lost hiker. The hiker mistakenly follows a game trail, then follows a creek downstream, only to end up in a swamp.

times they'll try to climb to the highest point in the area to get a good view of their surroundings. Sometimes, they'll remember the folksy advice to walk downhill until they find a creek and then follow that downstream to civilization. This doesn't always work out.

Figure 14 shows an example of goal-directed behavior in trying to find safety. A hiker is walking along an established trail, which is also used by animals in the region. At a certain point, the animals veer off onto their own trail to find water. If the hiker mistakes the game trail for the hiking trail, she veers off and follows this for some distance. As the trail approaches the water source — a creek — the game trail disappears, and the hiker is left wondering what to do. The hiker backtracks, but the main trail might not be easy to identify, so the hiker again follows the game trail. Realizing that she is lost, the hiker decides to follow the creek downhill, thinking it leads to civilization, only to find that it drains into a large swamp, at which point she is thoroughly lost.

At the start of chapter 2, I described two girls who were lost in the fog in Nantucket Sound. I couldn't understand why their kayaks were found many miles from shore, and in an area where strong currents developed. Tom Leach, the harbormaster in Harwich Port, developed a theory about what happened. There is a buoy two miles offshore that marks a shoal that large boats should avoid. The buoy has a long pipe that emits a low growling noise from wave action. I was familiar with the buoy's sound and location from many paddles in the area. Tom guessed that when the girls got disoriented in the fog, they paddled toward the sound of the buoy, thinking that this would mark the entrance to a harbor and safety. In reality, they were paddling farther out into the Sound.

With the possible exception of fugitives and despondent personalities, most individuals who are lost don't want to stay that way. Many actively pursue a strategy that they believe will get them out of trouble. Their behavior appears to fall into a number of recognizable categories. Ken Hill cataloged some common strategies that lost people use to try to find their way to safety.[13] These strategies are deduced from his interviews with people after they were found and are clustered around certain kinds of behavior:

> **Random traveling:** Much like a simple random walk, random traveling follows a loopy path. School-age children will often do this.
>
> **Route-traveling:** This involves following a trail or track to find civilization. The lost person follows a game trail, an abandoned track, even a perceived path with less brush, in the belief that it will lead to civilization.
>
> **Direction traveling:** When a person is able to establish and maintain a direction of travel using some means of orientation, such as a compass, he assumes that if he walks long enough in one direction he will eventually come upon something helpful. Hill reports that

this strategy is rarely effective. Often, the person will become so focused on traveling in a straight direction that he will walk right across a road without noticing it. In other cases he may find himself far deeper in the woods than where he started out.

Route sampling: The lost person tries to establish and maintain a base of operations, such as a trail junction, then moves along different trails or tracks or in expanding distances. Hill claims that this can be effective, particularly if the individual is willing to backtrack to the central location.

Direction sampling: This is similar to route sampling, except the person probes her environment in different directions, using some means to maintain orientation along the way. Existing tracks don't have to be used.

View enhancing: The person will walk or climb to the highest spot that's easy to reach to survey the surroundings, with the hope of finding an obvious route out. Cell phone coverage is often better at a higher elevation because of a lack of obstructions.

Backtracking: Since the starting point of a trek is usually known, if a lost hiker or hunter can find his own track, he can follow it backward. This requires the skill to recognize and follow tracks.

The strategies used by the lost mimic how people live their lives; it can serve as a metaphor. Some people route sample throughout life, trying one vocation after another. Some barrel ahead along a well-worn path, oblivious to their surroundings. Some seem to just bounce around randomly. Others just want to get high and take in the view. This metaphor is nothing new. The philosopher and mathematician René Descartes advised a method similar to direction traveling as a way to lead one's life. In his *Discourse on Method, Part III, Maxims on how to be more resolute and to know oneself for proceeding rationally in life*, he gives advice to the reader on resolute living. The traveler through life should "[follow] the example of travelers who, when they have lost their way in a forest, ought not to wander from side to side, far less remain in one place, but proceed constantly towards the same side in as straight a line as possible, without changing their direction for slight reasons."[14]

According to Descartes, an endpoint is guaranteed, "for in this way, if they do not exactly reach the point they desire, they will come at least in the end to some place that will probably be preferable to the middle of a forest."

However, at least for lost persons, the strategy described by Descartes hasn't been shown to be the most effective.

The goal-directed behavior of lost people has another parallel in territorial exploration. Many of the strategies employed by explorers in a new territory mirror the conduct of those who are lost. Lewis and Clark practiced route traveling up the Missouri River to gain knowledge of lands to the northwest. They used view enhancing to gain a pathway through the Rocky Mountains, and on his return from Oregon, Lewis sent parties in different directions to gain more knowledge of the western lands: an example of direction traveling.

In our times a number of strategies can help persons who are lost. I present two here. Both are predicated on the notion that a search-and-rescue operation will be looking for the lost person, and he is simply aiding that operation. He should leave a note with a friend or family member indicating the likely time of return. If word doesn't reach the friend by an agreed-upon time, authorities should be contacted and a search operation initiated. If he is traveling on foot, he should leave an imprint of his boot tread in aluminum foil and put it in a car at the trailhead. Trained trackers can find and follow the boot print.

In my case I carry a fanny pack that has a roll of bright orange surveyor's tape and an indelible marker. If it appears that the best chance of walking out of a situation is following a particular path, I will take it but will leave pieces of tape tied at eye level, so that each piece is visible from the previous one. On the tape I note the time, date, and direction of motion. If I stop, I can start a fire with matches or a lighter I'm carrying with me that will help a searcher find my location. Ideally, the fire would be in a clearing on a prominent feature, such as a hilltop.

If a searcher crosses my path, he only has to follow the line of survey markers to find my position. This is effectively the same technique Hansel and Gretel used in scattering bread crumbs to find their way through the woods. It is a relatively simple technique but requires some advance preparation.

A more likely scenario occurs when a backpacker, hiker, or hunter carries a cell phone. While cell phone coverage can be sporadic in remote locations, a user can reach distant towers through a line-of-sight communication if he is on a high promontory. Conservation of battery power is key, so the cell phone should always be powered off when not in use. It should only be turned on when it's likely to be in a line-of-sight range of communication. The communication should be done entirely through texting, as this uses far less power than voice transmission. The lost person should leave an "SOS" message giving information about his condition, any emergency needs he may have, such as medicine, water, or warm clothing. The transmission should give an estimate of his of location and a time when he will check back in. Often, authorities can use the information received to find the location of the transmitted signal and will relay back instructions via a text message.

THE HUBBARD-WALLACE-ELSON EXPEDITION

A classic tale mixes elements of exploration and lost-person behavior, along with a generous helping of self-denial: the infamous Hubbard-Wallace-Elson journey into uncharted reaches of

Labrador in 1903. These three explorers engaged in many of the kinds of behavior described by Hill for lost persons.

In 1903 Leonidas Hubbard, Dillon Wallace, and George Elson set out by canoe from a Hudson's Bay Company post in Labrador, determined to find Lake Michikamau. At the time, the only written records of Michikamau were from Hudson's Bay Company official John McLean and Canadian geologist A. P. Low. Low had traversed large sections of Quebec and Labrador on mapping expeditions for the Canadian Geological Survey and documented his findings in an extensive report.[15] In one of the maps accompanying Low's report was a sketch of a lake named Grand Lake and a river called the North West that formed an unexplored route to Michikamau (Figure 15). Low reported that the North West River flowed out of Lake Michikamau and into Grand Lake, but there was nothing to fill in the blank on the map other than some dotted lines. In his report Low went on to describe the annual caribou hunt of local Montagnais Indians from Michikamau during the month of September.

At the time geographic information about Labrador was still sketchy, but the holes in the maps were rapidly being filled in. This was in the era of Teddy Roosevelt, and Peary's trek to the North Pole, when some of the last unexplored regions of the planet were fodder for articles describing adventurous journeys. Hubbard hoped to make his reputation with his expedition to Michikamau via the North West River and then follow the annual caribou hunt. He planned to publish an article describing his exploits in a magazine called *Outing*. He convinced the editor to give him a generous advance to cover expenses for the trip to Labrador and to outfit the expedition.

Hubbard recruited a friend, Dillon Wallace, to join him. While on a camping trip in November 1901, Hubbard gave his recruiting pitch: "'Think of it, Wallace, a great unknown land right near home, as wild and primitive today as it has always been! I want to see it. I want to get into a really wild country

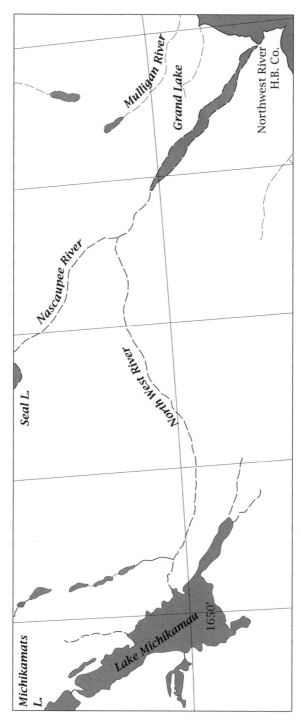

Figure 15 A portion of A. P. Low's original map, focused on the region of Grand Lake and the North West River route to Lake Michikamau. Rivers and lakes not explored but described to Low appear as dotted lines. Adapted from a map published by National Resources Canada © Department of Natural Resources Canada. All rights reserved.

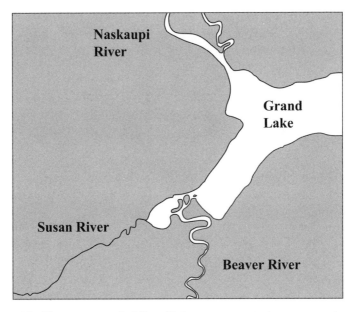

Figure 16 The western end of Grand Lake as represented on an accurate map. Hubbard ascended the Susan River mistakenly, rather than taking the Naskaupi River (Low's North West River) as he had planned. Adapted from map published by National Resources Canada © Department of Natural Resources Canada. All rights reserved.

and have some of the experiences of the old fellows who explored and opened up the country where we are now.'"[16] Wallace agreed to accompany him. Hubbard also hired George Elson, a half-white, half-Cree from the Hudson Bay as a guide. A. P. Low had employed local guides, but Elson had never been to Labrador before. They brought provisions but mainly planned to live off the land, shooting game and catching fish. In *The Lure of the Labrador Wild*, Wallace wrote a memoir of his experiences on the expedition.

Hubbard, Wallace, and Elson left the Hudson's Bay Company post near the entrance to Grand Lake on July 15, 1903, which was quite late in the season for such an ambitious undertaking. As they paddled to the western end of Grand Lake, the first task was to find the North West River. A current accurate map in Figure 16 shows the western end of Grand Lake. Today the river flowing into the end of Grand Lake that Low called the "North West" is known as the Naskaupi. There are a number of rivers and streams entering the western end of the lake, but A. P. Low's sketch only showed one: the North West. Other inflows include rivers called the Beaver and the Susan; the latter might be more accurately characterized as a large brook.

Hubbard was already concerned about his late start and didn't want to waste time exploring the western end of the lake. He missed the entrance entirely to the North West River, which entered into a bay on the northwestern end of the lake. Instead, he headed directly for the far end of the lake. From there he began to ascend the Susan River.

The North West River was reported to be quite broad, but the Susan was shallow and narrow and required a lot of dragging and portaging of their canoe and shuttling of gear back and forth. This was a classic case of "bending the map" — the descriptions of the North West River didn't match what they saw on the Susan, yet they continued to toil upstream. Every so often they would find an old blaze, which reinforced the conviction that they were following the old Indian trail to Michikamau. The journey up the Susan can be likened to the route-traveling behavior described by Hill.

The Susan got narrower and narrower, eventually petering out. The group had to find a way forward and resorted to view enhancing to find their way. Hubbard and Elson left camp to climb to the highest hills around to scout a way forward. Even with years of experience in the wilds of the Hudson Bay, Elson was not immune to getting lost. As twilight fell Elson returned in an agitated state that spoke of woods shock. Wallace noticed this and related the following in his memoir of the trip:

> There was that in his tone that excited my curiosity; he seemed all of a sudden to have acquired an unusual fondness for my society. "What's the matter, George?" I asked.

"I've been about lost," he returned. "Come on and I'll tell you."[17]

This surprised Wallace, as he'd seen Elson navigate through miles of nearly featureless woods. He asked what happened. "'I just got turned 'round,' he replied. 'I didn't have any grub, and I didn't have a pistol, or a fishhook, or any way to get grub, and I didn't have a compass, and I was scared.'"

Wallace was curious about what Elson would have done if he hadn't found his way out. Elson thought for a minute and replied, "'Gone to the highest hill I could see,' he answered with a grin, 'and made the biggest smoke I could make at its top and waited for you fellus to find me.'" This technique has been used many times by lost persons to aid searchers in finding them.

The same evening Elson got turned around, Hubbard himself didn't return to their camp. This caused Elson and Wallace more consternation and illustrates the problem of finding a point in a two-dimensional landscape:

> "Hubbard will have a hard night out there in the bush," said George.
> "Yes," I replied; " I don't suppose we can expect him back now before morning; and when a man is lost in this wild country it's pretty hard to find a little tent all by itself."[18]

Hubbard found his way back to camp the next day. The party portaged over two lakes, then linked up with the Beaver River, which roughly paralleled the Susan. They didn't know that they were on the Beaver but found the remnants of Indian camps with teepee poles and caribou bones, leading them to think that they were on the trail to Michikamau. Taking the Beaver upstream, they came upon some lakes that were its source, with only a range of mountains beyond. Their food supply was running low. They were losing weight at an alarming rate. Their gear was deteriorating. It was late August, and the short Labrador summer was coming to an end. There was no path forward on a water route. One might think that they would have turned around at this point.

What did Hubbard decide? He felt that God would provide and proceeded to make a forty-mile portage over the mountains in the general direction of Lake Michikamau, somewhere off to the northwest. As a general strategy this is classic direction traveling: they knew that they were moving in the direction of the large lake, and if they maintained a more or less constant heading they would come upon its shore.

After forty miles of hauling the canoe and what remained of their gear, they came upon a large lake that was actually attached to Michikamau, although they didn't know it at the time. They paddled to the far shore. Again view enhancing, Hubbard and Elson climbed to the highest hill, and in the distance they could see Michikamau stretching off to the horizon. Before they could explore any more of the region, a gale hit that kept them windbound for two weeks. Their food supply dwindled to eighteen pounds of pea meal, and there was virtually no game or fish to be had. Finally, Hubbard threw in the towel, and they decided to make a dash back to the Hudson's Bay Company post, backtracking as much as possible.

Their physical condition continued to deteriorate, and they suffered from acute hunger pangs as the weather turned to a mixture of snow and rain and temperatures dipped below freezing. They eventually made it back to the Beaver River, where the going was relatively easy, paddling downstream. Hubbard's physical condition was alarming, however, and he could no longer steer the canoe. When they reached the point where they entered the Beaver, both Elson and Wallace tried to persuade Hubbard to just continue downstream on the Beaver, which would be far easier than the fight they knew they were facing on the Susan. Hubbard stubbornly refused and insisted that they backtrack on their old route, despite his alarmingly weakened condition.

Abandoning their canoe, the three men tried to make it to the mouth of the Susan on foot, racing against time and the elements. Before long, Hubbard was incapable of walking any farther. Wallace and Elson left him in a tent with a supply of firewood. Wallace found an old sack of moldy flour they had cached on the way in and tried to return to Hubbard while Elson continued on toward Grand Lake to find help.

Wallace couldn't locate Hubbard's tent, as the snow was laying up, obliterating landmarks. He wandered back along the path of the Susan, hoping for rescuers, hearing the voice of his dead wife encouraging him onward. Eventually, Elson did find help and rescued Wallace, but Hubbard was found dead in his tent. The last words in his diary were, "I think the boys will be able, with the Lord's help, to save me."[19]

Navigational habits can keep us from getting lost, even in dicey situations. These include a keen sense of direction and distance traveled on the segment of a journey. Of course, such habits have to be coupled to sound judgment, which evidently failed Hubbard. The biggest single navigational error, and Wallace himself admits this in *The Lure of the Labrador Wild,* was their failure to explore the northwest bay of Grand Lake. This was compounded by their belief that the narrow and shallow Susan River was the Naskaupi, despite the different description they had received.

In the process of searching for Lake Michikamau, Hubbard, Wallace, and Elson engaged in many of the classic "lost person" behaviors categorized by Hill: view enhancing, route traveling, direction traveling, all while traveling along a route that they mistakenly believed to be the route appearing on Low's map. Had they not persisted for so long and turned back while they still had sufficient provisions, Hubbard probably would have survived. As it was Hubbard gained fame but not in the way he intended.

4. Dead Reckoning

• •

IN CHAPTER 2 I described how we find locations on a mental map using a history of travels. This process, called *dead reckoning*, is probably the most common form of navigation and can be developed into a careful practice. Throughout the world, languages have words and phrases that describe spatial orientation and communicate information about journeys.

If you were on a desert island, how could you measure distances and communicate these to others? Readily available measures are on our body and in the environment. In ancient Egypt short measures were based on finger, hand, and arm lengths. The ancient Egyptian *cubit* is the length of the forearm. Longer units of distance were based on travel. The word "mile" comes from the Latin phrase *mille pacem*, which means "a thousand paces." Roman legionnaires kept track of the distance traveled by counting paces. A *pace* is the distance you cover when the same foot (left or right) hits the ground. Two steps — left, then right — equal one pace.

This practice of counting paces isn't limited to humans. At least one species of ant navigates by counting paces. Harald Wulf studied ant habits by gluing tiny stilts onto their legs or cutting off sections of their legs to change the length of their paces. In Wulf's study the distance the desert ants (*Cataglyphis*) traveled was directly related to how many paces they took.[1] Ants with stilts would walk past a target and ants with shortened legs would come up short in a way that was consistent with pace counting.

Another natural measure is based on travel time. If you were giving directions to a friend driving a car, you might say something like, "You stay on the interstate for two hours and take exit twenty-seven." This assumes that everyone travels at the same speed, but it is usually clear from the context how fast your friend moves. Distance traveled (miles) is speed (miles per hour), times time (hours).

The *hour* was and is one of the most widely used units of time. It has its origins in the ancient Egyptian use of rising stars to reckon the time of night. They used thirty-six bright stars in all, which would rise in turn just before the Sun at different times of the year. The passage of one night was associated with the passage of twelve of these bright stars, giving rise to the night's being divided into twelve hours. The day was likewise divided into twelve units. This scheme, dating from roughly the fifth century BC, became widely adopted, but it was far from universal. Medieval Saxons reckoned the length of day in *tides*, with eight tides in one day.

Three miles per hour was, and is, a good reference for human-powered travel. This is how fast most people walk on level ground, how fast a person can row a boat, and how fast an old sailboat can move on water. It is far from universal. You could be running or limping, but it is a reliable standard shared among travelers and widely understood. If you convert this speed per hour into a distance, it is three miles. In ancient Persian and Arab cultures, a *parasang* or *farsakh* was a unit based on this same combination of speed and time. The length of a *league* is close to the farsakh and is also based on the distance covered by a person in an hour. Longer distances can be reckoned in how many days or lunar months elapse over the course of a journey. Native Americans used these lengths of time as standards. A *marhalah* is a day's journey in medieval Arab reckoning and is equal to roughly eight farsakhs (twenty-four miles), or eight hours of travel.

These are still viable measures. If GPS devices and odometers vanished overnight, we could fall back on human measures instantly. When I tested a group of thirty students, they took 980 paces per mile, which is very close to the 1,000 originating from soldiers in ancient Rome! They also averaged a walking speed of three miles per hour, so the concepts of a mile, farsakh, and league are true to their origins.

Other distance units came from farming. A furlong is currently recognized as 220 yards long. It has its origins in the practice of plowing fields. Oxen can plow for only so long before they get fatigued. On the other hand, turning around a plow and a team is a time-consuming task, so there's an optimal length for plowing that takes these factors into account. The term "furlong" is derived from the Old English words for "furrow" and "long." A related distance, often taken as synonymous with a furlong, is the ancient *stadia*, which is sometimes related as being 125 paces, or one-eighth of a mile long. Ancient Greeks and Egyptians often quoted distances in stadia.

In practice it's important to know how fast you can travel under different conditions and not overestimate your pace. A healthy person can walk as fast as four miles per hour on smooth, level terrain at sea level. Walking speed will vary because of many factors: weather, a backpack, traveling companions, and altitude. On rough terrain, uphill, you might walk at a pace of about two miles per hour. On a steep slope at high altitudes, this may slow to one mile per hour. Mountaineers on high summits where there is as little as one-third as much oxygen available as at sea level will have an excruciatingly slow speed.

There can be a considerable range of walking speeds and pace lengths among individuals. If you watch pedestrians on a busy sidewalk, you will immediately see the differences. This makes standardizing pace lengths difficult. To deal with this, rulers and officials often employed professional pacers to measure distances between cities and towns.

Even with professional pacers, units differed from one kingdom to the next, causing confusion. Egyptian and Greek stadia differed. Likewise, the definitions of leagues and miles differed. Navigators had trouble interpreting information from different sources. Columbus in particular appears to have mixed Italian and Arabic standards for miles in his proposal to sail westward to the Indies. Nevertheless, some standardization began to emerge by the end of the sixteenth century, when the English Parliament defined a *statute mile* to be eight furlongs.

While many of the measures are based on walking speed and length of a pace, there was no clear standard for distances covered by a ship at sea. The speed of sailboats can depend on a number of factors: their length, width, load, and hull shape. One common technique to estimate the speed of a ship is the *log line*. A log or piece of wood with a long rope attached to it was thrown overboard and allowed to drift behind the ship. A sand glass was turned over the moment it hit the water, and the navigator would allow rope to run out as long as sand was running in the glass. Once the sand stopped, a seaman stopped the rope, hauled it back in and measured its length. Knowing time from the sandglass, he could figure his speed.

Another trick used by mariners has been timing a piece of flotsam the ship overtakes on the water's surface. You start counting when the bow (front) passes the flotsam and stop when it crosses the stern (rear). If a sailor knows the length of his boat and how long it takes to pass the flotsam, a sailor can figure his speed. In a variation on this called *pacing*, a sailor drops a piece of garbage at the front of the ship, walks toward the stern (back) at the same speed the garbage is passing by, and uses his walking speed as the boat's. This doesn't take into account current that can move *both* the vessel and the flotsam together in the water but will determine the relative speed of the boat against the water.

Sailors, particularly those in the British Royal Navy, devised a standard for log lines. They had a long rope with knots tied at regular intervals, along with a standard sandglass that could time an interval of thirty seconds. The log was thrown overboard at the moment the sandglass was turned. The rope ran out and a seaman counted the number of knots passing by. The Royal Navy devised the intervals between knots so that the number passing by in the timed interval represented the speed of the vessel in nautical miles per hour, called, appropriately enough, *knots*.

A major departure from human measures appeared with the *nautical mile*, which merges physical distances with changes in latitude and longitude. Positions on Earth get a unique pairing of the two angles, based on the angular distance from the equator (latitude) and the angular distance from an arbitrary origin of longitude called the prime meridian. The angles are traditionally divided into degrees. Each degree is further subdivided into sixty arc minutes (denoted by a prime symbol following it — e.g., 15'), and each arc-minute is further subdivided into sixty arc seconds (denoted by a double prime symbol — e.g., 15").

As Western European navigators began to employ the latitude–longitude system, the use of a distance system based on one arc minute of latitude began to emerge. By the end of the seventeenth century, it was well established. One arc minute of latitude is 1.15 statute miles, a length called a *nautical mile*. This is a natural measure of distance for mariners because it is close to the statute mile and makes it easy to readily convert angles of latitudes and longitudes into nautical miles, which are sometimes called *sea miles*.

A second major departure from human measures of distance was developed by the French Academy of Science in 1793, when the definition of a *meter* was proposed as a standard equaling one ten-millionth of the distance from the equator to the North Pole. This was the start of the modern metric system that bases standards on universally accessible measurements.

Visual Estimation of Range

An important skill in navigation is the ability to estimate distances to visible objects. The direction to an object is called its *bearing* and the distance is its *range*. Multiple sightings of landmarks can give you your position. We instinctively use visual clues all the time when we're walking around a room or outside in a familiar environment,

Figure 17 The angular size of an object in our visual field depends on its distance. By using a foreground reference of a known size (e.g., a finger), we can estimate the distance to a structure of a known size. Image credit: A. Scherlis.

but this instinct can be honed. Our perception of the distance to an object is based partly on the memory of the size of the object and the relative size it takes up in our visual field.

Figure 17 shows a more systematic way of estimating range. The viewer is looking at a line of telephone poles along a road that disappears into the distance. By holding up a finger at an outstretched arm's length, the viewer can compare the visual size of a telephone pole to the image of the finger. The farther away the telephone pole, the smaller it appears relative to the finger.

You can estimate the distance to objects if you know their physical size and can estimate their angular size. This is a matter of trigonometry (steps omitted), which gives a rough conversion that can be used to associate the angular size of a known object to its range. At a distance of one mile, an object one hundred feet tall will span about 1 degree of angle. This gives a conversion formula:

Range (miles) = size (feet) ÷ (100 [feet] × angular size [degrees])

As an example of how to use this, if you know that a lighthouse is two hundred feet tall, and it appears half a degree high, you can find its range (distance from you) this way:

4 (miles) = 200 (feet) ÷ (100 [feet] × 0.5 [degrees])

It's four miles away. The farther away an object is, the smaller the angle. The taller an object, the larger the angle it covers. This rule of thumb is accurate to within about 10 percent and is easy to remember. This estimate works reasonably well and is particularly helpful when you're moving and don't want to stop to use instruments and make calculations.

In order to take advantage of this, you need a way of measuring angles. Like distance measures, angular measures can be based on human features. Different configurations of the hand and fingers at the end of an outstretched arm will create different angular widths.

Figure 18 shows three possible configurations of a hand at the end of an outstretched arm. In this configuration, the width of a finger spans about 1.5 degrees. Three fingers span 5 degrees. The palm and extended thumb span about 10 degrees. A hand with the pinky and the thumb fully extended (not shown) spans about 20 degrees. In the example above, the two-hundred-foot-tall light-

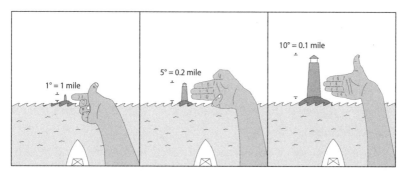

Figures 18a, 18b, and 18c Using fingers and the "One degree is one hundred feet at one mile" rule to estimate the distance to a lighthouse (this one is a hundred feet tall for illustrative purposes). Image credit: A. Scherlis.

house at a range of four miles visually appears to take up one-third of the width of a finger at the end of an outstretched arm.

The size of an object in a distance requires some prior knowledge, but one can make an educated guess with common structures. Sometimes it's easy: People are 5 or 6 feet tall, buildings are roughly 10 feet for each story, lighthouses can be anywhere from 50 to 250 feet tall. Church steeples can be 100 to 200 feet tall. The height of tall structures and mountains are often indicated on maps or are already known.

This technique of estimating angles with hands is by no means new. In the same way that measures of distance were named after human characteristics, angular widths were given names associated with parts of the hand at the end of an outstretched arm:

> *Ey-ass*   Caroline Islands term for hand width — 15
>   degrees[2]
> *Issabah*   Arabic term for finger width — 1.5 degrees[3]
> *Chih*   Chinese term for finger width — 1.5 degrees[4]

The use of fingers as an angular measuring scheme was employed by mariners to measure the height of stars on ocean voyages. This practice persisted well into the nineteenth century, as reported by Anglo-Indian scholar James Pinsep in an interview with an Arabian sailor in 1836: "He stretched out his arms, when I inquired about the issabah division, and placing his fingers together horizontally, counted with them the height of the polar star, just as I guess must have been the early and rude method of the Arab navigators."[5]

Pinsep reported that an issabah is 1°36′ (one degree and thirty-six arc minutes).

This technique works reasonably well up to one hand width, but for much larger angles it begins to break down.

Our ability to distinguish visual features on objects also depends on distance. Features become blurrier at greater distances. We are intuitively aware of the relation between visual distinctiveness and distance, but with more care we can improve on intuition. Below is a list of visual clues for distances given by author Harold Gatty, who wrote a number of treatises on navigation: [6]

- 50 yards — mouth and eyes can be distinguished
- 100 yards — eyes look like dots
- 200 yards — details of clothing can be distinguished
- 300 yards — faces can be seen
- 500 yards — arms and legs can be distinguished
- ½ mile — a person looks like a post
- 1 mile — trunks of large trees can be seen
- 2 miles — chimneys and windows can be distinguished
- 5 miles — large structures can be recognized
- 10 miles — very tall structures (water towers, church steeples) can be recognized

In my experience Gatty's list is overly optimistic and his distance estimates should be reduced by about one-third even for good viewing conditions. Depending on the lighting, the perception of distances can change dramatically. Details on front-lit objects are easier to discern, while they're harder to make out on backlit objects. As light levels drop toward sunset, details rapidly become obscured.

Our visual perception of distances can be strongly influenced by the entire visual field. This can lead to problems in unusual settings. Glaciers and large snowfields on mountains and in arctic regions where there are few visual clues other than ice, snow, and rocks can be particularly difficult. Mountaineers and arctic explorers have reported difficulties estimating distances, with both overestimation and underestimation common. In 1819–1820 William Edward Parry led an expedition to find the Northwest Passage.

His ship froze into the ice pack in the winter, and he noted the distortion of distance perceptions created by the environment:

> Parry noted that if he spotted a stone of more than usual size on one of the short walks he took from the ship, his eyes were hypnotically drawn to it and he found himself pulled in its direction. So deceptive was the unvarying surface of the snow that objects apparently half a mile away could be reached after a minute's stroll. In such a landscape it was easy for a man to lose his bearings.
>
> Pierre Berton, *The Arctic Grail*[7]

There are some unusual circumstances when there appear to be enhancements in the size of an object, although these may be an illusion. For example, it has been claimed that objects appear twice as large in the fog as they do normally.[8] Another visual field problem is the Moon illusion (Figure 19). When the Moon is rising near the horizon, people often perceive it as being quite large. If

Figure 19 The full moon with a foreground object (power lines) near the horizon.

you see this, hold up an index finger on your outstretched arm next to the Moon, and you'll see that it's only about half a degree in diameter. One evening my son came frantically to me, asking "Dad, Dad, what's the perigee [closest approach to Earth] of the Moon's orbit?" At first I scrambled to look up the answer; then I recalled that there was a full Moon out, and it was rising just about the time of sunset. I stopped and asked him, "Why do you want to know this obscure bit of information?" He told me that a friend had phoned him and told him that the Moon looked unusually large and figured it was much closer to the Earth. It just looked unusually large when juxtaposed against distant trees.

Not only are the rules above helpful in estimating distances for the traveler, they are used in warfare. When an infantryman or hunter aims a rifle, he has to compensate for the distance a bullet falls under the influence of gravity from his position to his target's position. The scope of a modern rifle has markings that are calibrated in *milliradians* or *mils*, equal to one one-thousandth of a radian (Figure 20); one radian is approximately 57 degrees. In the era of muskets, the precision of those guns was not as good

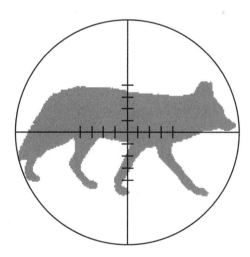

Figure 20 Illustration of a coyote sighted through a rifle scope. The gradations are in thousandths of a radian to aid the estimation of range. One radian is approximately fifty-seven degrees.

as that of rifles. The phrase "Don't shoot 'til you see the whites of their eyes," attributed to the Colonial commanders at the Battle of Bunker Hill, was an order to conserve ammunition until the British came within accurate musket range.

FINDING AND MAINTAINING DIRECTION

As we saw in chapter 3, there is a tendency for a lost person to move in an erratic, loopy path if he can't orient himself. When this happens the person makes little headway, and the average distance from the starting point may not change at all. To make progress on a journey, an individual must be able to establish and maintain some kind of spatial orientation. If a person is intent on direction traveling (i.e., moving in one direction), he can choose some likely but arbitrary direction and try to keep moving in a straight line. Better still are more permanent environmental clues; for example, a reference to the rising Sun or a prevailing wind.

The most straightforward way to combat the tendency to move in an erratic path is to find a close visual landmark, such as a tall tree, and one or more distant landmarks that line up with it, such as a mountain peak, then move toward the first landmark. When you're lining up landmarks, use the same eye for consistency. As you approach the first landmark, pick out another distant one that falls in line with the previous one and continue along in this way. Even if the footing is bad, as long as you've identified a landmark, you can turn your attention away temporarily and regain your bearings when you look up again. Often, caravans in the Sahara Desert travel in single file, so the last person in line can see if the leader of the caravan is deviating to one side or another, and the last person can shout changes in direction to keep the path straight.

As with distances, languages contain clues about how people established orientations in the past. Since most travel is along the surface of the Earth, we're not talking about "up" or "down" but rather direction on a horizontal plane. Most cultures partitioned directions into at least four *cardinal* (east, west, north, and south) points. This is perhaps related to the symmetries of the human body of front–back and left–right. In many cultures there is often a primary axis of orientation determined by something in the environment, such as the direction of prevailing winds, then a second axis perpendicular to that. The existence of names of cardinal points implies that cultures had the means and need to orient themselves and establish directions.

Some social scientists speak of languages that can only describe *absolute* spatial orientation, such as "north" and "east," as opposed to *relative* orientation, such as "front" and "back."[9] The names of the cardinal points in most languages are invariably derived from the names of environmental features. For example, the word for "east" is often associated with the direction of the rising Sun.[10] In Latin *oriens* means "east" and is derived from the direction of the rising Sun; it is also the origin of the word "orientation." *Occidens* means "west" in Latin and is, of course, the direction of the setting Sun. The primacy of the word for "east" for the direction of the rising Sun and the word for "west" for the direction of the setting Sun dominates many of the languages in Europe and Asia, including ancient Sanskrit. Latin, Greek, Sanskrit, German, and many other languages in Europe and south Asia are descended from a single language that linguists call Proto-Indo-European. But even in non-Indo-European languages, such as Hebrew, Arabic, Chinese, and Hawaiian, east and west are also associated with the rising and setting Sun.

In Chinese the symbol for "east," *dong*, is descended from a combination of two symbols: The Sun rising in the east is represented as the Sun behind a tree. Thus a tree, *mu* 木, is placed with the Sun, *ri* 日, to form east, *dong* 東. The modern form is a stylistic evolution from the older form, but its original shape is largely retained: 东.

The moments of sunrise and sunset are the easiest part of the day to get one's bearings because these represent approximate east and west. East is the preeminent direction for many, with good reason. People generally travel during the day and sleep at night. When travelers wake in the morning, the Sun is rising in the east, serving as a natural reference for the next day's travels.

Terms for "north" and "south" show more variation among cultures. If you face the rising Sun, the north is on your left and the south is on your right. In many Indo-European languages, the name "north" or *nord* is thought to derive from the Proto-Indo-European word *ner* for "left" or "below." The word "south" or *sud* (French) or *Süden* (German) is related to the Sun, which reaches its highest point in the sky at midday in the south in the Northern Hemisphere. In Hawai'i the western sky is preeminent; this is where the Sun is said to enter a new realm so it can return for the new day. In Hawaiian the word for "south" is related to left and "north" to right.

North and south can be derived from other meanings, sometimes rooted in visual references. In Latin "north" is *septentriones* or "the seven oxen," from the stars of the Big Dipper, which looks like a plow. It eventually became synonymous with "north." Similarly, in Greek one word for "north," *arctus*, means "bear," as the Big Dipper was identified as a bear. "South" in Latin is *meridiones* or midday — when the Sun is halfway across the sky in the south.

In other languages direction names come from local geographic features. In ancient Egyptian "south" and "north" are related to "upstream" and "downstream" on the Nile. For the Puget Salish tribe in what is now Washington State, the word for "east" literally means "from up river."[11] In other languages it can be related to winds (Greek, Native American languages). It can sometimes come from weather, in which the word for "north" is derived from the word for "cold."

As mentioned in chapter 2, winds can be used for orientation.

Wind was an important reference of the Netselik Inuit. You might worry if the wind shifts direction, but in many cases the wind direction was associated with important qualities of the wind itself. For example, another Greek word for "north," *boreas*, comes from the god of the north wind, which was distinctly cold and dry, while the western wind, *zephyrus*, was warm and damp, announcing spring.

The names for directions between the cardinal points, such as northwest and southeast, differ from culture to culture and often hint at navigational practices. To reach Iceland, Greenland, the Faroe or Shetland Islands, the Norse would sail along the coast of Norway until they reached a known landmark that served as a jumping-off point. In Old Norse the points between north, south, east, and west were identified in terms of their relationship to the coast of Norway. For example, northeast was "north-in-direction-of-land," *landnordr* (land-north), and southeast was "south-in-direction-of-land," *landsudr* (land-south). Northwest and southwest became "away-north," *utnordr*, and "away-south," *utsudr*, respectively (Figure 21).

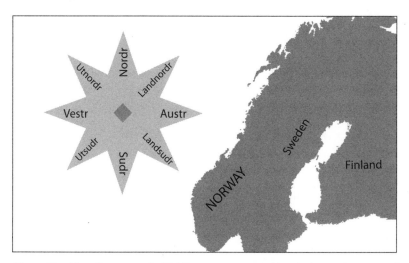

Figure 21 Direction names in Old Norse were named in relation to the coast of Norway, which served as multiple points of departure for voyages to Iceland and Greenland. Image credit: A. Scherlis.

It's common to find an eight-fold division of directions. The Tuscarora tribe from coastal North Carolina had eight compass points and used the characteristics of wind to fill in the points between the cardinals. British surveyor John Lawson documented the habits of the Tuscarora in his account *A New Voyage to Carolina*: "They have names for eight of the thirty two points and call the winds by their several names, as we do; but indeed more properly, for the North-West Wind is the cold wind; the North-East the wet wind; the South the warm Wind; and so agreeably of the rest."[12]

Further subdivisions of the angles around the horizon are common. Navigators in the Pacific Islands use the positions of rising and setting stars, and also winds for orientation on the open ocean. Here, one can see sixteen and even thirty-two-fold divisions of the directions toward the horizon. In principle the process of subdividing angles could continue indefinitely, but a thirty-two-fold division is a width of 11.25 degrees, which is about as good as a person can do in eyeballing angles without instruments. The 11.25 degrees also corresponds to the approximate width of a fist at the end of an outstretched arm, which may explain a widespread use of a thirty-two-fold division.[13]

Before the use of the magnetic compass for orientation, the Sun, stars, and winds were often the primary clues to finding directions. In some cases ocean swells were regular enough to establish directions, such as in the Marshall Islands, where "wave piloting" was common. On land there are many more clues to direction that can be employed on overcast days, not the least of which are known human paths.

The most straightforward means of navigation is, of course, asking directions, something of a dying art in the modern era. When we give directions, it would ideally be a series of orientations, landmarks, and distances, perhaps phrased as travel time. We can parse the sentence, "You travel west for three miles until you reach a cliff, then travel south for two miles to reach your destination." If we dig back to the origin of the words, this roughly translates to "You travel in the direction of the setting Sun for three thousand paces until you reach a cliff, then travel in the direction of the midday Sun for two thousand paces to reach your destination." This list becomes the basis for navigation by dead reckoning.

Practices in Dead Reckoning

Most of the time people practice some form of dead reckoning without consciously realizing it, but it can be developed as a methodical practice. Throughout history, this was the most common technique on land and sea. Dead reckoning is simple and easy to describe.

If you walk through a dense, flat forest on an overcast day or sail through the fog, you try to maintain motion in a constant direction and keep track of distance covered. Even with this practice there is an inherent uncertainty of position that grows as every hour of travel goes by. You only know your speed and direction with limited precision. When a landmark is sighted and you are able to fix your location with more precision, the dead reckoning clock gets reset, and the traveler starts the process again.

Three critical pieces of information are needed to carry out dead reckoning: (1) a starting location, (2) a knowledge of the direction you are moving in at all times, and (3) a knowledge of how far you've traveled.

Some common terms in dead reckoning are:

> *Point of departure*: start of a journey from a known location
> *Heading:* the direction of travel
> *Bearing:* the direction from the observer to an object
> *Fix*: an unambiguous determination of location

There is a Persian proverb that says, "Fortune is infatuated with the efficient." This is an astute observation for many human activities, but it is particularly true when one is engaged in dead reckoning through an unfamiliar territory. Below, I describe some common practices.

An important skill is the ability to constantly update your location on a map, whether mental or physical. This involves comparing all available clues to expectations. To this end the initial fix for a point of departure must be absolutely certain. Having found the point of departure or a waypoint, you then determine the direction of travel — a heading. When you do this try to ascertain multiple ways of maintaining orientation while traveling, at least a primary way and a backup. As an example, these could be the use of a magnetic compass and observing wind direction. Compasses can be lost, the wind direction can change, but each can be used to cross-check the other.

When you are moving, find a landmark that lies in the intended direction of motion. Walking with your head down, looking at the compass or your feet, is far less accurate than taking a sighting to a landmark in the intended direction of travel or, better still, finding a set of landmarks that line up on your heading and moving toward them. When you reach the first landmark, find another landmark that coincides with your heading and move in that direction. On water, if you lack a magnetic compass, wind direction or stars form a natural compass.

As you move keep track of the number of paces, or number of strokes of the canoe or kayak paddle, and keep track of time if you have the means. Try to make an estimate of your speed so you can regularly update your mental or physical map for your position.

1. If you are traveling on or in sight of land, keep track of landmarks as you pass them, and note the bearings in the direction of these landmarks. Oftentimes, if a person is lost, he can backtrack to the point of the most recent fix and figure out how he went wrong.

2. When changing direction, you need to take a new fix to determine your new position. Make a note of how far you traveled on the previous bearing, and proceed as before. Look backward in the direction from which you traveled, so that on your return journey you can be accustomed to what you'll encounter.

3. On the open ocean periodic fixes may require the observation of stars or the Sun. Often, these aren't observable because of bad weather, and you must rely on dead reckoning in between fixes until a clear observation is available.

4. Don't hold conversations with companions while moving through difficult terrain unless they directly pertain to the task of navigation. Conversations can become a major distraction and make it easy to lose track of landmarks. Hold conversations during rest breaks.

5. If you have a physical map, mark your progress on the map. If you don't have a map, use a piece of paper to indicate landmarks and progress, including turns (Figure 22).

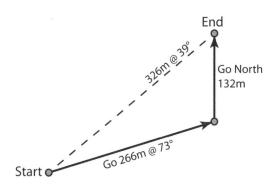

Figure 22 On a map or paper, keep track of different legs of your travel. The geometric addition of multiple legs of a journey will give you a "distance made good." Image credit: A. Scherlis.

Lewis and Clark used dead reckoning in their early-nineteenth-century expedition across the Northwest of the United States. In their journals they noted each leg when they altered direction to keep track, then put together the segments to arrive at a new position at the end of the day. Periodically, when the weather permitted, they would then take celestial observations to obtain a fix that could be compared to their dead reckoning. Shown below is a snippet from the journal of Lewis and Clark with one day's dead reckoning. The first column gives their direction of travel, the second gives the distance on that leg of the day's journey, and the third shows comments used to compare landmarks to positions deduced from dead reckoning.

Bearing	Miles	Comments
South	1½	To a point on the Stard [starboard or right] Side
S 28° W	2¼	To a rock resembling a tour [tower] in the Stard [Starboard] bend
S 10° E	1¾	To the Stard. point passing a rapid
S 60° W	¾	To a tree in the Std [Starboard] bend rocks & rapds [rapids] all the dist[ance]
South	¾	To some bushes on a Lard [larboard or left side] point passed a large Creek at 1/2 a mile on the Lard. Side which we called Shield's Creek [named after John Shields, whom Lewis and Clark sent to explore this river]

Table 1: Portion of Lewis and Clark's journal.

On land, particularly in urban or coastal environments, we frequently have to make detours to avoid difficult areas. These can include swamps, lakes, cliffs, dense brush, large buildings, or an island. If you can see to the other side of an obstacle, such as across a lake (Figure 23), you can find a clear landmark on the other side and make your way around the obstacle. Using a box pattern, where there is a set of two displacements perpendicular to travel and one parallel to travel, you can keep track of the distance made good while avoiding the obstacle. In the case of a cliff or hill where a sighting across the obstacle is obscured, a boxlike path still can help a person retain an orientation.

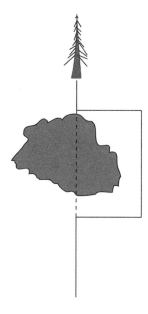

Figure 23 When detouring around an obstacle, such as a lake, sight a landmark on the far side, and move in a box pattern to establish a reliable distance made good.

The accuracy of dead reckoning depends on experience and the ability to make careful observations. Over time, people get better at it, but this is mainly through trial and error. In most cases for dead reckoning, the precision of a heading can be within 5 or 10 degrees at best, and this is with a compass. If wind direction is being used as a direction estimator, it might be precise to only 20 degrees. Estimates of speed are rarely accurate to better than within 10 percent, because changes in the environment will affect travel. If you travel on foot, variations in terrain and fatigue can affect walking speed. A sailor won't notice changes in the vessel's speed because of subtle changes in wind speed and direction, even while he is periodically using a log line to estimate speed.

This concept of a fixed percentage of uncertainty in direction and travel speed is not universal, but like the simple random-walk model in the previous chapter on lost people, it serves as a kind of baseline description for uncertainties inherent in dead reckoning. It tends to work reasonably well and is a conservative assumption.

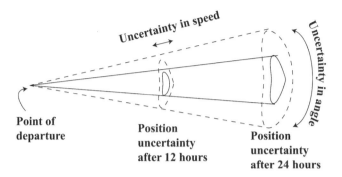

Figure 24 How fixed-percentage uncertainties in speed and heading contribute to a range of possible positions when relying solely on dead reckoning to obtain a position.

Assuming that the uncertainties in direction and travel speed are a fixed percentage, the uncertainties in heading and speed only accumulate with time, resulting in an ever-growing uncertainty in position as the journey progresses. If you travel at 3 miles an hour with a 10 percent uncertainty, after twelve hours, your position is 36 miles from your starting point with an uncertainty of 3.6 miles. After twenty-four hours that uncertainty grows to 7.2 miles. The same accumulation happens for uncertainties in direction as well (Figures 24 and 25).

Navigators can gain on the inherent imprecision that grows over time from dead reckoning. Anthropologist David Lewis undertook long-distance voyages similar to those attributed to Pacific Islanders. The settlement of Hawai'i and the northern island of New Zealand are believed to be the result of long-distance voyaging over distances of two thousand miles. Lewis set out to recreate these voyages using indigenous navigation techniques, including stars, winds, and ocean swells for orientation. After voyages of two thousand miles or more, he made landfall to within about thirty miles, a precision of within a percent, much better than the 10 percent number I used above.[14] Lewis speculated that he was subjected to many random factors that could have taken him off

Range of possible positions after 12 hours

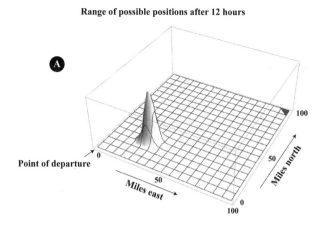

Range of possible positions after 24 hours

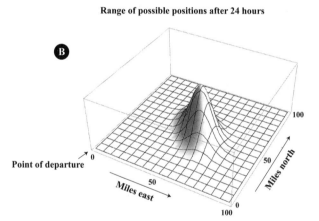

Figures 25a and 25b   A sailor sets off in a northeasterly direction at three miles per hour with a 10 percent uncertainty in speed and a 10-degree uncertainty in heading. The two figures show the probabilities of the sailor's location after 12 and 24 hours.

course, like winds and current. He concluded that, over the course of a very long voyage, the factors tended to cancel each other out, leading to a much better precision than my nominal 10 percent.[15]

The Polynesian Voyaging Society, founded in 1973, recreated a number of long-distance journeys using indigenous techniques. In 1976 and 1980 they made round-trip journeys between Hawai'i and Tahiti, twenty-four hundred miles in total. In all four voyages the knowledge of their position was correct to within sixty to one hundred miles throughout the voyage, which, again, is substantially better than the nominal 10 percent margin I mentioned above. The navigators on these voyages used the position of stars to find latitude and were sailing on an approximate north–south line.[16] Longitude was also relatively precise and didn't benefit from position fixing of any kind.

A likely scenario on a long voyage is an alternation of periods where positions are determined by dead reckoning, followed by a fix of some kind that reduces the position uncertainty substantially. After the fix the journey proceeds, with position again being derived from dead reckoning until the next fix. By alternating these two, the navigator can limit the growth of position uncertainty. In this case the uncertainty does not grow infinitely with time.

The precision of dead reckoning depends, in part, on skill and experience. While experienced navigators may do as well as 10 percent, I find that people who are freshly taught dead reckoning rarely achieve that level of precision. I taught a group of a hundred students how to estimate distances by counting paces and measuring angles using their hands. When I gave them the task of walking a mile from one spot to another landmark and estimating their relative position, their precision was within 25 percent, not 10.

Uncertainties in dead reckoning dictate certain habits of travel. If you're trying to reach a tiny speck in a trackless wilderness using dead reckoning, the odds of successfully reaching it rapidly go to zero the farther it is located from the starting point. However, if you know that your goal lies along a river, a road, a coastline, or another linear feature, it makes the problem much easier. Find a heading that will intersect the road, a river, or a coastline, but one that is deliberately offset from the goal. This allows you to take into account any uncertainty in dead reckoning. If you make a deliberate offset in your heading, you know that when you hit the road or coast or river you only need to travel in a given direction

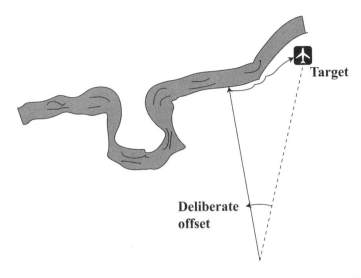

Figure 26 Using a deliberate offset to reach a target. Rather than trying to find the airfield on the river directly, steer deliberately to the west of the target and follow the river to the east.

along the river, road, or coastline until you find your destination (Figure 26). This effectively reduces the problem of finding a point in a two-dimensional plane to first finding a line in a two-dimensional plane, then finding a point along that line.

I've had to use this technique more than once. When kayaking off the coast of Maine, I was enveloped in a thick fog. I wanted to reach the easternmost point of an island, then turn south to paddle to my destination. If I'd paddled directly to the eastern point, there was a good chance I would have missed it and paddled right out into the open ocean. I made a deliberate offset to guarantee that I would reach the north coast of the island. Once I found that, I followed the coast to the east until it turned south and used that as my point of departure for the final leg of my journey.

A variation on the use of a deliberate offset is finding a way to enlarge the effective size of the target. Pacific Islands navigators would sail in the direction of an island and try to observe birds that flew thirty miles out from land to feed during the day, then follow them back as they returned to their nests when evening approached.

Navigation in snowy terrain is a particular problem because everything is white. Storms can limit visibility to a few feet, and the appearance of terrain shifts constantly. When Roald Amundsen led his Norwegian team to the South Pole in 1911–1912, he cached food and fuel for the return trip in a series of depots spaced seventy miles apart. Route finding near the poles has its own peculiar challenges. Amundsen laid a road of markers along his route, consisting of flags, hummocks and even trash at intervals of three to eight miles. In each he wrote the bearings and distances to the next signpost. To mark the depots themselves, he laid a picket of black flags at half-mile intervals perpendicular to the direction of travel, up to a distance of five miles, bracketing the depots (Figure 27). Each marker had a piece of paper with the direction and distance to the depot itself. In building the picket Amundsen effectively expanded the size of the target depot to take into account the imprecision of dead reckoning. His choice of five miles is consistent with the sort of uncertainty one expects after dead reckoning seventy miles.[17]

It is often possible to get a position fix when two landmarks line up with each other. This creates something called a *line of position*, which means that you have a position that lies somewhere along

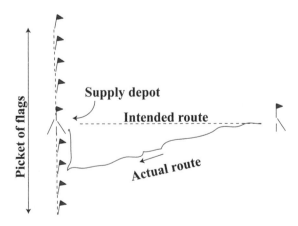

Figure 27 Amundsen's picket of flags bracketing his supply depots.

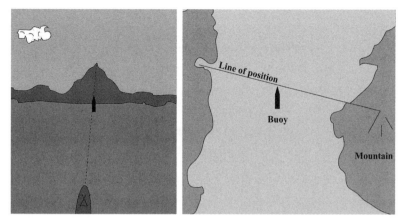

Figures 28a and 28b A line of position is formed when two distant objects line up. This can be extrapolated and a fix can be obtained from the intersection of this with another line of position or a one-dimensional feature, such as a coastline.

a line connecting those two features. In Figure 28 the kayaker is located somewhere along a coast and sights the peak of a distant mountain lining up with a buoy. On a map the kayaker knows that he is located somewhere on a line connecting those two features. If he knows that he's close to the coast, the intersection of this line of position with the coast gives him a position fix. Variations on this theme involve using a line of position from two objects, and some visual impression of range, or the intersection of two lines of position, which also can create a position fix.

The great benefit of natural bearings is that it can be done quickly, and without the need for consulting a magnetic compass. This technique can be used in any environment. In an urban setting, for example, you can use high-rise buildings and other landmarks as a way of establishing a line of position.

All of these techniques are matters of habit. Reading about them can be a curiosity, but they need to be practiced. For example, there are exercises that can be done with an hour or two of free time. Try just counting steps sometime when you have to walk somewhere. Notice the wind direction and the position of the Sun in the sky. Find the height of a building or other tall landmark near where you live, then estimate your distance to the building from a location by stretching out your hand and measuring its angular height. Walk to the building, and count your paces. See if your distance estimate from the angular height and the pacing agree with each other. Many of the suggestions in this chapter are based on long trial and error over generations and didn't arise from pure thought: they were the result of practice.

5. Urban Myths of Navigation

• •

SPATIAL ORIENTATION IS critical to navigation. From the origins of words for the cardinal points (east, west, north, and south) in different languages, you can see how major environmental features were historically used for orientation. "How-to" books and outdoors magazines often have chapters or articles with descriptions of clever ways to establish orientation using nontraditional means. These often seem folkloric or like urban mythology. There are far too many of these to catalog here, and frankly, most work only under limited circumstances. However, I choose to trace the origins and histories of three curious methods of orientation that are sometimes found in the "how-to" literature: moss on trees, church orientation, and satellite dishes. It is not my intent to convince the reader to regularly use these techniques for orientation. Rather, I wish to examine their utility as navigational aids and convince the reader that this kind of aid is only helpful with familiarity and specialized knowledge.

A familiar adage says that moss grows only on the north side of trees. There is a foundation for this concept, but you must be careful. In the midlatitudes in the Northern Hemisphere, the path of the Sun makes an arc that swings across the southern sky. This illuminates the southern side of trees, drying out the bark, while the north side stays in the shade. Moss tends to grow in dark damp areas, so the northern sides of trees in the Northern Hemisphere, on average, are conducive to better moss growth than the southern side. By the same logic you should expect moss to grow on the south side of trees in the temperate parts of the Southern Hemisphere.

If you look at a tree trunk with green things growing on, it may be any of three distinct organisms: lichen, algae, or moss. People often mistake lichen and algae for moss. Lichen is a symbiotic organism of fungus and algae, coming in many kinds. The most common has a light blue-green color and produces flat, flaky disks. Other lichens that grow on trees can have a more filamentary structure. Some can be bright yellow, orange, or red in hue. The north- or south-facing character of lichens is not as pronounced as for moss or algae, and they can often be found growing all around a tree trunk. *Pleurococcus* is a very common genus of algae and is found on rocks, tree trunks, and soil. It is light green and, like moss, tends to prefer dark, moist habitats. In this respect *Pleurococcus* can be used like moss to determine a north–south growth pattern.

While *Pleurococcus* is a single-celled organism, moss is a plant that has distinct structural components. Unlike many plants, however, moss lacks a vascular system to transport water and has to rely on the environment to supply water, which is why it grows preferentially in a moist, shaded habitat. Mosses are usually darker in color than *Pleuroccus*, and a close inspection of moss reveals the structural plantlike components.

People should be suspicious about the moss-on-the-north-side folklore because many factors create a dark, damp environment. If trees are on the north side of a hill, their bases are perpetually shaded and the base of their trunks can be fully ringed with moss. In the middle of a dense forest where the floor is covered in shade, there is little distinction between the north and south side of a tree for favorable growth conditions. Prevailing wind directions and windbreaks also play a significant, if not dominant, role.

The growth of moss depends on the tree itself: the kind of bark, age, species, and health. The bark of oak trees is rougher than most, and more moss grows on them more than on other species. Birches and beeches have smoother bark, and you're less likely to find moss at their bases. Sick trees will rot and have a damper surface than healthy ones. If a tree is old, you can often find a small mound of dirt at the base that is heaped up more on the north side than the south side (in the Northern Hemisphere). As moss grows at the base, it eventually dies, leaving dirt, and more moss grows on top of that mound.

If you're able to look at the average characteristics of a number of trees with moss on them, or if you can look around at the location of a particular tree and are adept at assessing the local conditions favorable to growth, it can provide a crude directional indicator. It takes some experience to pick out the right trees and average over the local conditions.

In my experience the location of moss on a tree is a potentially useful direction indicator, but it takes time to learn and is subtle. If a neophyte is dropped into the woods, I seriously doubt he or she could exploit this trick. But my aim here is not to convince you that "moss grows on the north side of trees" is a helpful adage but rather to trace some of the history of this piece of folklore.

The first descriptions of moss as a direction indicator appear to come from reports in the late seventeenth and early eighteenth centuries on the area that is now North Carolina. These were based on contact with Native Americans in that region, who used moss as a directional indicator when the sky was overcast. John Lawson, the British surveyor-general of North Carolina, was commissioned to report on the conditions of the Carolinas. In *A New Voyage to Carolina*, he wrote a detailed account of the tribes in that area. Most of his contact was with the Tuscarora tribe, who at that time inhabited the coastal regions of what is now North Carolina. Lawson relates the Tuscaroras' use of moss as a direction-finding guide: "They are expert travelers, and though they have not the use of our artificial compass, yet they understand the north-point exactly, let them be in never so great a wilderness. One guide is a short moss that grows upon some trees, exactly on the north-side thereof."[1] Other eighteenth-century descriptions of North Carolina include similar references to the use of moss by Native Americans as a direction indicator.

Toward the end of the eighteenth century and at the start of the nineteenth century, the description of moss as a directional indicator found its way into treatises on plants:

> It may not be irrelative to observe in this place, that it
> is to the presence of fungus or moss on the north side
> of trees, that travelers and Indians direct their course
> through unfrequented forests.[2]
>
> > Robert Walsh, "Plant Physiology," a paper
> > appearing in *The American Quarterly,* 1837

Similar advice on moss also appears in encyclopedias, books of advice to young readers, and dictionaries in that era.[3]

Before the Civil War, antislavery tracts appeared that described the Underground Railroad. Missionaries from the Anglican Church wrote many of these pieces. Some painted comparisons of the plantation (field) slaves with local Native Americans, indicating the transfer of these skills: "Almost all plantation slaves know the north star; like the Indians, are peculiarly observant of Nature; the various changes of the weather, their habits and wants compel them to notice objects that may be of use to them, which pass unobserved by others. Thus the Indian can travel the pathless forests without compass, and when the north star cannot be seen, he can find his way by observing the moss on the trees which is always on the north side. The slave whose heart, like the needle of the compass always points to liberty at the north, learns the pole star, and takes it for his guide."[4]

Other writers speak of this skill without reference to Native Americans:

> The Abolitionists in the Slave States, (for there are
> many, though they cannot advocate their principles),
> very kindly give the slaves information as to the direc-
> tion of Canada, and the Free States through which they
> must pass to reach it, also the names of the most impor-
> tant rivers, all which information the slaves remember.
> With this instruction alone, the slave starts for Canada,
> the North Star is his guide. By it he knows his course.

> When the clouds intervene, and thus obscure the flick-
> ering light of this "beautiful star," Nature has a substi-
> tute. A smooth soft substance called moss, which grows
> on the bark of the trees is thicker on the north side of
> the tree and thus serves as a guide northward, till the
> heavenly guide again appears.[5]

Some writers noted that this skill was particular to the plantation (field) slaves and not shared by domestic (house) slaves, who
would not be exposed to the same lore or have a chance to observe
natural phenomena as much. After the Civil War much of the
discussion on moss as a direction-finding tool is absent, but many
references are made to moss as an annoyance on the north side of
fruit trees and roofs.

The concept of camping and wilderness travel as a vacation
or hobby slowly emerged in the latter half of the nineteenth
century. References to direction finding with moss began to
appear at this time. One early tract on wilderness travel, *The Art
of Travel* by Francis Galton, was directed at a European audience and dates from 1855. Here is the earliest mention of the
use of moss in Europe: "Bearings by moss — The moss that
grows strongest on the north side of firs and other trees, in the
latitude of Europe, gives, as is well known, a clue by which a
course may be directed through a forest. For, looking on the
surrounding mosses of trees, much more moss will be observed
in some one direction than in any other, and that moss, lying as
it does on the north side of the trees, is of course due south with
reference to the observer. And as he walks on, and fresh trees
come constantly in sight, he is able to correct any slight error of
direction into which the peculiarities of the particular trees may
at first have led him."[6] Galton indirectly hints that moss may not
be infallible but that the use of multiple trees can be reliable.
This is the first time that cautions on moss as a direction indicator begin to appear in the literature.

During the presidency of Theodore Roosevelt, interest in outdoor vacationing blossomed. Horace Kephart was a prolific author on the subject of outdoor skills; many of his articles appeared in the popular *Outing* magazine. In one of these articles, published in 1903, he addresses the use of moss in an article, "How to Tell Direction in Forest and on Prairie":

> It is a common saying of woodsmen that "moss grows thickest on the north side of a tree." This rule is fairly reliable in flat, timbered country, in northern latitudes, but not in mountainous regions, nor in swamps, nor in the damp forests of the South.
>
> Moss grows best where there is continuous moisture. It is intolerant of sunlight. Where the land is fairly level, but not swampy nor subject to overflow, and the winds do not differ materially in dampness, the moss grows thickest on the north (shady) side of the tree, and the south side is graced with the largest and longest limbs.[7]

Here, more cautions appear on the use of moss, but it's largely a positive endorsement of the technique. *Outing* and a competing magazine, *Forest and Stream*, continued to be quite popular until the outbreak of World War I, when the nation's attention was diverted. During the war, however, field manuals for officers and privates alike had advice on direction finding from moss.

At the end of World War I, a U.S. Army officer reported a statistical study of moss on 107 trees and concluded that moss was not at all favored on the north side of trees. This was reported in 1919 in an article, "Does Moss Point North?" *Outing* editors Caspar Whitney and Albert Britt countered by saying that the study didn't allow for obvious conditions that would distort the appearance of moss, nor the knowledge of woodcraft that an experienced outdoorsman might possess:

These are the things that a woodsman would entirely ignore in seeking a sign of direction from the moss on trees.

Instead, he would single out for examination the straight-shafted old trees of rather smooth bark, knowing that in such cases there would be fairly even lodgment of moisture all around and that the wet would evaporate least from the north and northeast sides of the tree. Hence the moss would preponderate there.

The woodsman would expect to find such difference more pronounced on the edge of thick forests than in their densely shaded interior. He would give special heed to the evidence of trees that were isolated enough to get direct sunlight throughout the day. Those in the shade of cliffs or steep mountains so that they could catch the sunbeams only in the morning or afternoon would be ruled out of court.

Such a woodsman does not take into account all kinds and conditions of trees. Among a hundred trees he might examine only ten but those ten would be more trustworthy for his purpose than their ninety neighbors. That is woodcraft.[8]

Fervor for outdoor activity waned in the 1920s and '30s, but many Boy Scout manuals in that era continued to advise on the use of moss, with some cautions. The general commentary on this technique becomes increasingly skeptical throughout the twentieth century. By the 1960s it is almost entirely dismissed. In a 1968 Scout publication, *Boys' Life*, this advice is given: "It's said that moss grows on the north side of tree trunks. True, but it often grows on every other side, too, and you can't depend on it."[9]

Finally, the last blow came in *Garden* magazine in 1979: "For real life hikers without benefit of compass there is the old rule of thumb: The moss grows on the north side of the tree.

Unfortunately for hikers, the rule is false, and in two ways at that."[10] Not only do these authors list cautions similar to those in the *Outing* articles, but they also claim that hikers will confuse algae and lichens for moss.

Ironically, while the moss folklore was slowly being demolished in the twentieth century, during the same time period another moss legend was emerging: Harriet Tubman. She fled slavery from the Brodas estate in Maryland 1849, finding refuge in Philadelphia. Tubman later became a famous conductor on the Underground Railroad, helping fugitive slaves find freedom. In 1943 Earl Conrad published a biography of Tubman, where we can find the first reference to her use of moss: "In addition to the fundamental requisites of travel on the Road, such as arming with a pistol or other firearm, and the possession of funds and a knowledge of routes to the North, there were countless tricks of travel, and Harriet was acquainted with most of them. The most elementary knowledge, of course, was to be able to recognize the North Star, for all travel was done at night, and this star was the main reliance. Most conductors knew the trick of looking for moss on the north side of trees, and by this being able to determine a northerly direction."[11]

Conrad cast the knowledge of using moss as a generic skill employed by conductors on the Railroad, which mirrors the commentary of British missionaries before the Civil War. By the 1970s, however, the skill was directly attributed to Tubman. In *Nine Women: Portraits from the American Radical Tradition*, Judith Nies writes of Tubman: "She had only two means of determining her way: following the North Star when the weather was clear; feeling the moss, which grew on the north side of tree trunks when it was cloudy."[12]

The most interesting aspect of this quotation is the emergence of the use of touch to find north using the moss on trees, which, to my knowledge, only made its appearance in literature on Tubman. Fictionalized accounts of Tubman's life appeared around this time that amplified her use of moss until it appeared to become commonly accepted wisdom.

Now, I don't mean to cast doubt on the possibility that Harriet Tubman used or even felt moss on the bottom of trees to find north. It could well be the case, based on the antebellum reports of the British missionaries. We simply don't have concrete evidence that she practiced this technique. The principal observation is that the legend of Tubman's use of moss grew in one genre of literature (women's studies), while at the same time its utility was being demolished in another genre (outdoor).

Church Orientation

At first glance the concept of urban navigation might seem strange. After all, you can ask for directions if you're lost in a town. But people often get lost in unfamiliar populated areas. There are usually local clues that can help a lost person regain orientation in a town or city. In many regions of the world, the orientation of religious structures can be used as a navigational aid. Not only can the orientations act as approximate guides, but they also illustrate the human need to create connections to the world through structures. Many mosques, temples, and churches are constructed with specific alignments. Here I focus on a northern European tradition, the roots of which are not fully understood.

The earliest orientations of structures in northern Europe are associated with Neolithic culture, circa 2000 to 4000 BC. Many gravesites and stone arrangements show astronomical alignments that were probably part of rituals. Chambered tombs from that era have openings that point toward the rising and setting Sun at the winter solstice.[13] The orientations are not precise enough to call them "ancient observatories," but they nonetheless indicate an awareness of the sky. Stonehenge is the most famous. The modern interpretation of the alignments of Stonehenge has shifted a number of times over the course of decades, drawing in part on new archaeological finds. The current wisdom is that a lone "heel" stone in Stonehenge to the northeast of its center was part of a more elaborate structure that marked the rising of the Sun and the extremes of the Moon's orbit at the summer solstice.

Traditions don't have to be associated with a "meaning" in a commonsense fashion. They may simply be carried out because someone before us did the same thing, and he or she was copying someone else without understanding why. There is a sense that the imitation of ritual will revitalize an old but poorly understood magic. One example of this is the Saxon practice of copying the burial practices of people from the Neolithic era.[14] At the end of the Saxon epic *Beowulf*, a destitute man finds and steals from a burial mound full of treasures from a long-dead civilization. A dragon guarding the tomb is aroused by the theft and begins torching the local villages. When Beowulf enters the tomb to slay the dragon, he is mortally wounded; his people cremate his body and create a lavish burial mound full of treasure, in imitation of the practice of the lost civilization.

As in the legend, the earliest Saxons in England buried their dead in mounds, imitating the Neolithic practices. We may never know the motives of the Neolithic peoples who built the stone rings and burial mounds; the chain of history from that period has been broken, and we are left with only archaeological remains and speculation. Yet these remains have an undeniable power over the human imagination.

In northern Europe medieval churches and their associated burial grounds show a distinct east–west orientation. The origins are obscure, but they appear to be rooted in a tradition of prayer while facing east practiced by early Christians. One of the oldest declarations of this practice comes from the Christian author Tertullian, circa AD 200: "*Cum ad orationem stamus, convertimus ad orientem.*"[15] This translates to "When we stand to pray, we turn to the east." Subsequent writers tried to justify this tradition from

Biblical passages, including the location of Eden in the east in Genesis 2:8. There is also the association of Jesus as the "light of the world" with the rising Sun. The origin of this practice cannot be traced to a single passage or authority, however.

If, as Tertullian suggests, Christians pray to the east by tradition, church orientation would follow as a natural consequence. This doesn't appear to be the case in the oldest known churches; many were built on the foundations of ancient Roman basilicas. But in northern Europe in the early Middle Ages, eastern orientation emerged as a practice and persisted until the Reformation and later. The church naves in that era run roughly west to east, culminating at an altar to the east. Church graveyards also had an east–west orientation, and bodies were placed with their feet pointing toward the east. These alignments weren't perfect or universal; often, local geography, such as hills and roads, would force other alignments, but this practice still seems to have been common.

Are these alignments reliable as an aid to navigation? Much like moss on trees, these are only approximate, and the explanations for shifts from a pure east–west orientation are probably the most curious. Assuming the eastward orientation for churches was widespread from the eighth century on, we must ask, "How did the medieval architects and masons determine east in the construction of churches?" Without a compass, the only available means of orientation was the Sun. In northern Europe the bearing to the rising Sun depends on the time of year. At the latitude of Stonehenge, 51 degrees north, the Sun rises approximately 40 degrees north of due east at the summer solstice and 40 degrees south of due east at the winter solstice. This is a huge range. At the vernal and autumnal equinoxes, the Sun rises due east. If an architect were determined to create a perfect east–west alignment, the position of the rising Sun at one of the equinoxes would be the best solution.

Let's suppose the church builders created their alignments from the bearing to the rising Sun on random days. What kind of distribution of churches would this create? The rising Sun's position seems to come to a standstill at the solstices and moves the most rapidly during the equinoxes. If the builders chose random days to carry out their orientation, there should be a preference for the direction of the rising Sun on the solstices, which is not seen.

In a survey of twenty Saxon and twenty Norman churches built in the Middle Ages, the average orientation is consistent with due east, with approximately 70 percent of the churches lying within 14 degrees of due east.[16] This is a distribution that shows a purposefulness of true east, as opposed to the direction of the rising Sun on random dates.

Another concept offered for church orientation is the tradition of naming churches after saints. In this scheme the altar would point in the direction of the rising Sun on the day of the saint's feast. A quick perusal of the feasts of saints shows that these dates are randomly distributed throughout the calendar. If consecration dates were employed for church construction, the orientations would tend to favor the direction of the solstices, which again isn't borne out by data. A more obvious factor in church orientation is real estate: local forces can and will dictate construction possibilities.

Like moss, church orientations are something of a statistics-and-guessing game. In Figure 29 I show a map of two churches in the town of Bradford-on-Avon in western England. St. Laurence church is one of the few intact surviving Saxon churches. The church across the street, St. Laurence, is contemporary. The orientation of the Saxon-era church is 50 degrees, or 40 degrees north of due east. The azimuth of the contemporary church is 63 degrees, somewhat closer to due east. We know very little about the Saxon church; written evidence suggests a construction date of AD 705, while stylistic elements point to a date around 1000. With two churches and twice as many rationalizations for church orientation, what can we conclude? From the statistics of Saxon church orientation, St. Laurence is an outlier: very few point that far north. We can run through some possibilities.

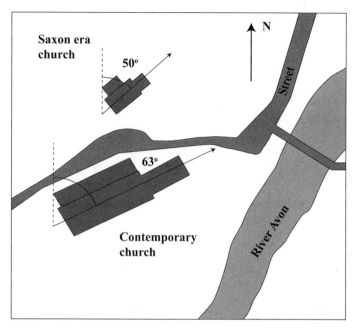

Figure 29 The alignments of a Saxon era church (circa AD 700) and a contemporary church in Bradford-on-Avon, England, show only an approximate east–west orientation.

1. Summer solstice: The azimuth of the rising Sun on the solstice at this latitude is 50 degrees, which neatly coincides with the orientation of St. Laurence. The only problem with this hypothesis is that there is a fairly steep hill to the east of the church, making it impossible to observe the Sun directly at the horizon.

2. Date of the saint: The date associated with St. Laurence is August 10. The azimuth of the rising Sun on that date at this latitude is 63 degrees, which neatly coincides with the orientation of the contemporary church.

3. Due east: the churches both have a rough east-facing orientation, but the large deviation from true east is uncommon.

4. Local conditions: The Avon River turns through a steep valley where the town is located. The hillside and

local roads parallel the river, so this may have been one of the determining factors in the construction.

My opinion is that the churches at Bradford-on-Avon were oriented to face approximately east, but the local conditions were important factors (option number 4 in the list above).

Salisbury Cathedral is only a few miles south of Stonehenge. Completed in 1258, it stands as a magnificent example of High Gothic architecture. The town of Salisbury is located on a plain, so there are no major physical impediments to construction that would force a particular alignment. The nave is oriented to an amazing precision of better than 1 degree toward true east. It is so accurate that you can calibrate your compass with the cathedral.

Salisbury Cathedral shows a distinct purposefulness on the part of the builders. Although aided by the flat ground surrounding the cathedral, it clearly took substantial care by the builders to achieve this precision. Since magnetic compasses were barely being used in Europe when construction commenced on the cathedral in Salisbury, the architects most likely used the Sun.

This tradition of east-facing churches is roughly limited to northern Europe, particularly during the Middle Ages. Churches on the other side of mountain barriers in Italy and Spain don't show such purposefulness in alignment. There is a modest tendency for an eastward alignment in some North American churches built in the eighteenth and early nineteenth centuries, but that appears to have died out, and local conditions tend to dictate orientation.

Satellite Dishes

Affordable satellite TV emerged at the end of the twentieth century. One key factor was the development of small, inexpensive fixed-satellite dishes. The very first satellites thrown into orbit circled the Earth in ninety minutes. To track these, expensive dishes had to be equipped with motors that would allow them to follow the motion of the satellites as they crossed the sky overhead. Although ham radio operators could catch the passing signals of such satellites as the early *Sputnik*, it was not a system capable of reliably transmitting TV signals to a household.

The modern system takes advantage of an important property of Newton's theory of gravity: the farther the orbit from the surface of the Earth, the longer the orbital period. While *Sputnik* took ninety minutes to orbit the Earth, the Moon, being much farther away, takes twenty-eight days. Somewhere between *Sputnik*'s orbit and the Moon's orbit, there's a sweet spot where a satellite placed into orbit will go around the Earth in precisely twenty-four hours, the length of a day. This is called a *geosynchronous orbit* (also called a *geostationary orbit*). Any satellite put into such an orbit will appear to hover at a single point over the Earth's equator (Figure 30). From the point of view of a person on the ground, the stars

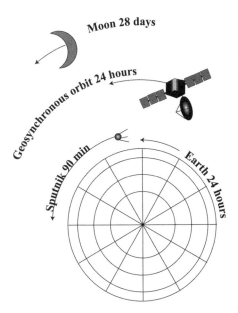

Figure 30 Principle of geosynchronous satellites. The time it takes for the satellite to complete its orbit is precisely twenty-four hours, the same as the length of a day. Each satellite hovers over a single location high above the equator.

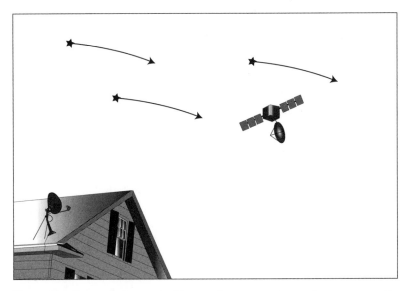

Figure 31 Geosynchronous satellites as they appear to an observer on earth. Satellite dishes on houses are oriented toward the satellite providing the local service. The satellites appear to hover at one spot in the sky, unlike the stars, which move.

and the Sun will trace arcs through the sky once a day, while the geosynchronous satellite will always appear in the same place in the sky (Figure 31).

The homeowner installing a TV dish must orient it to face the provider's satellite. The good thing about this technology is that once it's installed nothing more needs to be done. Unless it is disturbed, the dish always points toward the satellite's position in the sky. Most providers suggest that a professional do the installation, but for the astronomically minded it is not such a big chore. The main trick is finding the location in the sky where the satellite is located. Fortunately, there are websites available where the homeowner can put in the location of his provider's satellite and his latitude and longitude. The website will do a calculation that will tell him the compass bearing and the angle to use above the horizon to orient the dish.

As with churches it seems unlikely that a person would be truly lost if she is within sight of a satellite dish. On the other hand, if a person is disoriented in a strange. deserted neighborhood on a cloudy day, satellite dishes could help to regain orientation. Different providers will use different satellites, so you have to know the location of your particular provider's satellite. How does this bear on urban navigation? It's a matter of reverse engineering. If you see a satellite dish and know the provider from the logo printed on the dish, you can figure out the location of the satellite and use this as a direction indicator.

The use of satellite dishes as urban navigational beacons began to hit online blogs on the web in the early 2000s and spread in much the same way that the moss trick gained notoriety in eighteenth-century literature. Here's one example: "John writes: 'I have discovered a new urban navigation technique. If you're ever unsure as to which direction you're heading. Take a look at the satellite dishes. They're always pointing SW. It's like the moss growing on the north side of the tree!' Does this only work in northeastern North America?"[17]

In answer to John's question: this only works in northeastern North America if you're using a satellite provider named DirecTV. There are hundreds of geosynchronous satellites in orbit, many of which are obsolete. In the northeastern United States, the principal DirecTV satellite is called DirecTV 6 and orbits above a longitude of 110 degrees W. If you think of a pillar high in the sky, this satellite connects with a spot in the Pacific Ocean a few thousand miles off the coast of Peru and is on a meridian that runs north through the Rocky Mountains.

In Boston, at a longitude of 71 degrees, a dish set to this DirecTV satellite should be oriented at a compass heading of 219 degrees, or toward the southwest. Dishes in other parts of the world need different orientations to take into account different locations and different satellites. Much like moss and churches, specialized knowledge is required to use the dishes as a navigational aid. A group of my students walked around Boston with a compass and measured the orientation of a group of about twenty satellite dishes from the same provider; they found that the mean orientation of the local DirecTV dishes was indeed 219 degrees and agreed with the precision of their compass work. So it works!

Say you are about to travel to an unfamiliar city and want to use satellite dishes for orientation. It's quite easy to look up providers on the web, then use online calculators to find the orientation of the common dishes.[18] Let's take Casablanca, Morocco, as an example. It's located at latitude 34 degrees north and longitude 7.6 degrees west. A likely satellite that would be heavily used is called Nilesat 101, which hovers over the equator at a longitude of 7.0 degrees west. It carries such channels as Fox Movies Middle East, Al Jazeera, Al Jazeera Sport, and MTV Arabia. A satellite dish for this popular provider should be oriented due south and have an elevation angle of 51 degrees in Casablanca.

Another popular provider is carried on Hot Bird 8, which hovers over the equator at a longitude of 13 degrees east. A dish pointing toward Hot Bird 8 in Casablanca should be oriented at 34 degrees east of due south. If you get lost wandering around the city, an inspection of the many satellite dishes and knowledge of the popular providers can help you orient yourself.

All of these sources for orientation — moss, churches, and satellite dishes — give only approximate directions but can be used with some understanding and skill. Regarding satellite dishes, I often wonder what an archaeologist in the distant future would think of these funny dishes pointing toward the same spot in the sky. Would they view it as some worship of a pagan deity in the same way we wonder about the motivations of the builders of Stonehenge? Perhaps.

6. Maps and Compasses

· ·

YOU MAY RECALL that in chapter 2, I described the difference between route knowledge and survey knowledge. Route knowledge is largely one-dimensional, describing paths and connections. Survey knowledge is inherently two-dimensional and allows shortcuts and an understanding of the physical relationship among many landmarks. Mammals, including humans, appear to have survey knowledge once they've had a chance to explore their environment.

How is survey knowledge communicated? We can talk about geographic features and their relationships, but this tends to convey only a portrayal of routes; that is to say, if someone describes a region to you, she may say, "Turn left at the fork in the trail, go three miles, and you'll come to a spring." You won't gain insight into possible shortcuts or how multiple landmarks relate to each other. On the other hand, if you were handed a graphic representation of a region — in other words, a map — it would convey a large amount of information at a glance that would otherwise take a long time to describe with words. Language cannot economically convey the kind of two-dimensional information that is readily captured in maps.

The oldest known maps appear on the walls of caves and carved into rocks. Some findings date from roughly 12,000 BC and show outlines of local topography and fauna: mountains, rivers, fields, and animals. The oldest known map in Western Europe (Figure 32) was recently reported from archaeological work led by Pilar Utrilla from the University of Zaragoza in Spain. Her team spent fifteen years deciphering a set of stone etchings found in a cave in northern Spain. It is dated to around 12,000 BC and shows local geographic features carved into stone, including a river, mountains, the cave entrance, and paths.[1]

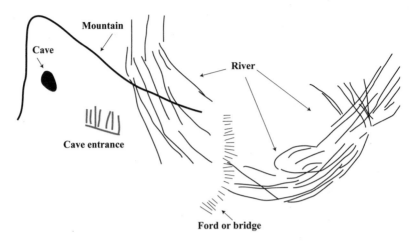

Figure 32 Outlines of geographic engravings found on a paleolithic map and possible interpretations, based on findings by researchers at the University of Zaragoza, Spain. The estimated age is 13,660 years. After P. Utrilla et al., A palaeolithic map from 13,660 calBP: engraved stone blocks from the Late Magdalenian in Abruntz Cave (Navarra, Spain), *Journal of Human Evolution*, 57:2 (2009): 99–111. © (2009), with permission from Elsevier.

Some early *petroglyphs* — stone carvings — show arrays of dots that look like constellations in the sky. Others show patterns of huts or arrangements of housing.

When we look at a modern map, we expect two broad features: *orientation* and *scale*. A map is a miniature representation of some region. The features shown on an accurate map should have the same relative distances that they have in the mapped area. The scale of a map is the conversion factor needed to calculate the true distances between shown features from their relative distances on the map. So, for example, on a map with a scale of one inch to a mile, two features separated by half an inch on a map are half a mile apart in reality.

The *orientation* of the map is a matter of convention. Today we hardly give it a second thought and assume that "up" is north on a map and positions more to the left are west and more to the right are east. This is a visualization that we practice where the mind's eye is hovering over the landscape with the head pointing north. This wasn't always the case.

One of the earliest examples of a map showing a global orientation is a clay tablet excavated at Yorgan Tepe in Iraq. It dates from about 2300 BC and depicts a region bounded by two ranges of hills and a plot of land belonging to an owner named Azala. The map is oriented with east at the top.[2]

In chapter 4, I mentioned a large number of languages that put a primacy on the direction east being oriented by sunrise. While it may have started in the Middle East, the tradition of orienting a world map with east at the top persisted as a Christian tradition throughout the Middle Ages. Often, the Garden of Eden was depicted at the top of medieval maps. In Genesis 2 Eden was described in the east: "The Lord God planted a garden toward the east, in Eden; and there He placed the man whom He had formed."

In many medieval maps an image of Jesus often appears at the top. As we saw in chapter 5, the orientation of Christian prayer to the east was incorporated in the construction of Northern European churches.

Many medieval maps follow a "T-O" format. Here, the world is represented as a circular disk, where all known land is encircled by ocean. Figure 33 shows the features of a T-O map that has the visual appearance of the letter "T" inside an "O." There is an encircling ocean that forms the O. The T is formed by the waters of the Mediterranean as the vertical part of the T, with the horizontal formed by the Tanais (Don) and Nile Rivers. The horizontal branch of the T divides Asia from Europe and Africa, and the vertical branch divides Europe from Africa. Jerusalem is usually shown at the center.

These maps had varying degrees of detail. In some cases they are quite simple and mainly symbolic, while in other cases they began to approach a more literal depiction of the known parts of

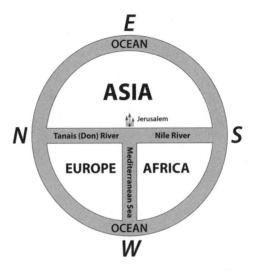

Figure 33 Structure of T-O maps common during the Middle Ages in Europe. These were oriented with east at the top. Image credit: A. Scherlis.

the Earth. One of the most detailed depictions of geographic locations in a T-O map is found in Hereford Cathedral and dates from *circa* AD 1300. The T-O maps of that era were not intended for navigation. Rather, they cataloged known geographic regions of the Earth in a figurative fashion.

A different mapmaking tradition emerged in ancient Greece. One of the earliest descriptions of a map appears in the *Iliad* by Homer, which dates to roughly the eighth century BC. In Book 18 of the *Iliad*, Homer describes a shield forged for Achilles by Hephaestus, the god of fire and metallurgy. It consists of five concentric rings, with the innermost depicting the Sun, Moon, and stars. Surrounding the cosmos are two cities, one at peace and one at war. The next two rings depict many forms of human activity, including agricultural scenes. The outermost ring was an encircling river, Oceanus. Flat circular maps of the world with an encircling ocean are characteristic of early Greek representations.[3]

The concept of a spherical Earth is attributed to the era of Pythagoras, in the sixth century BC. In the third century BC, Eratosthenes first measured the Earth's circumference by compar-

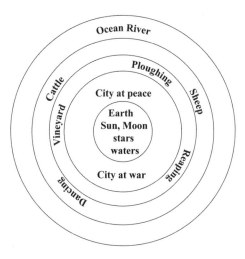

Figure 34 Five layers of the shield made for Achilles by Hephaestus as described in the *Iliad*.

ing the differences in the lengths of shadows cast in two Egyptian cities, Alexandria and Aswan, at the summer solstice. Knowing the distance between the two cities, he could calculate the radius of the Earth. The concept of a spherical Earth enabled a rapid growth in understanding of astronomy and geography in ancient Greece.

As Greek culture developed, geographic knowledge expanded. The conquests of Alexander the Great in the East and the travels of the navigator Pytheas of Massalia (Marseilles) in the fourth century BC expanded the horizon. Pytheas navigated north to an island he named Thule, possibly modern Iceland. He was the first to notice the change in the length of day with latitude and the differences in climate as he traveled from south to north. He also proposed that the Earth rotated about a fixed axis.

Some five hundred years later, Greek knowledge of geodesy and astronomy culminated in the work of Ptolemy of Alexandria. While others before him developed latitude and longitude as measures for positions on the Earth, he systematized these concepts and developed tables of positions of major cities and islands.

One key idea that evolved from the time of Pytheas to the time of Ptolemy was the concept of *climes*. A clime is a strip of latitudes that circles the Earth. In this scheme the physical characteristics of a clime are the same everywhere in the world, including the kind of people you meet. This is not without reason. As you travel north from the equator toward the Pole, the average temperature drops, the kinds of trees and vegetation change, and even the inhabitants' skin color shifts. The demarcation of different climes was cataloged by the length of day at the summer solstice in each zone. The farther north you travel, the longer the day lasts at the solstice.

In broadest terms Ptolemy outlined a world with a northern temperate zone that encompassed the inhabited regions of Europe, Asia, and Africa. An illustration of the kind of map associated with Ptolemy's geography is shown in Figure 35. The region around the equator was considered too hot for people to live and was referred to as the Equatorial Torrid Zone. South of the equator laid a region that was called the *Antipodes*. This literally means "opposite feet," as a person on the other side of the globe would be upside down relative to a person in the Northern Hemisphere. There also was a southern temperate zone. North of the northern temperate zone was a frigid zone, which was uninhabitable, and there was a corresponding southern frigid zone.

The model of climes, along with the rotation of a spherical Earth about a fixed axis gave impetus to maps that depicted north as up or down. The strips of climes create an east–west symmetry that would make it logical to represent east–west horizontally.

Another feature of Ptolemy's geography was the idea of the *oikoumene*: the inhabited world. The only inhabited regions known to ancient Greece were Europe, Asia, and North Africa. The farthest point west was marked by the position of a place called the Fortunate or Blessed Islands. Historians often associate the Fortunate Islands with the present-day Canary Islands. The oikoumene spanned 180 degrees around the globe. An encir-

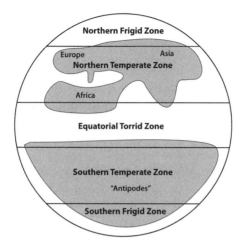

Figure 35 Areas of the world as expressed by Ptolemy and furthered by Arab scholars. Northern and southern temperate zones were bordered by an equatorial torrid zone and frigid zones near the pole. Image credit: A. Scherlis.

cling ocean surrounded the inhabited world, as in the design of Achilles's shield and in the later T-O maps, described above.

After the time of Ptolemy, Western Europe plunged into the anarchy of the Dark Ages. Where order existed, it was enforced by the Catholic Church, which established rigid canons of belief. Muslim cultures in the Middle East, Persia, and North Africa, however, flourished with great importance accorded to the legacy of Greek inquiry. Much of Ptolemy's work was translated into Arabic, along with the works of other scholars, such as Euclid. Pioneering work in astronomy, geography, and mathematics was furthered by such scientists as the Persian polymath al-Bīrūnī. Tables of the latitude and longitude of major cities were generated for the purposes of astrology and figuring out how to face Mecca.

Many Arab maps of the world from AD 800 onward had an emphasis on the region around Persia and the Arabian Peninsula. Many of the Arab world maps retained the Greek concept of an encircling ocean and the concept of climes. At the pinnacle of their empire, the Arabs dominated the trade on the ocean routes between East Asia, India, and the Arabian Peninsula, although there is no evidence that they ventured far beyond the Strait of Gibraltar (the Pillars of Hercules) into the Atlantic.

The major navigational feats in medieval Western Europe were travels of Irish monks to Iceland and the subsequent Norse colonization of Iceland, Greenland, and the short-lived settlements established in North America (Vinland). No maps depicting these regions from the period of Norse voyaging survive, if one discounts the controversial Yale Vinland map.[4] Most of the surviving large-scale maps from the early Middle Ages are the T-O maps. The only maps from that era that approached the ideal of survey knowledge were smaller-scale maps depicting local towns and environs.

THE COMPASS AND PORTOLAN CHARTS

In the thirteenth century the magnetic compass first appeared in the Mediterranean. Its origin as a navigational tool is still debated, but its rapid and widespread adoption by Western Europeans and Arabs during this period is undisputed. For the first time a mariner could set a course by using the compass, as opposed to a reliance on natural indicators such as wind or the Sun. This could be done in the fog, on overcast days, in practically any condition, and it improved the reliability of navigation. If you wanted to sail from one location to another, you only needed a compass heading, and you were set. The use of the compass in the West was accompanied by the sudden appearance of a new kind of map that historians call a *portolan chart*. Unlike previous maps, these were drawn primarily for navigation. Many call these the first true maps because they display a consistent orientation and relatively accurate scale.

In portolan charts landmarks and ports along the coast were represented in great detail. Figure 36 illustrates the major features. Overlaying the chart is a spiderweb-like network of *rhumb lines*. A rhumb line is the path a person or vessel follows if a constant compass heading is kept; that is to say, the orientation of a course

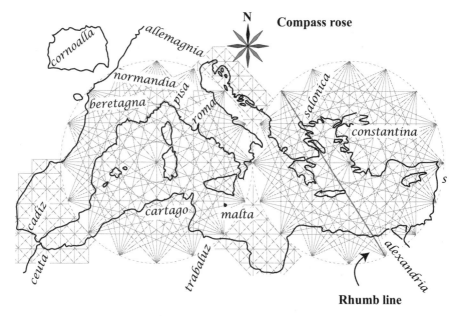

Figure 36 Representation of a portolan chart characteristic of the thirteenth through early sixteenth century. These charts had an accurate scale and were covered with a network of rhumb lines showing sailing directions.

with respect to north is fixed. The name *rhumb* comes from the Spanish *rumbo*, meaning "'course." Compass roses also first appeared on these charts and were segmented into eight, sixteen and even thirty-two compass points.

The *Carta Pisana* is the oldest surviving portolan chart and dates from roughly 1275. It is drawn on sheep vellum and depicts the Mediterranean, parts of the North Atlantic coast, and the Black Sea. In contrast to the contemporary T-O maps, the Carta Pisana displays the Mediterranean much more accurately and has north at the top, along with many rhumb lines.

A map folio dating from 1375 by cartographers Abraham and Jehuda Cresques portrays Europe, the Mediterranean, and North Africa with precise attention to scale and a heavy use of grids of compass lines that allowed navigators to set courses from one port to another. Many maps from this period also depicted fanciful islands from legends, as if to cover all possible bases. The Cresques' chart shows a river of gold flowing through the central Sahara, south of the Atlas Mountains.

The appearance of portolan charts also corresponded to a rise in commerce between the Mediterranean, the Baltic, and the North Sea. Groups such as the Hanseatic League in the Baltic and city-states such as Genoa and Venice in the Mediterranean controlled much of the trade in Western Europe. The Baltic, Mediterranean, and North Sea have different characters, and to cover the much wider range of trade, the folios containing the charts had detailed navigational advice for mariners.

For example, while the tides in the Mediterranean are small, the tides in Northern Europe and the British Isles are substantial, which mariners had to take into account. The timings of the tide varied from port to port but were predictable if a navigator knew the phase of the Moon. Many of the navigation folios from this era had tables for each port that gave the time of high tide relative to the position of the Moon. This kind of information would be critical for a Mediterranean navigator making his way to a port on the North Sea.

Compass Principles

The Chinese and Greeks knew about magnetism circa 500 BC. The name is derived from a stone called magnetite, which was found in the region of Magnesia in Thessaly. Magnetite could be fashioned into lodestones that displayed a mysterious attraction to each other and to iron. When an iron bar was drawn across a lodestone, it acquired magnetic properties. Although well known, the magnetic properties of iron, nickel, and magnetite were not used in practical applications for centuries.

A compass magnet is called a dipole (Figure 37), meaning that it has two poles: north and south. This is more commonly called

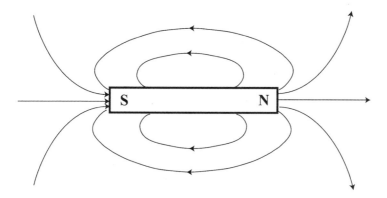

Figure 37 A magnetic dipole with a north and south magnetic field. The field lines point from north to south.

a bar magnet. By convention if we take a dipole and let it swivel freely in the horizontal plane, then let it settle, the north end aligns in the direction of the northbound magnetic lines of the Earth. We draw magnetic fields as continuous lines beginning on the north pole of the magnet and ending on the south pole. The arrows in Figure 37 indicate the southbound direction of the field lines. Magnetic poles of the same kind (north-north and south-south) will repel each other and unlike poles (north-south) will attract each other; that is to say, a north pole will repel a north pole, while a north pole will attract a south pole.

If a magnetic dipole is put in a uniform magnetic field, it will experience a *torque* (Figure 38) that causes an object to rotate. A playground seesaw is a good example of torque in action. The seesaw rotates about a fixed point, and the weight of the children on each end exerts a force at some distance from the point of rotation. In making a compass we put a pivot point in the middle of the magnet that allows it to rotate freely. The torque it experiences from the Earth's field will cause it to oscillate back and forth like a seesaw, eventually settling down pointing along the magnetic field lines.

A magnetic compass needle rarely points directly toward the geographic North Pole. Rather, it lines up with the local magnetic

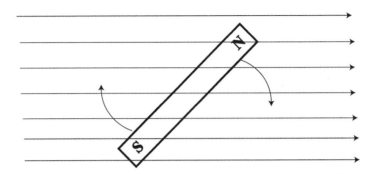

Figure 38 The effect of a uniform magnetic field on a dipole is to create a torque that will make it rotate so that the north pole of the magnet is aligned with the field lines.

field, which can display a substantial variance from true north. We talk about magnetic *variation* or *declination* as the difference between "true" and "magnetic" north. The term *magnetic declination* is used mainly for land maps, while nautical charts use the term *magnetic variation*. Figure 39 shows a recent world map of magnetic variation from the U.S. Geological Survey.[5] In the map you can see where the lines of variation converge on the magnetic poles.

In my home city of Boston, a compass needle will point roughly 16 degrees west of true north, while a magnetic compass needle in the Pacific Northwest will point 16 degrees east of true north. If you go tromping around in the woods with a compass and don't pay attention to the local variation, you might make a substantial error in your bearings. Fortunately, present-day magnetic surveys of the Earth are unusually comprehensive and easy to obtain.

Magnetic variation arises because the field is created by the flow of molten metals deep inside the Earth. This flow is not completely regular and gives rise to changes in the field, depending on location. In addition to the geographic variation, there is a variation over time called *secular variation*. This is again due to the vagaries of the flow of molten material deep inside the Earth.

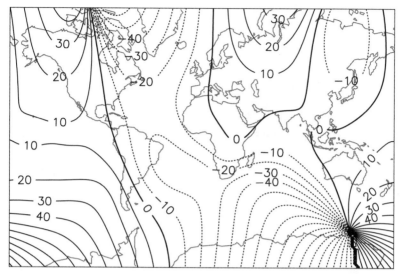

Figure 39 Map of the magnetic variation in degrees of the Earth's field from true north. The lines correspond to contours of equal magnetic variation. Solid lines are variations to the east of north, dotted are variations to the west of north. The lines concentrate near the north and south magnetic poles. This map is for the year 2000. Courtesy of the U.S. Geological Survey.

Before the days of magnetic surveys, one regular duty on shipboard was a measurement of the local magnetic variation so that a ship could hold its course throughout the day. This was done at sunrise and sunset. Armed with a table of the angles of sunrise and sunset, a navigator could compare the bearing to the Sun against the compass and adjust accordingly. These variations would be duly noted in logbooks.

The Earth's magnetic field has both a vertical and a horizontal component; that is to say, the field can be described as a line parallel to the Earth's surface (horizontal) and a line perpendicular to the Earth's surface (vertical). We almost always use horizontally oriented compasses, but close to the magnetic poles, the horizontal component becomes very weak, and the field is mostly vertical.

The guts of a compass are made from strips of iron placed underneath a *compass card* (Figures 40 and 41), which has different angles laid out with respect to north. Some of the earliest

Figure 40 Compass card and housing from a dry-card compass from the nineteenth century. Photograph courtesy of the Collection of Historical Scientific Instruments, Harvard University.

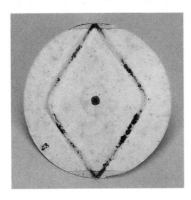

Figure 41 Magnetized iron wires on the back of a compass card to orient the compass to Earth's magnetic field lines, dating from the seventeenth century. Photograph courtesy of the Collection of Historical Scientific Instruments, Harvard University.

compasses were dry compasses, meaning that they rotated freely in air on a bearing. One of the drawbacks of a dry compass is that it can take a long time to settle and bounces around in a rocking boat. To combat these problems the compass dial was later placed in a damping fluid to allow it to settle more rapidly.

The earliest compass cards were called "roses" because they had the appearance of the opening of flower petals. The design was based on the basic geometric scheme of the successive bisection of angles. Starting with the cardinal points of north, east, south, and west, one can subdivide further into points containing northeast (NE), southeast (SE), southwest (SW), and northwest. Next would come north-northeast (NNE) and east-northeast (ENE). A further subdivision would be north-by-east (NbE) and northeast-by-north (NEbN). Most compasses in the era of Western European exploration consisted of thirty-two or sixty-four compass points.

The development of celestial navigation drew heavily from techniques of astronomy, which used a degree system for dividing up the horizontal angle or *azimuth*. The word "azimuth" comes from the Arabic *assumut*, which means "the ways." The use of points, as opposed to numbers, to describe horizontal directions persists to this day in many sailing communities.

By convention the azimuth in the horizontal plane starts at 0 degrees pointing due north, then proceeds clockwise toward east (90 degrees), south (180 degrees), west (270 degrees), and back to north (360/0 degrees). The traditional division of the azimuth into 360 degrees comes from the Babylonians, who used a base-sixty number system. The use of 360 has a number of advantages: it is very close to the number of days in the year (365.24), and it is divisible by a large number of factors: 2, 3, 4, 5, 6, 8, 9, 10, 12, and so on. As the use of celestial navigation became more common, the compass rose system slowly fell out of favor, and angles were widely employed. Many nautical compasses, however, still carry both a degree and a rose system.

Theories of the Earth's Magnetic Field

The origin of Earth's magnetic field was a mystery in the Middle Ages. Given that lodestones exerted forces on each other, it was perhaps reasonable to guess that there was a giant magnet somewhere that made the compasses point north. One of the earliest philosophers to attempt to explain this was the nobleman Petrus Peregrinus in the thirteenth century. During that era the Earth was viewed as the center of the universe, and it was thought that the cosmos rotated around it. Peregrinus believed that there was a universal sympathy between earthly forces and the cosmos. This universal sympathy caused the iron in the compass to seek alignment with the axis of the celestial sphere.[6]

Over time, the Earth was no longer regarded as the center of the cosmos, so Peregrinus's universal sympathy concept seemed implausible. The next wrinkle was the belief in a giant mountain of lodestone at the North Pole that caused a worldwide attraction for compass needles.

Magnetic variation was at first not appreciated by mariners in the Mediterranean. The variation was close to 0 degrees in Southern Europe and North Africa. Closer to the north magnetic pole, variations become larger, and values taken from Northern Europe showed larger and obvious variations from true north.

In 1541 Swedish mapmaker Olaus Magnus from Lund took measurements of magnetic variation from the Netherlands and Gdansk, Poland. He drew a set of great circle lines pointing to a location he deduced to be the magnetic mountain near the North Pole. From the intersection of the two lines, he figured that the great magnetic mountain must exist some distance from the Pole, off the north coast of Siberia.

Mapmaker Gerardus Mercator (1512–1594) is best known for his method of projecting a circular Earth onto a flat map, which has become known as the Mercator projection. Mercator produced a map in 1569 that showed two magnetic islands in the Arctic. There were reports of a zero magnetic variation on the island of Corvo in the Azores and also in the Cape Verde Islands. In Regensberg, Bavaria, the variation was 17 degrees west of true north. By taking the intersection of a great circle line from Regensberg and the 0-degree lines from both Cape Verde and Corvo, Mercator concluded that two magnetic islands generated the Earth's field. On his map, he placed these in the Arctic Ocean between the meridians of California and Asia.

The problem with the constructions of Mercator and Magnus is that they resulted from misconceptions about magnetic variation. Just because a compass shows a misalignment with true north at some location does not mean that a great circle can be extended directly to a magnetic mountain.

Soon the theories of magnetic mountains disappeared, and the concept of a giant bar magnet inside the Earth took hold. The precise nature of this dipole was unknown, but this view was more realistic. Our present-day understanding of the Earth's magnetic field has evolved substantially since the era of magnetic mountains: The interior of the Earth is heated by the decay of radioactive elements and heat left over from its formation. At the center of the Earth is a solid inner core consisting of iron and nickel. The inner core is very hot, approximately 9,800 degrees Fahrenheit, and would be in a molten state if it were not for the extraordinary pressure surrounding it, over three million times the pressure on the Earth's surface. A liquid outer core surrounds the inner solid core (Figure 42). The liquid in the outer core is also rich in iron and can flow freely. Surrounding the outer core is the *mantle*, which is liquid but more viscous than the outer core. The mantle extends up to the outermost layer called the *crust*, which is the solid outer layer that we live on.

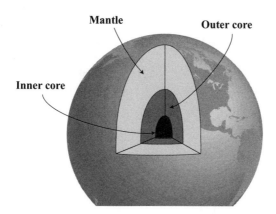

Figure 42 Interior of the Earth, showing the solid inner core, fluid outer core, and mantle.

Fluid flow in the outer core creates the Earth's field. Molten iron and other elements at the surface of the inner core are heated and become less dense. Like a hot air balloon, the heated fluid is more buoyant than the surrounding fluid. It rises and slowly cools as it approaches the mantle. When the fluid cools, it falls again toward the center of the Earth. This process repeats itself over and over again, giving rise to currents of fluid that circulate from the inner core to the mantle and back again. These are called *convection cells*.

An additional factor is caused by the Earth's rotation. This is called the Coriolis effect and is the result of the fluid rotating more quickly around the Equator than near the poles. The combination of the convection in the outer core and the Coriolis effect gives a kind of twisting motion to the fluid in the outer core. There's a strange interplay between the magnetic field created by the fluid flow and the flow itself. These two feed on each other and conspire to create a reasonably stable version of a bar magnet (Figure 43).

The flow and field are not entirely stable. The north and south magnetic poles wander several miles each year. Currently, the magnetic North Pole is moving northward at about twenty miles per year. Because of the complexity of the effects creating the field,

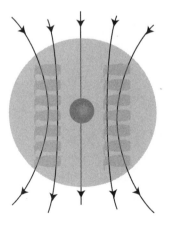

Figure 43 Dynamics of the fluid flow in the outer core give rise to a self-sustaining effect that generates the Earth's magnetic field.

only supercomputer models can really capture the dynamics of the generation of the field, in a process that is termed a *geodynamo*.

The reversal of the Earth's field is known to happen from time to time and is documented by magnetic fields frozen in stripes of the spreading Atlantic seafloor. The stable periods between reversals can be as short as a hundred thousand years or as long as several million years. The actual duration of the reversal itself may be quite short and has been the inspiration for a number of apocalyptic scenarios in fiction and screenplays. Computer simulations have been able to capture reversals, which roughly confirms that the models are reasonably accurate.

Using a Compass

Although the practical use of a compass is a skill regularly taught to Boy Scouts, use of the compass is rapidly becoming a lost art. Lives have been lost through the inability to use compasses in emergency situations, yet they're one of the cheapest and most reliable ways of providing an orientation reference.

The practical use of compasses can be found in many handbooks, so I won't repeat much here. If you know the local magnetic variation and correct for it, a handheld backpacker's compass can provide directions to a precision of no better than about 5 degrees. For better precision one would need a military quality compass or a surveying instrument like a transit. While traveling on foot, a magnetic compass can be used to establish and maintain a heading, based on a point of departure and a destination marked on a map. It is a good practice to use the compass to sight an object in the direction of travel, then move toward the object. Once that landmark is reached, sight something else on the same heading. If you walk with your eyes glued to the compass, it's easy to go astray.

If you're unsure of your location but can find two landmarks on a map, it is possible to use a compass to determine your position

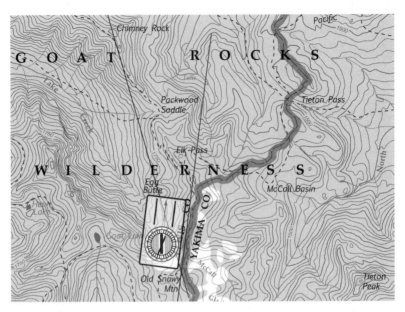

Figure 44 Triangulation using two lines of position. The first line of position comes from a sighting of a marked location called Chimney Rock. The second comes from a sighting of the promontory just south of Elk Pass (at lower right of the "R" in "WILDERNESS"). The compass in the figure is aligned with the second sighting. The intersection of the two lines of position gives a fix for the observer. Base map after USGS topographic map, courtesy of the U.S. Geological Survey.

through a process called *triangulation*. You use the compass to find the bearings to the two landmarks. The bearing to each landmark can be used to find a line of position. If you imagine yourself at each landmark, there is a *back bearing* or *reciprocal bearing* to your position that is 180 degrees opposite your bearing. Each of the back bearings can be used to create a line of position to where you are located.

Figure 44 illustrates the process of triangulation on a map of a region in Washington State called the Goat Rocks Wilderness Area. A hiker is located at an unknown location, but he knows it is somewhere on that map. He wants to find his position. First, he sights a landmark that he can identify called Chimney Rock, which is visible on the map. He uses a compass to find that Chimney Rock has a bearing of 330 degrees true north, after correcting for magnetic variation. He finds that the back bearing from Chimney Rock is 330° − 180° = 150°. This means that a hypothetical observer on Chimney Rock would look toward the hiker and find a bearing of 150 degrees. This establishes one line of position from Chimney Rock extending in a direction of 150 degrees. The hiker is somewhere along this line.

To find his position the hiker must find another line of position. In Figure 44 the compass is shown oriented along a bearing to a promontory that is just a bit south of the location marked Elk Pass on the map. The promontory is located just at the end of the bottom right-hand corner of the letter "R" in the word "WILDERNESS" printed on the map. The hiker takes a sighting with his compass on the promontory and reads a bearing of 10 degrees, again corrected for variation. In this case the back bearing from the promontory is 10° + 180° = 190°. Another hypothetical observer looking toward the hiker from the promontory would see him at a bearing of 190 degrees. The hiker then draws a second line of position from the promontory with a line of 190 degrees. The intersection of the line of position from Chimney Rock and the line of position from the unnamed promontory is the location of the hiker.

The precision of triangulation with compasses is usually limited by the precision of the angles measured and the geometry of the bearings. There is an inherent uncertainty because each line of position is only as accurate as the compass bearing. If the bearing is accurate to within 10 degrees as a typical value, there is a range of possible lines of position on which the hiker may be located.

Likewise, a second bearing will yield a range of possible lines of position (Figure 45). The intersection of the two swaths of possible lines of position gives a diamond-shaped region, inside of which the hiker is located. If the two bearings are very close together in angle, the diamond is elongated. If, on the other hand, the two bearings are nearly perpendicular, the diamond of uncertainty is more limited in extent. As a general rule of thumb, it's a better

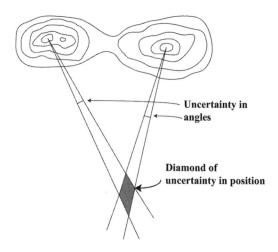

Uncertainty in angles

Diamond of uncertainty in position

Figure 45 Diamond of uncertainty generated from the uncertainty in angles from two bearings used in triangulating a position.

practice to choose two landmarks separated by approximately 90 degrees to get a good position from triangulation.

SURVEYING AND MAPMAKING

Most of us use maps to navigate, but someone has to create the maps in the first place. Until the seventeenth century maps were largely based on dead reckoning reports of travelers, sometimes supplemented with latitudes from celestial observations. In 1533 the Dutch scientist Gemma Frisius (1508–1555) published a scheme of mapmaking based on triangulation. Frisius's scheme employed a set of control points on hills or platforms that could be seen from a distance.

The scheme developed by Frisius is the basis of much of present-day cartography. In many ways it resembles triangulation with compasses. Where the hiker in the above example uses a map with known locations and compass directions to find his position on the map, the mapmaker has to establish a control network that is also based on triangulation. Figure 46 illustrates the basic principle. Let

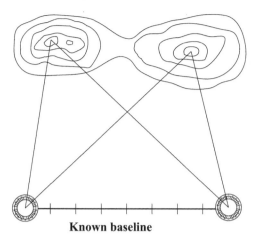

Known baseline

Figure 46   Identification of the location of two mountain peaks using triangulation from control points along a known baseline.

us suppose a mapmaker wants to find the distance between two mountain peaks. The mapmaker chooses a pair of control points to make the measurement. The distance between the control points is measured precisely, often with calibrated chains or some other means. The two mountain peaks are then sighted from the two stations and the angles with respect to true north are noted. The mapmaker can then take this information and use it to calculate how far apart the two peaks are separated and their distances and bearings from the control points.

While Figure 46 shows the basic principle, the process of creating a network for cartographic measurements is a bit more involved. A cartographer will choose a starting point where the latitude and longitude are precisely determined using astronomical observations. The control network is then extended through the construction of a baseline of a known length and bearing from the starting point. Surveyors then set up a network of control points at locations that are visible from at least two other points.

Figure 47 illustrates a control network. From each of the control points, surveyors will measure the angles to other visible stations, building up the network through observations. At the far end of

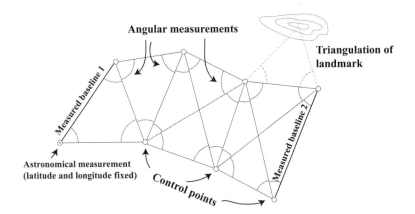

Figure 47 A control network established for the creation of accurate maps. A baseline is extended from a point of known latitude and longitude to a second measurement station. The angles between stations are measured with precise instruments to create the network.

the network, the cartographers create a second baseline. An analysis of the data from the network can be used to predict the length of the second baseline. The comparison of the calculated to the measured lengths of the second baseline will give the cartographers some idea of the accuracy of the measurements and the control network.

Once the control network has been set up, the surveyors can proceed to fill in information such as the location of prominent landmarks, mountain peaks, coastlines, and roads. This information can then be the basis for a map where lines of latitude and longitude are superimposed.

The Danish astronomer Tycho Brahe published an early map that employed Frisius's technique of a network of control points. His network ran from Copenhagen north to Kronborg, Sweden, a distance of 115 miles. On average he achieved accuracies of within a few percent by combining triangulation with precise celestial observations, a substantial improvement over the portolan-style maps.

Following Brahe's work was the first modern measurement of the curvature of the Earth carried out using triangulation.

Dutchman Willebrord Snel Van Royen (also known as Snellius) used the technique to create a network from Alkmaar in northern Holland extending 85 miles to the town of Bergen Op Zoom in the south. His measurement of the length of a degree of latitude, published in 1621, was correct to within about 3 percent of the present-day value and became a standard for nautical navigation of that time.

Surveys of coastlines were and are carried out using triangulation to create marine charts. The main difference between land-based surveying and marine charting is the use of anchored ships as part of the network. In addition to measuring features on land, the ships would make soundings to chart the depth of the seafloor.

One of the most impressive mapping feats was the Great Trigonometric Survey of India conducted by the British in the nineteenth century. This survey had control networks thousands of miles long extending the length and width of the subcontinent, with countless numbers of control stations creating a vast interlocking network that was used to establish locations that were precise to better than a centimeter. It was during this survey that the highest mountain peaks had their altitudes measured through sightings in the distance. Mt. Everest, which bears the name of the leader of the survey, was simply a distant peak that was cataloged in the field as part of the effort. Only after the data were analyzed in London was it realized that it was the highest mountain on Earth.

A portion of the trigonometric survey is in Figure 48, showing a sample of the control network and long-distance shots of mountains in the Himalaya range. In the figure the control network looks like a spiderweb of crossed lines, while the shots of the distant mountains are the triangles extending beyond the control network and terminating on the high peaks.

Present-day mapping techniques include the use of GPS receivers and precise electronic techniques, including laser range finding. Technology works only up to a point, however, as physical locations

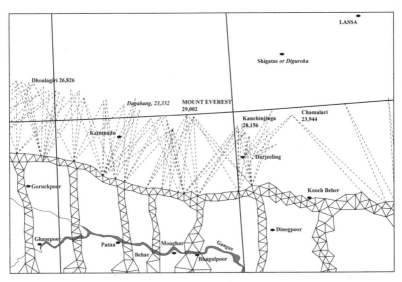

Figure 48 Northeast portion of the Great Trigonometric Survey of India. Note the network of control points and the shots of distant mountains in the Himalaya Mountains. The heights and spellings are reported as they appear in the original master map. From Clements Markham, *A Memoir on the Indian Surveys* (London: W. H. Allen and Co., 1878).

have to be mapped and recorded. Oftentimes, handheld GPS devices in the field can give users a false sense of security. Shoals will move; positions of prominent features, particularly in nonindustrialized countries, may not be accurately recorded on maps or in any database. Although a GPS receiver may give someone a position to an accuracy of less than a meter, this is no guarantee that the features on a map appear anywhere close to that precision.

7. Stars

. . . .

IT SEEMS THAT humans have always seen patterns in the sky. The earliest known star maps may be found in the Lascaux Cave in southwestern France, dating roughly from 18,000 to 15,000 BC. These paintings show clusters that resemble the Summer Triangle and Pleiades (described below). We don't know when was the earliest use of stars for navigation, but we do know that the Pacific Lapita people ventured from the Bismarck Archipelago northeast of New Guinea, over large expanses of ocean to settle remote islands. These voyages commenced around 1300 BC. The Lapita are believed to be the ancestors of many of the present-day Pacific Islanders.

By 800 BC the Lapita had made several large leaps across the open ocean from New Caledonia to Fiji, Tonga, and Samoa. The navigational strategies of the Lapita are unknown, but many of the Pacific Islands cultures that descended from the Lapita use navigational schemes based on stars, suggesting a common origin from the earliest voyaging. In the *Odyssey* Homer describes Odysseus steering with stars as his guide. Although his description is phrased in poetic terms, it is quite practical and demonstrates the use of a kind of celestial navigation in European waters as early as 800 BC.

Stars helped create calendars to mark the passage of the seasons and were used as means of fortune telling. By the time of the Babylonians, astrology was a well-established art linking movements in the heavens to affairs on Earth. The stars governed daily life and guided the traveler at night. The relationship between the heavens and Earth is often viewed as a spiritual bond, as captured in chapter 38 of Job:

> Canst thou bind the sweet influences of Pleiades, or
> loose the bands of Orion?

Canst thou bring forth Mazzaroth in his season? or
   canst thou guide Arcturus with his sons?
Knowest thou the ordinances of heaven? canst thou set
   the dominion thereof in the earth?

                              Book of Job 38:31–33

In this passage the meaning of the word "Mazzaroth" is a bit obscure, but many biblical scholars have taken this to mean the passage of the stars over the course of a year, using them as a calendar. The book of Job dates from roughly 400 BC but is probably based on earlier texts.

With the advent around 500 BC of the concept of a spherical Earth by the Greeks, tables of latitude and longitude marking major cities could be created. Star maps of the heavens used a similar system and allowed for a further development of astronomy and astrology. When Western Europe passed into the early Middle Ages, the torch of science was picked up by Muslim scholars, who furthered the arts of astronomy and mathematics. Circa 1100 AD this knowledge began to filter back into Europe, with translations of works from Arabic to Latin. The modern era of celestial navigation began with Portuguese efforts in the fifteenth century to adapt astronomical tables for use on shipboard.

In our times people rarely, if ever, notice stars. Part of this is due to light pollution near urban areas. Part of it has to do with stars' lesser importance because we have calendars, accurate watches, and GPS units. While few modern humans can name a single star, let alone navigate by them, every so often a contemporary story emerges of someone who still has the skills.

Marc Aubriere worked for the French intelligence agency (Direction Générale de la Sécurité Extérieure, or DGSE) in Mogadishu, Somalia. In July 2009 he was kidnapped by a radical Islamist group, and his jailers took his shoes to discourage an escape. Dubbed a "French James Bond" by the media, he eventually escaped. He waited for a night when the wind was high, to mask any sounds. The guards had neglected to bolt the lock on his cell door, and on August 25 he silently slipped out, emerging onto the deserted streets of Mogadishu at midnight. Initially, he was confused by the darkened streets of the world's most dangerous city but managed to regain his sense of direction by using the stars to find his way. In a Radio France interview, he said, "I walked for five hours through the night to reach my destination, guided by the stars."[1,2]

NAMING THE STARS

Unlike the Sun, Moon, and planets, stars don't move in their positions relative to one another in the sky, making them ideal for navigation purposes. The main issue for any navigator is the ability to recognize stars rapidly and efficiently, then translate the observations into information about position and heading. Below I describe some of the more common groupings of stars, particularly those visible from the Northern Hemisphere. For most people the first exposure to stars is in easily recognized constellations, but the view is limited to a small patch of the sky. With some effort a more global view can emerge that incorporates stellar locations and their relation to places on Earth. People familiar with the stars can take a relatively brief look at a patch of the sky and immediately recognize what they see and can orient themselves rapidly. It is akin to seeing a familiar scene and instinctively knowing directions. As with any other skill, it takes practice. In many cases the names of and myths about stars and constellations aid the learning process.

THE BIG DIPPER, DUBHE, ARCTURUS, AND SPICA

The Big Dipper (Ursa Major) is the most universally recognized star grouping. With seven stars of the same color and magni-

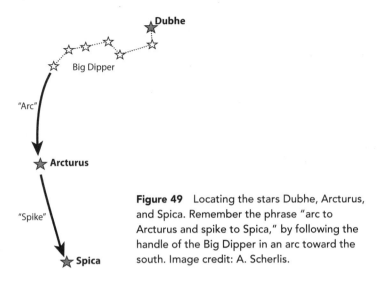

Figure 49   Locating the stars Dubhe, Arcturus, and Spica. Remember the phrase "arc to Arcturus and spike to Spica," by following the handle of the Big Dipper in an arc toward the south. Image credit: A. Scherlis.

tude, it stands out in the night sky. The association of Ursa Major with a bear is fairly common. Both the Greeks and many Native Americans saw the Big Dipper as a bear. Others saw it as a plow or a drinking gourd; the ancient Romans called it *Septentriones* or seven plow oxen. In Hindu tradition it's called *Saptarishi* or the "seven sages." Note the common Indo-European word for "seven": *sept* and *sapta* in both the Latin and Sanskrit. The word *septentriones* became synonymous with "north" in Latin. In England and Ireland the Big Dipper is often called the Starry Plough. In parts of Scandinavia it's sometimes called "Karlsvognen," meaning "Charlie's Wagon."

On the upper far tip of the ladle is the star Dubhe. The handle of the Big Dipper serves as a pointer to two bright stars farther south: Arcturus and Spica (Figure 49). These stars are often visible just after sunset toward the west. A jingle used to locate these stars tells you to follow the handle of the Big Dipper to Arcturus and Spica:

Arc to Arcturus and spike to Spica

This helps locate the stars, which are by far the brightest in this part of the sky, and their identification is nearly foolproof if you remember this association. Arcturus is part of the constellation Boötes, the plowman, named possibly because he "drives" the plough of the Big Dipper. Spica is the brightest star in Virgo, the Virgin. The region of the sky west of Spica and Arcturus is relatively lonely, with few bright stars.

In Figure 49 and most of the illustrations of constellations that follow, west is to the right, and east is to the left, with north at the top. At first thought, this convention might seem backward, but it allows for a presentation of the relative positions of the stars as seen by an observer on the ground, looking up at the sky with his or her head oriented to the north. This is the normal convention for star charts.

THE SUMMER TRIANGLE, SCORPIO, AND ANTARES

The association of stars with the seasons is widespread for patterns seen just after sunset. In the Northern Hemisphere the Summer Triangle (Figure 50), lying to the east of Arcturus and Spica, is high in the sky just after sunset in the summer months of July and August, spanning a large arc in the heavens. The western star in the triangle is Vega in the constellation Lyra, the lyre. Vega has a bluish-white hue and is used by astronomers as a standard for stellar brightness. Deneb lies to the east of Vega, and slightly to the north. Deneb is the tail of Cygnus, the swan (Figure 51). Cygnus is often called the Northern Cross, where with Deneb is at the head of the cross. Altair in Aquila, the Eagle, is the southern tip of the Summer Triangle.

If you look to the south and west of Altair, you can find the constellation of Scorpio. In the middle of the tail of Scorpio is a bright orange-red star, Antares. Scorpio is one of the more prominent signs of the zodiac. When the Sun passes through Scorpio, it is headed toward its most southerly point in the course of the year. Like the Summer Triangle, Scorpio also appears in the summer months of June, July, and August.

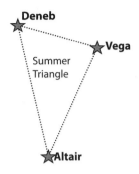

Figure 50   The Summer Triangle and Antares in the constellation Scorpio. Image credit: A. Scherlis.

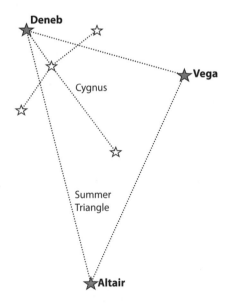

Figure 51   More detail of the Summer Triangle, showing Cygnus, the Swan, also known as the Northern Cross. Image credit: A. Scherlis.

While the stars of the Summer Triangle and Antares should be visible even in large cities at night, if you get away from light pollution, out in the country, you'll see the Milky Way arc across the northern sky toward Scorpio.

Winter Stars

In contrast to empty regions of the sky west of Arcturus and Spica, the winter sky is rich in stars. Just after sunset in the winter in the western sky lies a cluster of seven stars, all blue in tint, that form a bright foggy patch. These are the Pleiades or Seven Sisters, all mythical daughters of the titan Atlas. The story of the Seven Sisters relates that Orion, the hunter, pursued them. The Greek poet Hesiod advises the would-be sailor to stay ashore when the Pleiades appear:

> But if desire for uncomfortable sea-faring seize you;
> when the Pleiades plunge into the misty sea
> to escape Orion's rude strength,
> then truly gales of all kinds rage.
> Then keep ships no longer on the sparkling sea,
> but bethink you to till the land as I bid you.[3]

In one account, the gods placed Taurus, the Bull, to the east of the Pleiades to protect them from Orion. Figure 52 shows the relative positions of the Pleiades, Aldebaran, and Orion. The eye of Taurus is the red-orange star Aldebaran, which appears in a legend about the discovery and settlement of Hawai'i. As related by ethnographer Bruce Cartwright, the legend of Hawai'i-loa mentions Aldebaran as a guiding star: "Let us steer [the wa'a] in the direction of Iao [the eastern star], the discoverer of land. . . . There is land to the eastward, and here is a red star, Hoku Ula [Aldebaran] to guide us, and the land is there in the direction of those big stars which resemble a bird."[4]

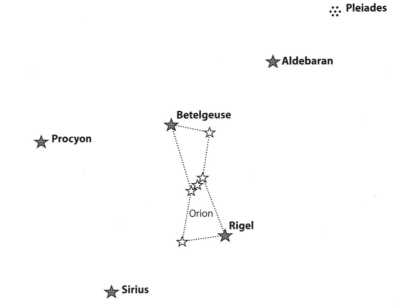

Figure 52 The relative positions of the Pleiades, Aldebaran, Betelgeuse, Rigel, Procyon, and Sirius in the winter sky. Image credit: A. Scherlis.

Orion is also significant to the navigator: it lies across the celestial equator, an imaginary line that divides the sky into a northern and a southern hemisphere. The right-hand star in the belt rises due east and sets due west for all observers, regardless of their location. The upper left-hand shoulder is the red giant star, Betelgeuse, and the lower right-hand star is Rigel, which forms Orion's western foot.

Following to the east of Orion are his two faithful hunting dogs, Canis Major (big dog) and Canis Minor (little dog). Sirius, in Canis Major, is the brightest star in the sky. Its rising just before the Sun in the morning represented the season of the flood of the Nile to the ancient Egyptians. Many ancient Greeks and Romans thought that the combination of the heat of the Sun and Sirius's appearance was responsible for summer, hence the name "dog days." Procyon is the bright star in Canis Minor and appears to the northeast of Sirius. Figure 52 also shows the positions of Procyon and Sirius relative to Orion.

Spring Stars

Moving on farther to the north and east from Orion's dogs, we encounter the constellations of Gemini and Leo (Figure 53). The appearance of Gemini and Leo in the sky after sunset signal the coming of spring. Gemini — the twins — is recognizable as the twin stars of Castor and Pollux, which have roughly the same coloration. The more southern of the pair, Pollux, is a bit brighter.

East of Gemini is Leo, the Lion. By far, the brightest star in Leo is Regulus, which has a white-blue coloration. Figure 53 shows the relative positions of Regulus, Pollux and Procyon. Pollux is almost due north of Procyon, so a line drawn from one to the other creates a good north-south line for figuring directions. South of Regulus is one single bright star, Alphard in the constellation Hydra. This is in the region with few bright stars, and the name, meaning "the solitary one" in Arabic, reflects this.

Once we swing eastward again from Regulus, we run into Arcturus and Spica again, completing the circle of the seasons in the evening sky.

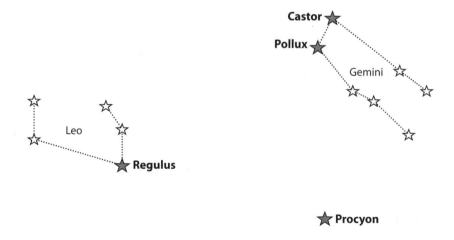

Figure 53 Leo, Gemini, and Procyon. Image credit: A. Scherlis.

Polar Stars

Moving back north again, we can revisit the Big Dipper. Probably the most famous way in recent history of finding north is the use of the Big Dipper to point to Polaris, the North Star (Figure 54). If you follow the end of the ladle of the Big Dipper through Dubhe, it points to a star — Polaris — on the tail of the Little Dipper, Ursa Minor. Polaris wasn't always the pole star. The Earth's axis wobbles, causing a phenomenon known as the *precession of the equinoxes*. In the time of Columbus, Polaris orbited around the North Pole at a distance of about 4 degrees. When Homer composed the *Odyssey*, Polaris was just a minor star that was 11 degrees away from the Pole and inconsequential. There are few if any references to Polaris in antiquity; it only begins to appear in the Western age of discovery in the fifteenth century. Now Polaris is not precisely at the north celestial pole but is approximately forty-four arc minutes away.

Cassiopeia is on the other side of Polaris from the Big Dipper. Cassiopeia was the mythological queen of Ethiopia in Greek mythology. The distinctive constellation looks like the letter "W" when it is close to the horizon but orbits around the North Pole and in the upper part of its orbit appears as the letter "M." Arab cultures identified this cluster as a kneeling camel, while Laplanders identified it as moose antlers. The brightest star in Cassiopeia is Schedar, which forms the eastern bottom corner of the "W."

In northern latitudes the Big Dipper and Cassiopeia are both *circumpolar*, meaning that they orbit the North Pole and never set below the horizon. The circumpolar nature of the Big Dipper and its distinctive shape figure in sailing directions, as it can be seen at night any time of the year above latitudes of 35 degrees.

In the *Odyssey* Homer writes about using the Big Dipper for directions. At the opening, Odysseus has been held captive for seven years by Calypso on the island of Ogygia. Calypso finally releases Odysseus, who builds a raft and follows the sailing directions of

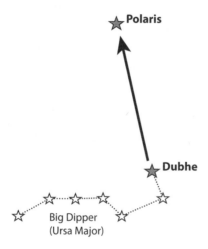

Figure 54 Finding Polaris from the Big Dipper and the location of Schedar in Cassiopeia. Image credit: A. Scherlis.

Calypso to the island of the Phaeacians, considered by many to be modern-day Corfu. In this famous passage the voyage is sung:

> The wind lifting his spirits high, royal Odysseus
> spread sail — gripping the tiller, seated astern —
> and now the master mariner steered his craft,
> sleep never closing his eyes, forever scanning
> the stars, the Pleiades and the Plowman late to set
> and the Great Bear that mankind also calls the
>   Wagon:

she wheels on her axis always fixed, watching the
   Hunter
and she alone is denied a plunge in the Ocean's baths.
Hers were the stars the lustrous goddess told him
to keep hard to port as he cut across the sea
and seventeen days he sailed, making headway well;
on the eighteenth, shadowy mountains slowly loomed
the Phaeacians' island reaching toward him now
over the misty breakers, rising like a shield.[5]

At the latitude of Corfu, the Big Dipper is circumpolar, meaning that it orbits the north celestial pole and never sets. Hence she
"wheels on her axis" and is "denied a plunge." The head of Ursa
Major faces roughly east and to Orion the Hunter. Calypso's directions called for him to keep Ursa Major to his port (left), which
means that he sailed east for seventeen days.

Three Maps

Stars exist in a three-dimensional space. Often, dimmer stars are
farther away than brighter stars. Nonetheless, at first glance, we
visualize stars in the sky as existing on the surface of a dome.
For the celestial navigator, this works well enough, although for
astronomers, the true three-dimensional nature of the location of
stars can be very important, particularly for closer stars.

Like maps, real or mental, we can visualize star locations by
placing them on a kind of map. The correspondence between star
maps and locations on the planet are what celestial navigation is
all about. To understand this we need to know about three kinds
of maps: one centered on the observer, one for the Earth, and one
for the sky. Figure 55 shows some of the stars mentioned above in
a more global setting. You can see that Orion orbits over the equator, while Castor and Pollux orbit at higher latitudes, and Polaris

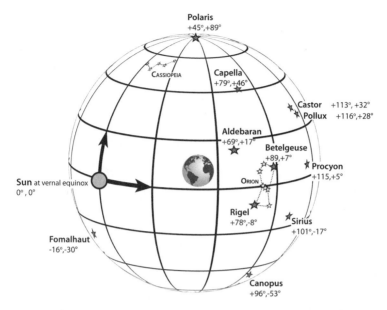

Figure 55 The relative locations of some of the major stars and clusters described above relative to positions on Earth. Image credit: A. Scherlis.

remains fixed over the North Pole. The figure shows two of the three maps in action: a map of Earth and a map of the sky.

1. **Observer-based** (Figure 56): Altitude and azimuth are the names of the two coordinates. These identify the location of objects in the sky, based on what you see from your vantage point on the ground. The job of the navigator is to figure out her position on Earth from this local view of the sky. The heavens look like a big dome with stars plastered on the surface of it. The observer-based system is a way of expressing the position of the stars on the surface of this dome.

Even though it's really curved, the Earth seems locally flat, like a big plane extending out to the horizon. The *zenith* is the point directly above you, the observer. *The azimuth* is the angle going around the horizon clockwise, starting at north as 0 degrees and

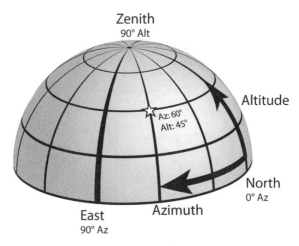

Figure 56 Observer-based system for stars, where the angle from north along the horizon is called the azimuth and the angle above the horizon is altitude. Image credit: A. Scherlis.

extending through 90 degrees as due east, 180 degrees as due south, and 270 degrees as due west; then it wraps around to 360 degrees or 0 again at north. The angle above the horizon is called the *altitude* of the star, and runs from 0 degrees right at the horizon to 90 degrees when the star is directly overhead at the zenith.

Lines of constant azimuth converge at the zenith. As a consequence stars right at the zenith don't have a well-defined azimuth and don't help in direction finding, while stars close to the horizon can be used as a kind of natural compass. Many traditional navigators in the Pacific Islands used, and still use, stars near the horizon for their direction finding when at sea. Stars near the zenith can be useful for latitude determination, and there's some belief that the Pacific Islanders used this method in the era of long-distance voyaging before contact with the West.[6]

The *meridian* is an imaginary arc in the sky that runs from the horizon at due north (0 degrees)

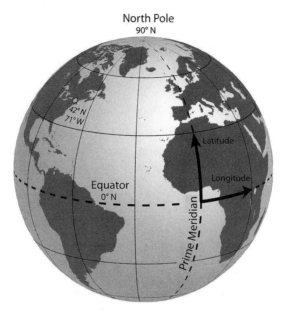

Figure 57 Lines of latitude and longitude. The equator is the zero of latitude and the prime meridian is the zero of longitude. Boston is indicated at 42° N, 71° W. Image credit: A. Scherlis.

through the zenith and back down to the horizon at due south. As a result of the Earth's rotation, celestial objects will be seen to move from east to west, rising somewhere east of the meridian, passing through the meridian at their highest altitudes, then setting somewhere west of the meridian. The moment when a celestial object passes this imaginary line is called the *meridian passage.* The name comes from the Latin *meridies*, literally meaning "midday," for the passage of the Sun past due south.

2. **Earth-based** (Figure 57): Latitude and longitude are the two coordinates identifying positions on the surface of the Earth. One of the goals of using the stars for navigation is the determination of a *fix*, an unambiguous location that can be compared to a map or a table of latitudes and longitudes.

Lines of latitude encircle the Earth extending east–west, with the equator being at 0 degrees. Positive latitudes extend north from the equator, and latitudes to the south of the equator are designated negative. The North Pole is at latitude of +90 degrees, or 90 degrees N. The South Pole is at −90 degrees or 90 degrees S. The lines of latitude are sometimes called *parallels*, as they never converge to a single point but run parallel to each other.

Lines of longitude run north–south, crossing the equator and lines of latitude at right angles. The convention of dividing the world into an Eastern and Western Hemisphere comes from Ptolemy's description of an occupied half of the world stretching to the east from a prime meridian or zero of longitude. Ptolemy's choice for a zero of longitude was the westernmost point of land known in his time, which he named the Fortunate Islands and were likely the present-day Canary Islands. His system used only positive values of longitude for locations in the occupied half. For people living in the Western and Eastern Hemispheres, we use negative values for the west and positive values for the east, starting from the prime meridian. The values go from −180 degrees to +180 degrees or 180 degrees W to 180 degrees E.

The choice of the zero of longitude or prime meridian was arbitrary. Arab and medieval tables of latitude and longitude had a number of different prime meridians, which created confusion. Even into the nineteenth century, different countries used different locations for their prime meridians, but this was finally settled in 1884 at the International Meridian Conference, using the British Royal Observatory in Greenwich as the ultimate origin of longitude. Lines of longitude converge at the poles in the same way lines of azimuth meet at the zenith for the observer-based system.

While a navigator in many ways could carry out a precise determination of latitude using the altitude of stars, a precise determination of longitude proved to be a more elusive goal. The Earth rotates once every twenty-four hours, and the navigator at some latitude will see the sky pass overhead looking more or less the same, regardless of her longitude. This remained a major navigational problem until the widespread adoption of the chronometer as a means of finding an absolute time reference.

3. **Celestial maps** (Figure 58): Declination and celestial longitude are the two coordinates identifying positions on a map of the sky. These coordinates reuse the latitude-longitude system from terrestrial maps but with slightly different names. If we projected every object in the sky directly onto the face of the Earth at some instant in time, each of these would have a value for latitude and longitude that we could call out. If everything in the sky was fixed with respect to the Earth, we could just go by the names of the latitudes and longitudes of the stars in the sky and be done with it, but since the Earth rotates about its axis and orbits the Sun, we need to give the stars their own map, which mirrors as closely as possible the latitude-longitude system.[7]

   There are obvious equivalents with the terrestrial map. Hovering over the Earth's equator is the *celestial equator*. Hovering over the Earth's North and South Poles are the *north celestial pole* and the *south celestial pole*. The angles of stars north and south of the celestial equator are the equivalent of latitude and go by the name of *declination*. We could very well call it "celestial latitude" if we liked, and this term would probably

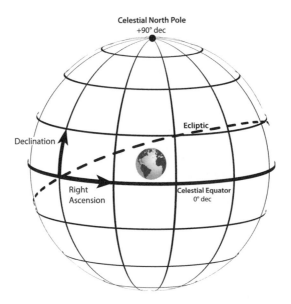

Figure 58 Celestial coordinates. Declination is the equivalent of latitude on earth, and celestial longitude is the equivalent of longitude on earth. The north and south celestial poles are directly over the Earth's North and South Poles. Image credit: A. Scherlis.

be more descriptive, but "declination" is a historical convention. Like lines of latitude, lines of declination are a set of parallel circles that run east–west in the sky. As with latitude, the celestial equator has a declination of 0 degrees and runs north to +90 degrees at the north celestial pole and –90 degrees at the south celestial pole.

The declinations of major stars allow a navigator to pair the stars with places on Earth. A star with the declination equal to the latitude of a place on Earth will cross its zenith once a day. Knowing the declination of a major star, a navigator can keep track of his latitude from its altitude at its meridian passage.

I identify the equivalent of longitude with a *celestial longitude*. This has the same properties as the terrestrial longitude and runs from +180 degrees to –180 degrees or can be designated east or west celestial longitude.

> The equivalent of the prime meridian for star maps, like terrestrial maps, is arbitrary.

The Earth's orbit around the Sun and the tilt of its axis create an apparent motion of the Sun against the fixed background of stars. This path is called the *ecliptic* and moves from a high declination of 23 degrees N at the summer solstice to a low declination of 23 degrees S at the winter solstice. It crosses the celestial equator twice, once at the spring equinox and once at the autumnal equinox. By convention we take the celestial prime meridian to be the point where the Sun crosses the celestial equator at the spring (vernal) equinox and measure celestial longitudes from there as a reference. This celestial version of the prime meridian is called the *vernal equinoctial colure.*[8]

Figure 59 shows a mapping of the major navigational stars I've described above and a few more in the Southern Hemisphere. The plot also indicates the rough outlines of familiar constellations, along with the path of the ecliptic. The plot shows the stars on a rectangular grid. You should bear in mind that the lines of longitude all converge at the North Pole and South Pole, so there is a distortion in the relative positions of the stars as you look toward north and south declinations.

Figure 60 is the same map as Figure 59, but with the outline of the Earth, with the latitude corresponding to declination and longitude corresponding to celestial longitude. The Earth's rotation can be thought of as the outline of the Earth moving from right to left underneath the chart of the stars once every twenty-four hours. You'll note that the Earth itself has to be displayed as a mirror image of the way we would normally view it, to correspond to the map of the stars as we view them from the ground.

The mirror image of the Earth in Figure 60 illustrates a curious aspect of how we create and imagine maps. Maps are created as we visualize a scene, so a map of the Earth has the appearance of what

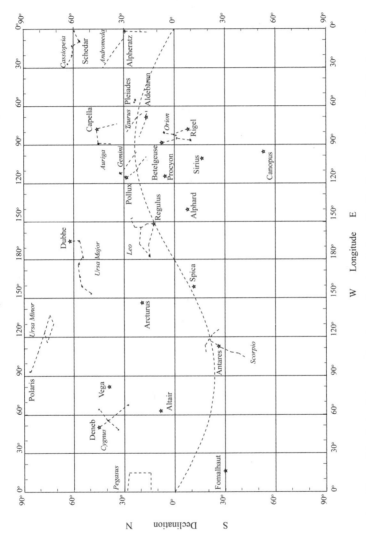

Figure 59 Celestial map of major navigational stars. Declination is on the vertical axis and celestial longitude is drawn to correspond to terrestrial longitude. The path of the Sun over the course of the year (ecliptic) is shown as the dashed line.

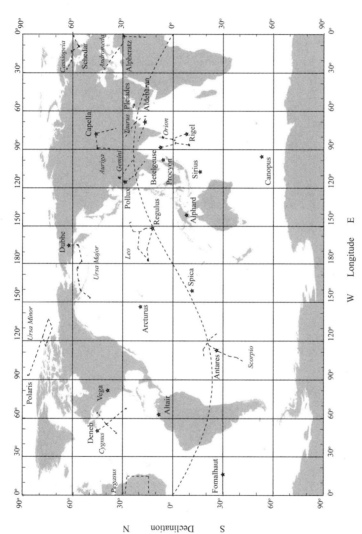

Figure 60 Map of the Earth overlaid on the star map so that latitude on Earth corresponds to the declination of stars and longitude corresponds to celestial longitude. The zero of celestial longitude is taken to coincide with the prime meridian (e.g., noon in Greenwich on the vernal equinox).

we might see as we hover in the sky looking down with our head oriented to the north. A map of the heavens has the appearance of what we might see in the sky by looking upward with our head oriented to the north. If we wish to overlay a map of the Earth against a map of the heavens, one of them has to have west and east flipped around, as if viewed in a mirror. For the representation of Figure 60, I chose to flip around the image of the Earth, as the outlines of the continents are more familiar to most people than the images of the stars in the sky.

In Appendix 1 I list the declinations and celestial longitudes of the major navigational stars in the chart. Along with them I list the latitudes and longitudes of some major points on Earth that correspond to the locations on the celestial map. The locations I've used are arbitrary and subject to my own whims. In Figure 61 I show on a map of the Earth the locations I've chosen in making this correspondence. As I describe below, some Pacific Islands navigators associate star positions with the islands that pass beneath their zenith. In this way they've effectively memorized a kind of chart akin to Figure 60, where star positions are associated with positions on Earth. Note that it's exceedingly rare that a memorable position on Earth lines up in *both* latitude *and* longitude with the corresponding coordinates of a major star in the sky.

On the overlay map and in Appendix 1, I've chosen separate memorable locations on Earth. The first is to give a correspondence between that position and a star's declination, and the second to give a correspondence between that position and a star's celestial longitude. In a number of cases, the memorable locations are the same, as two stars may be at the same declination or celestial longitude. For example, Procyon and Pollux both line up at a celestial longitude of approximately 115 degrees E, corresponding to the longitude of Beijing.

The alignments between longitudes on Earth and on the overlay of Figure 60 happens once a day at different times, depending

Figure 61 Places used in the star–Earth associations listed in appendix 1.

on the date. On the vernal equinox, when the Sun has a declination of 0 degrees and a celestial longitude of 0 degrees, this overlay map (Figure 60) corresponds to noon at Greenwich, UK.

PATHS OF THE STARS IN THE SKY

The navigator can use stars in two fundamentally different ways. On one hand, stars can be used for orientation, acting as a natural compass. On the other hand, a more sophisticated use of stars is as a means of fixing position on the Earth. In either case it's important to understand the motion of stars throughout the course of a night, as this will be different for observers at different latitudes.

The motion of the stars can be confusing at first, although it all stems from the Earth's rotation and orbit. If you sit directly at the North Pole, it is like being at the center of a giant merry-go-round. The paths of all stars form complete circles overhead. If you sit at the equator, stars will form arcs high in the sky. In the midlatitudes, the paths are some mixture of partial arcs and full circles. Because of Earth's rotation all stars move at a rate of 15 degrees per hour in the arcs. This corresponds to 1 degree every four minutes.

To create a map of the stars for the observer, one needs to know the mathematical formulae that convert the star positions to local coordinates. The Arab mathematician al-Jayyani solved the problem of mapping the motions of the stars for observers on the ground at different locations in a treatise called *The Book of Unknown Arcs on a Sphere*, published around AD 1060. If you understand where you are on the face of the Earth and what stars you are following, there are equations that allow you to predict what you will see on your local dome in the sky. For the navigator, this can be turned around: by observing the altitudes of stars on your local dome in the sky, and knowing which stars you're looking at, you can find your latitude. With a clock, you can also find your longitude.

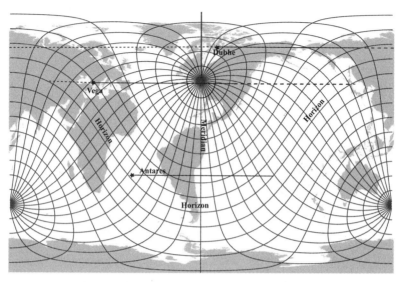

Figure 62 Motion of Antares, Vega, and Dubhe from sunset until sunrise on the night of May 31–June 1 for an observer in Boston. Lines of constant azimuth converge at the zenith, while lines of constant altitude encircle the zenith. East is to the left in this figure. The horizon (altitude = 0 degrees) is indicated. Base image credit: D. Ragozzine.

Let's say you are in Boston (42 degrees N) on the night of May 31 to June 1. What do you see in the sky? As an illustration Figure 62 shows the lines of altitude and azimuth and the local horizon for an observer along with the paths that the stars Antares, Vega, and Dubhe take over the course of that night. The star paths are lines passing over through constant latitudes: solid when visible at night, dashed lines during the daylight hours. In Figure 62 the horizon is shown mapped onto the star map. The zenith is the point where the lines of azimuth converge to the observer in Boston. The lines encircling the zenith point are lines of constant altitude. The local meridian is indicated as a north–south line. For the date and time interval of the figure, Antares is seen to rise in the southeast (left), increase in altitude, cross the meridian, decrease in altitude, and set in the southwest (right).

Figure 63 shows the same paths as in Figure 62, but from the point of view of an observer on the ground, looking up into the

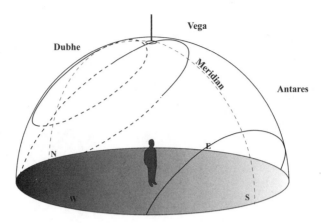

Figure 63 Paths of Antares, Vega, and Dubhe for an observer on the ground, corresponding to Figure 62. This is shown for the time between sunset and sunrise on the night of May 31–June 1 in Boston.

sky. Antares rises in the southeast, arcs over the meridian toward the south and sets in the southwest. Vega rises in the northeast and nearly passes over the zenith at its meridian passage. Dubhe is circumpolar and can fully orbit the north celestial pole without setting.

If the you look to the east, you'll see stars rising and setting at an angle of 90 degrees minus your latitude for stars right on the equator, like those in Orion's belt. Figure 64 shows what this would look like where the stars paths have been made into streaks to aid the eye. This is the view you would get with a camera recording the star paths over the course of an hour.

If you look northward toward the celestial pole, you will see the circumpolar stars orbiting the pole, as in Figure 65. There is something called a *polar distance*, which is simply 90 degrees minus the star's declination. You can find your latitude by measuring the smallest altitude for a circumpolar star. In this case your latitude is the smallest altitude plus the star's polar distance.

One circumpolar star that's easy to use for most observers in the Northern Hemisphere is Polaris. Although it is approximately forty-four arc minutes away from the true north celestial pole, let

Figure 64 Paths of setting stars in the sky looking toward the west in the Northern Hemisphere. Stars on the equator rise and set at an angle of 90 degrees minus the latitude.

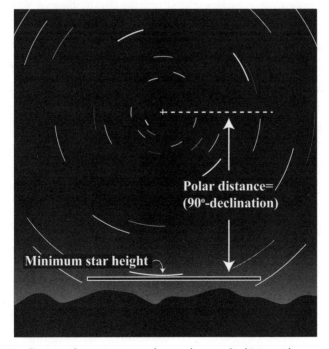

Figure 65 Circumpolar stars as seen by an observer looking north.

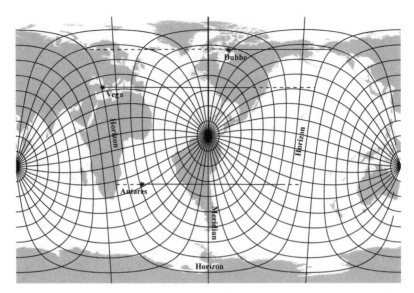

Figure 66   Paths of Antares, Vega, and Dubhe for the same period as in Figure 62 for an observer in Caracas, Venezuela. Base image credit: D. Ragozzine.

us for now take it as being at a precise declination of 90 degrees. That being the case, the altitude of Polaris above the horizon would directly give the latitude of the observer. If Polaris is relatively low in the sky, you could estimate your latitude by using the hand at the end of an outstretched arm to find its altitude. Looking back on chapter 4, you may recall the names given to the angular widths of various configurations of fingers and hands by Arab and Caroline Island navigators.

Now, let's say we magically transport you to Caracas, Venezuela, on the same night. What would you see? Figures 66 and 67 show the corresponding versions of Figures 62 and 63 for an observer in Caracas (10 degrees N).

In Caracas, which is near the equator, there will be far fewer, if any, circumpolar stars. Dubhe is no longer circumpolar. Vega makes its meridian passage well to the north of the zenith, and Antares arcs much higher in the sky. At 10 degrees N, you might barely be able to see Polaris. From Figure 67 you might be able to see that paths of stars seem to intersect the horizon at nearly right

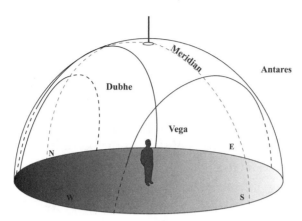

Figure 67 Paths of Antares, Vega, and Dubhe for the observer in Caracas for the same period as in Figure 66.

angles. For observers between 20 degrees S latitude and 20 degrees N latitude, the stars will rise and set at nearly the same azimuths for all locations, making this belt of latitudes ideal for a concept called a *star compass.*

DIRECTION FINDING: STAR COMPASSES AND STAR PATHS

The oldest navigational use of stars may have been for orientation at night. If you're comfortable with the star positions and their paths at different times of the year, the orientation comes naturally. However, it is a skill that requires practice and doesn't come from reading a book. The more ambitious reader can try to memorize the association of stars with places on Earth listed in Appendix 1, then go out and actively name stars and sight the cardinal points of east, west, north, and south to build up intuition. The precise values of declination or celestial longitude aren't as important as an understanding of the relative positions of stars.

As an example of the use of stars for orientation, we can return to the story of Marc Aubriere mentioned at the beginning of the chapter. We don't know exactly how he oriented himself with the stars at night, but I can take an educated guess. He was kidnapped by the Islamist group al-Shabaab from the Sahafi Hotel in Mogadishu. Al-Shabaab controlled the neighborhoods on the east side of Mogadishu at the time, while the government palace and most European interests are located in the western zone of the city. At 10 degrees N, Mogadishu has the same latitude as Caracas; the paths of stars in the sky will be the same as in Figures 66 and 67.

If I emerged from an al-Shabaab stronghold at midnight, I'd want to head west in the city to find safety. Probably the easiest way of heading west would be to notice the constellation of Cassiopeia to the north, which would be high in the sky at that time of year. By keeping Cassiopeia to his right, he would have been able to keep moving westward. A more sophisticated use of stars would be to use the setting azimuths of some of the brighter stars at that time of year. Altair would set 9 degrees north of due west and would make a good beacon as a kind of star compass. Pacific peoples used stars as direction-finding beacons this way.

In chapter 2, I described how navigators in the Caroline Islands used an *etak* system to mark the passage of their voyages from one island to another. The primary basis of this system was a star compass based on the rising and setting azimuths of stars. Near the equator the rising azimuths of stars are very close to 90 degrees minus their declinations and change little as long as you voyage in a range of latitudes between 20 degrees N and 20 degrees S. Figure 68 shows an example of a star compass. Dubhe rises in the north-northeast and sets in the north-northwest. Betelgeuse and Rigel rise straddling due east and set straddling due west. A navigator could keep a canoe on a constant heading if he could keep track of the rising and setting positions of the major stars.

As long as a star is relatively close to the horizon, it can serve as a direction indicator, but when it begins to climb into the sky, its azimuth becomes less discernable. Pacific Islands navigators

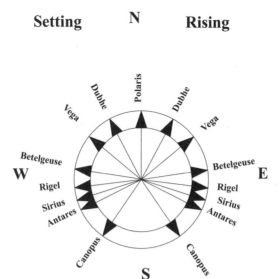

Figure 68   Star compass using rising and setting positions of major stars.

solved this problem with the concept of *starpaths*. A starpath is a sequence of stars that all rise and set at the same azimuth. A navigator doesn't need to memorize all the stars in a given path. As long as he can identify a major star in that path when it rises, he can then identify another star at the same azimuth and track that, then find another star as that one rises. This allows for a reliable star compass without the memorization of an impossibly large number of stars. Sir Arthur Grimble, a long-time resident of the Gilbert Islands, reports an interview with a navigator named Biria who could name 178 stars, constellations, and nebulae and report their relative positions at different times of the night on different days of the year. One hundred seventy-eight stars is on the high side for most Pacific Islands navigators, but it shows the lengths some will go to in their practice.[9]

As long as a navigator is restricted to voyaging in a zone between 20 degrees N and 20 degrees S latitude, the star compass concept works rather well. Arab sailors, plying the waters of the Arabian Sea and the Bay of Bengal also employed star compasses.

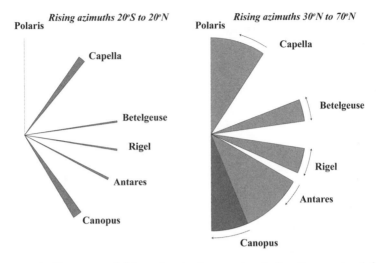

Figure 69 The range of rising azimuths for stars bracketing the equator (left) and stars between 30 and 70 degrees north latitude.

If, however, you venture beyond this band of latitudes bracketing the equator, the whole idea rapidly breaks down. Figure 69 shows star compasses for two ranges of 40 degrees of latitude. In the zone bracketing the equator, there is some modest spread in the azimuths of the rising positions of major stars, and you could sail from the southern reaches of Tahiti to Hawai'i with a single star compass without difficulties.

On the other hand, if that range of 40 degrees spans 30 degrees N to 70 degrees N rather than the equator, the system is useless unless you are sailing due east or due west. On the right-hand side of Figure 69, I've drawn the range of rising azimuths for Betelgeuse, Rigel, Canopus, and Capella. Canopus will be visible at 30 degrees N but will rapidly dip below the horizon. Antares will be visible at more northern latitudes than Canopus, but it, too, will disappear. Capella will make the transition from rising at some point in the east to becoming circumpolar. Even Rigel and Betelgeuse, which bracket the equator, will have rising azimuths that change substantially with latitude.

Star Pillars and Zenith Stars

If your position has a latitude equal to the declination of a major star, that star will pass over your zenith at some time over the course of twenty-four hours. As a crude way of estimating position, this might be able to give you an estimate of latitude to a precision as good as half a degree. To exploit this trick you need some means of determining vertical. This could be the mast of a ship or a weighted line hanging down. If a star is within about 5 degrees in declination of the observer's latitude, the navigator can lie on his back as the star is approaching its highest point in the sky, called the *meridian passage*. He notes the angular distance from his vertical reference and keeps checking it. At some point it will make a minimum angle with respect to the vertical reference, which the navigator then notes. By knowing whether the star passes north or south of the zenith, and by what amount, the navigator can subtract or add the minimum angle from the star's declination to find his latitude.

This use of a *zenith star* to find latitude has been discussed by a number of authors as a technique that may have been used by Pacific Islanders voyaging long distances. An island can be identified with a star that moves directly over its zenith. Arcturus is frequently called out as the zenith star for Hawai'i, as it has a declination close to the latitude of the southern tip of the Big Island. Figure 70 illustrates the concept of a zenith star, which creates a correspondence between the celestial map and the terrestrial map. At some point during the day or night, a star with a declination equal to the latitude of an island will pass through the zenith of that island. A navigator could find an island by sailing north or south until a star associated with a specific island appeared at the zenith, then sail east or west until he encountered the island. The zenith position of a star could be found by sighting along the height of the mast.

Although Pacific Islands navigators don't appear to use zenith stars at present, there are some hints that they were used in earlier times. Tahitians had a song about star pillars holding up the dome of the sky. One version was recorded in 1818 of a woman named Rua-nui who lived on the island of Bora Bora. The stars were later identified with their equivalents in English by the aid of one Baora'i, who was counsellor of Bora Bora, A transcript of the song was published in the *Journal of the Polynesian Society* in 1907. In the chant she names a number of the major stars and associates qualities to them. The song reveals a knowledge of star positions that portrays a kind of celestial map in a lyrical fashion.[10]

In her song Rua-nui calls Antares the "entrance pillar of the dome of the sky." With a declination of 26 degrees S, Antares would be at the zenith of the southernmost islands known to the Tahitians: Rapa Iti in the Austral Islands. Spica is the "pillar of perfect purity," and Dubhe is the "upper side pillar" and would form the northeastern and northwestern points of a star compass. Alphard is "a red star that flies in the open space south." As I described previously, Alphard is located in an underpopulated stretch of the sky, so the description is apt. Finally, Rua-nui calls Polaris "the pillar to fish by, in the boundary of the sky."

The last reference to Polaris is curious; it is not observable from Bora Bora, which is well south of the equator (16° 30' S). The North Star is only visible from latitudes above roughly 6 degrees N. Farther south than this latitude, Polaris is too close to the horizon to be seen until it disappears altogether. Bora Bora was quite isolated from the West until the late nineteenth century, so knowledge of Polaris would have had to be associated with distant voyaging, perhaps trade with Hawai'i, where it would be visible. Calling it the "boundary of the sky" is a sensible description from the perspective of culture that spans latitudes of 22 degrees N (Kaua'i, Hawai'i) to 26 degrees S (Austral Islands).

Author David Lewis describes interviews with Ve'ehala, a navigator from Tonga who speaks of a *fanakenga*, or a star that "points down toward an island" or its "overhead star." There are stars associated with specific islands, their fanakenga stars. In the

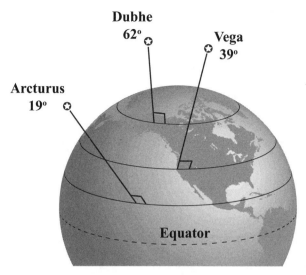

Figure 70 Concept of a zenith star, where each star will be directly at the zenith of a place on earth with a latitude equal to the star's declination. At some time during the night, Arcturus will be directly over Hawai'i (after Burch, *Emergency Navigation*, 45).

discussion Ve'ehala relates that the use of fanakenga is no longer practiced and he learned it from an old woman, who in turn had learned it from chants. Ve'ehala could describe the rough concept, and it sounds like a zenith star, but it was not being actively used at the time of the interview.[11]

If the navigational target is a distant archipelago a navigator wishes to reach, a zenith star can give useful information. Consider a voyage from Tahiti to Hawai'i, a distance of some two thousand five hundred miles. Arcturus, at 19 degrees N latitude, would be at the zenith of the southern end of the chain. A navigator from Tahiti could sail due north until Arcturus just passed over his zenith to the south, then turn and sail west at constant latitude. The Hawaiian Islands form a wide enough target that this strategy, called "running down latitude," would be sufficient, and a navigator would surely intercept this at one of the islands.

Interviews with present-day Pacific Islands navigators by such authors as David Lewis indicate the prevalence of star compasses, but they find no evidence of the present use of zenith stars. On the other hand, around the time of the earliest Western European exploration of the Pacific, much of the long-distance voyaging had already died out. By that time the inhabitants of isolated locations such as Easter Island had already lost contact with their nearest island neighbors.

Geoffrey Irwin, a professor of archaeology at the University of Auckland, discusses the systematic loss of contact of distant Pacific Islands peoples in his book *The Prehistoric Exploration and Colonisation of the Pacific*. Irwin reviews archaeological evidence and computer simulations of voyages to recreate the trajectory of the rise and decline of long-distance voyaging of Pacific Islanders. Irwin argues that the ancestors of the present-day inhabitants may have used more sophisticated strategies for navigation than are presently in use. If this is the case, the concept of a fanakenga (overhead star) or the chant about star pillars may be the faint echoes of a more daring period of navigation.[12]

8. The Sun and the Moon

· ·

WE'VE ALREADY SEEN the use of the Sun in navigation: the naming of "east" with sunrise and "west" with sunset in many cultures. Although its influence on the climate of Earth is profound, the motion of the Sun throughout the year is more complicated than the regular motion of the stars and presents more of a challenge for someone who uses it to navigate. The Moon is more complicated still, yet its motion presents a predictable pattern that ancient civilizations discerned. For the earliest navigators the Sun was probably most useful as a direction indicator, particularly in temperate climes. The length of day, the Sun's height at noon, and its rising and setting positions all give information that can be converted into the latitude of the observer.

On any day the Sun's motion creates a path through the sky that looks like that of a star. However, over the course of a year, it makes a full circuit around the celestial longitude and has a declination that varies between 23 degrees N and 23 degrees S. This path is the result of the tilt of the Earth's axis combined with its orbit around the Sun. Figure 71 shows the tilt and orbit at the equinoxes and solstices.[1]

Figure 72 shows the path the Sun takes in the sky at the equinoxes and solstices for an observer at the latitude of Boston (42 degrees N). At the spring and fall equinoxes, the Sun is directly over the equator, rising due east for every observer on the globe. The equinoxes were an occasion to find the latitude of cities by measuring the angle of a shadow cast by a stick at midday.

On any date, if the solar declination is known, there is an equation that gives the maximum altitude of the Sun at the time of its meridian passage:

$$\text{Alt} = 90° - |\text{Lat} - \text{decl}|$$

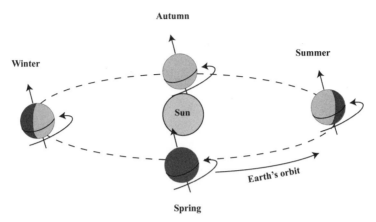

Figure 71 Passage of the Sun in the Northern Hemisphere looking south (after Burch, *Emergency Navigation*, 86).

Here "Lat" is latitude and "decl" is declination. The vertical bars enclosing the expression "Lat – decl" mean "take the absolute value of." This means you always subtract the positive sense of latitude minus declination to get the maximum altitude of the Sun for any declination.

The Sun reaches a declination of 23 degrees at the summer solstice. On that day its maximum altitude is high for observers in the Northern Hemisphere. For example, in Boston, at 42 degrees N, it is $90° - 42° + 23° = 71°$. On the other hand, at the winter solstice, the Sun's declination is –23 degrees, so the highest altitude is a meager 25 degrees above the horizon.

In Figure 72 some other features are apparent for an observer in Boston. While the Sun rises due east and sets due west at the equinoxes, it rises and sets substantially north of east and west at the summer solstice. At the winter solstice it rises and sets substantially south of due east and west. The lengths of the arcs are also quite different. The Earth rotates at a rate of 15 degrees in one hour, and the longer arcs produce a longer day in the summer. While these paths are descriptive, they don't convey the bigger effect. The amount of sunlight that reaches the Earth's surface depends on the angle of the Sun's rays. When the Sun is shining

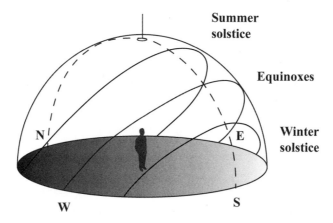

Figure 72 Path of the Sun in the sky as seen from the latitude of Boston (42° N) at the solstices and the equinoxes. The dashed line is the local meridian for the observer.

down from the zenith, directly overhead, it blasts a thousand watts of power per square meter onto the Earth. That's the equivalent of ten one-hundred-watt lightbulbs shining on an area the size of a small throw rug.

In the summer in the Northern Hemisphere, the Sun is shining more directly down on the Earth, and the surface catches a denser batch of rays. In the winter, when the Sun is lower in the sky, the intensity of light drops off dramatically. The combination of longer days and direct sunlight heats up the Earth in the summer, while the short days combined with the more rarified light allows the Earth to cool in the winter.

You might think that most people know "why it's cold in the winter and warm in the summer," but you'd be surprised. In 1987 filmmakers interviewed a group of twenty-three graduating Harvard seniors and faculty members. Twenty-one of the twenty-three gave the wrong answer. Most offered up the explanation that the Earth is closer to the Sun in the summer (it is a little closer in January). The producers of the film suggested that these misconceptions point to significant flaws in the way science is taught. The problem is perhaps not just a problem in our educational system,

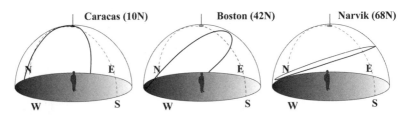

Figure 73   Path of the Sun during the summer solstice at three latitudes: 10, 42, and 68 degrees north.

but because people are oblivious to the Sun's position and path in the sky, it has ceased to have meaning in their lives. If it is presented as a disembodied fact, there is no reason to link the passage of the Sun to the seasons.

On the other hand, if it is observed on a regular basis, it acquires a tangible meaning. During the spring the Sun rises higher in the sky, and the quality of light changes rapidly with each passing day. In the summer when the Sun is at its highest and the days are longest, light produces strong contrasts and bright colors. In the autumn days rapidly become shorter, and the Sun drops in the sky. In the winter the Sun is low in the sky, and the character of lighting is dim and in short supply.[2]

The amount of solar warming is called *insolation*. The effect of insolation produces the change in temperature as one moves away from the equator toward the poles. The zone of latitudes where the Sun reaches the zenith at some time during the year is called the *tropics*, bounded by the Tropic of Cancer at 23 degrees N and the Tropic of Capricorn at 23 degrees S. These are named for the zodiacal signs where the Sun appears at its extremes in declination. The *northern temperate zone* stretches north from the Tropic of Cancer to the Arctic Circle, where the Sun becomes circumpolar. Likewise, the *southern temperate zone* extends from the Tropic of Capricorn to the Antarctic Circle. Figure 73 shows the path of the Sun in the sky for three latitudes: 10 degrees N (tropical), 42 degrees N (temperate) and 68 degrees N (polar). In the town of Narvik, Norway, the Sun

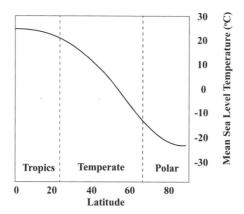

Figure 74 Average surface temperature of the Earth for different latitudes.

is circumpolar during the summer solstice, part of the land of the midnight Sun.

Figure 74 shows the mean average sea level temperature for different latitudes. The tropical temperatures are relatively constant around 26° Celsius (79°F). The effect of changing insolation with latitude is clear in the figure. As you move away from the equator in the temperate zone, the mean temperature drops about 1°C for every one hundred kilometers of travel toward the poles (1°F for every thirty-five miles). It levels out at –25°C (–13°F) in the polar regions. You may recall from chapter 6 the climate categories of the Greeks and Arabs, dividing the Earth into torrid, temperate, and polar regions.

Vincent van Gogh's paintings starkly show the contrasts of the character of sunlight between northern and southern climes. His early work in the Netherlands and Paris are dark and subdued, but his later paintings in Provence are bright and full of strong contrasts.

Figure 75 illustrates the passage of the Sun in the sky over the course of a day from the vantage point of an observer in the Northern Hemisphere. In this figure the Sun rises at 7:00 a.m. and sets at 5:00 p.m. The day is ten hours long in the figure, where the date is on the winter side of the equinoxes. The arc of the Sun moves 15 degrees every hour, and at the meridian passage it attains its highest elevation.

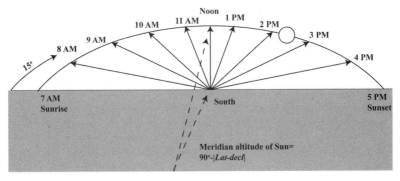

Figure 75   Passage of the Sun in the Northern Hemisphere looking south (after Burch, *Emergency Navigation*, 76).

As the Sun sets, shadows get longer and longer. As the Sun approaches the horizon, even the smallest blade of grass or grain of sand begins to cast a long shadow until the surface of the Earth itself casts a shadow. The transition between the light of day and the shadow of night is called the *terminator*. The terminator on the Moon is what we associate with the Moon's phases. Figure 76 shows the Earth and its terminator at the summer solstice. You can see that regions close to the North Pole will not fall into dark-

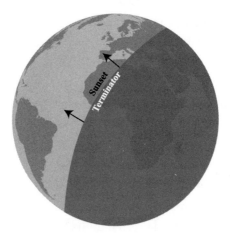

Figure 76   The shadow of the Earth forms the boundary between night and day and is called the terminator. The line perpendicular to the terminator is the direction to the setting (or rising) Sun. This shows the terminator at the summer solstice.

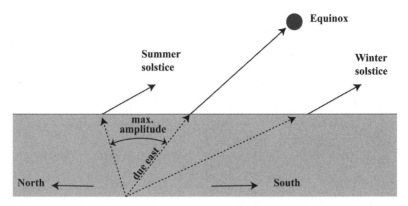

Figure 77 The paths of the rising Sun at the summer solstice, equinox, and winter solstice. The maximum amplitude is the largest swing north or south that the rising (or setting) Sun makes with respect to due east (or west). The Sun rises the farthest north at the summer solstice and the farthest south at the winter solstice (after Burch, *Emergency Navigation*, 79).

ness at all as the Earth rotates. The direction of the setting Sun is perpendicular to the terminator. You can see from the figure that direction to the setting Sun depends on the latitude of the observer at the solstice. At the equinoxes the terminator runs north–south along a meridian and the Sun rises due east and sets due west.

The position of the Sun at sunrise and sunset can be particularly helpful. A regular duty of navigators was to use the azimuth of the rising and setting Sun to measure the magnetic variation (difference between magnetic and true north). The angular departure from a due east rising is called the Sun's *amplitude*. Figure 77 shows the rising Sun at the four important times of the year: the two solstices and the two equinoxes. At the equinoxes the Sun rises due east. Each day after the vernal equinox, it will rise farther north of due east until it reaches its most northerly point at the summer solstice. After the solstice it will begin to rise more to the south until it again rises due east at the autumnal equinox. After the autumnal equinox it will rise farther and farther south of due east until it reaches its maximum southerly journey at the winter solstice, then will begin to move north again. At the solstices the amplitude is at its largest swing north or south during the year.

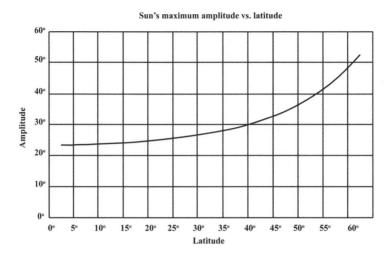

Figure 78 Sun's maximum amplitude varies with latitude.

The builders of Stonehenge placed emphasis on the summer solstice with an opening in a large embankment to the northeast. When one stands at the center of the complex, a heel stone in the middle of the opening points toward the rising Sun on the summer solstice. A number of other megalithic complexes show orientations toward the rising or setting Sun at the solstices.

The Sun's maximum amplitude will vary with latitude (Figure 78). In the extreme case of locations above the Arctic Circle or Antarctic Circle (67 degrees N or S) at the solstices, the Sun doesn't set, and there is no real sense of amplitude. The range of amplitudes is the smallest near the equator, equal to 23 degrees, and rises through the temperate zone. At Stonehenge, 51 degrees N, the Sun's amplitude is 40 degrees at the solstices, meaning it rises at an azimuth of 50 degrees (northeast) on the summer solstice and at 130 degrees (southeast) on the winter solstice.

You don't need a miniature Stonehenge to mark the positions of the rising and setting Sun. If you observe the sunrise and sunset from your house with respect to distant landmarks, such as trees, other houses, or hills, you can track the passage of the seasons. The Hopi tribe in Arizona used the positions of the rising and setting

Sun against distant landmarks as a way of monitoring the progression of the seasons. Anthropologist Alexander Stephen reported on this in his studies of the Hopi. According to Stephen, there was a chant of twenty nearly identical verses that describe the points on the horizon where the Sun rises or sets on the dates corresponding to when certain crops are planted and mature.[3]

LENGTH OF DAY

The length of day at any location depends on the date and latitude. If you live in Narvik, Norway (68 degrees N), in mid-June, the day is twenty-four hours long, while in mid-December it is perpetual night. If you live in Boston, the day can be as long as fifteen hours and as short as nine hours. If you live in Caracas, the length of day hardly varies from twelve hours. At the equinoxes the lengths of day and night are the same for all latitudes. Figure 79 shows the length of day at different latitudes at the summer solstice. Note that the curve is flat above the Arctic Circle, where the Sun is circumpolar at that time of year.

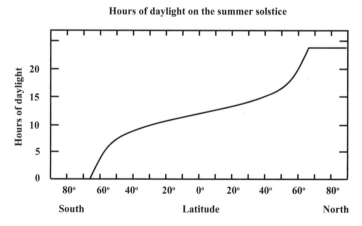

Figure 79 Length of day and latitude during the summer solstice. Beyond the Arctic Circle (67° N), the day is twenty-four hours long. At the equinoxes there is no dependence of length of day on latitude.

Timing the length of day to find latitude appears in a number of survival handbooks. During the Cold War the U.S. Air Force developed a fleet of long-range bombers, some of which are still in active use. The bomber crews have to be ready to fly to just about any place on Earth on short notice. Anticipating the possibility of a crew being forced to bail out, the Air Force trains the airmen in survival techniques and publishes a manual that includes directions on position finding without navigational equipment. Among a number of other techniques, the manual advises that length of day can be used to find latitude and has a full-page nomogram for that purpose. The technique assumes you have a watch and can mark the time of both sunrise and sunset, deriving the length of day. Once you have that, you look up the length of day for the particular date of the year, and the nomogram in the manual allows you to find your latitude.[4]

The technique of using the length of day to find latitude works reasonably well around the time of the solstices, when the Sun reaches its extreme declinations. At the equinoxes the technique is useless, as the length of night and day are the same everywhere on the planet. For the period of a month around the equinoxes, a small error in a measurement of the length of day will become magnified into a large error in latitude. For example, in one attempt to find latitude from the length of day in mid-October, I was off by over three hundred miles.

THE VINLAND SAGAS AND DAY-MARKS

The length of day and the path of the Sun appear as crude markers of the latitude of new lands discovered by the Norse. The "Vinland Sagas" is a term that refers to two texts, the *Saga of the Greenlanders* and the *Saga of Leif Eiriksson*. These were written several hundred years after the discovery of a new land they called Vinland, widely taken to be the northeast coast of North America.

The explorations themselves date from roughly AD 1000. The language of the sagas gives some insight into how the Norse used the path of the Sun to tell time.

For the Norse time reckoning and direction were intertwined. As seen in Figure 73 above, the passage of the Sun at Narvik, Norway (68 degrees N) is circumpolar at the summer solstice. In Iceland and Greenland the Sun has a low path in the sky, allowing its position to be compared to landmarks on the ground as a way of estimating time. While the ancient Egyptians divided up the day into twenty-four hours, the Norse used an eight-fold division of time. Here is a list of the Norse times and their rough translations (note that the symbol "ð" is sometimes translated from Old Norse as a "th" and sometimes as a "d"):

1. *miðnætti*   midnight
2. *ótta*   3:00 a.m.
3. *miður morgunn*   midmorning: 6:00 a.m.
4. *dagmál*   day-meal: 8–9 a.m.
5. *hádegi*   high day: Noon
6. *eykt*   3:00 p.m.
7. *miðaftann*   midevening: 6:00 p.m.
8. *náttmál*   night-meal: 9:00 p.m.

In Iceland a comparison of the Sun's position relative to landmarks as seen from the family farmhouse was one of the most common ways of reckoning time and formed a natural clock. The landmarks were called *day-marks* or *dagsmaurk*. A visitor to Iceland in the early 1800s writes of a mountain used by a family farm to sight midday:

> This mountain forms the meridian day-mark of the Grimstäd family. Few of the Icelanders being in possession of watches, the only Sun-dial they make use of is the natural horizon, which they divide into eight

equal points, called day-marks (*dagsmaurk*), avail-
ing themselves of certain peaks or projections of the
mountains; or, in the absence of these, they erect pyramids of stones on the corresponding heights. Most of
these kinds of pyramids have originally been raised by
the first settlers from Norway, and have been held in
repair from generation to generation; which circum-
stance will account for the difference of time between
the Icelandic computation and that in common use
with us.[5]

The path of the Sun at the latitudes in Icelandic settlements
gives positions at different times of the day that mirror the eight
divisions of the horizon into north (midnight), northeast (*ótta*), east
(midmorning), southeast (*dagmál*), south (*hádegi*), southwest (*eykt*),
west (*miðaftann*) and northwest (*náttmál*). The time of day is then
associated with a place, or *stað*. The noon-marking place name
would be *hádegistað,* the day-meal position would be *dagmálastað*,
and so forth. Place names associated with the day-marks are found
in Iceland, such as Miðmorgunshnjúkur (midmorning peak), or
Dagmálahóll (daymark hill).[6]

In the *Saga of the Greenlanders*, Leif Eiriksson raises a crew to
systematically explore regions west of Greenland. He finds a rich
land that he calls Vinland for the wild grapes that grow on its
shores. The saga describes the characteristics of the Sun in this
new land in midwinter. In the original it reads:

*Meira var þar jafndægri en á Grænlandi eða Íslandi. Sól
hafði þar eyktarstað og dagmálastað um skammdegi.*

One modern translation of this passage is: "The days and nights
were much more equal in length than in Greenland or Iceland. In
the depth of winter the Sun was aloft by mid-morning and still
visible at mid-afternoon."[7]

The literal translation of *skammdegi* is "short days," meaning the time around the winter solstice. A more literal translation of the second sentence is:

> The Sun passed *eyktarstad* and *dagmalastad* on the
> short days [in midwinter].

or

> The Sun passed the *ekyt*-point and the day-meal-point
> in midwinter.

or

> The Sun passed from the southeast to the southwest in
> midwinter.

In Iceland and Greenland there are very few hours of daylight in midwinter, but the implication of the above passage is that the new land had roughly six hours of sunlight around the winter solstice. Some writers interpret this statement very literally and assign a precise latitude for Vinland, which is probably not warranted. The author(s) of the saga seems to be drawing attention to the length of day and character of the Sun's path as seen in this new land. The Vinland Sagas also speak of voyages to Ireland, as far south as Dublin (53 degrees N). At the winter solstice Dublin has seven hours and thirty minutes of daylight, while Reykjavik, Iceland (64 degrees N), has only three hours of daylight. Presumably, the Norse sailors in the era of the sagas were already familiar with the dependence of the Sun's path on latitude.

THE SUN COMPASS

Shadows were employed for a large part of human history to mark time. Traditionally, the length of the shadow of a stick is used to mark the start of the Muslim midafternoon prayer, *Asr.* On the southern wall of medieval Saxon churches, *mass dials* were used to set the time of services. When the shadow of a stick reached a marking on the wall, the sexton would ring a bell to call the faithful. The path of a shadow can also be used for the purposes of orientation.

Figure 80 shows an overhead view of the paths of a shadow cast over the course of a day by a vertical stick at the solstices and the equinoxes. These are the paths that you would see if you laid a small pebble at the end of the shadow periodically over the course of a day. Calculating the path of a stick shadow is a trigonometric problem similar to those encountered in finding the paths of stars. At the equinoxes, however, the path of a shadow for all observers on the Earth will be a straight line. The only differences among varying latitudes will be the shortest shadow length. The closer

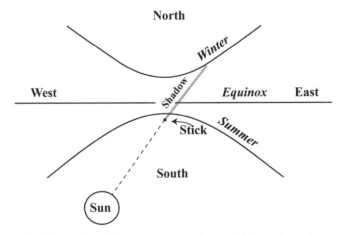

Figure 80 The paths of the shadows cast by a stick traced out by a stick at three different days of the year: the summer solstice, the winter solstice, and the equinoxes as seen from overhead. The paths are shown for 42 degrees north latitude.

the observer is located to the equator the shorter the length of the shadow becomes.

At times other than the equinoxes, the path traced out by the tip of the shadow assumes a more complicated shape called a *hyperbola*. A hyperbola is an open curve; it never closes in and touches itself. The curvature is the largest at the closest approach to the stick.

Figure 81 illustrates the principle of a Sun compass. By tracing the path of the shadow cast by a stick called a *gnomon*, a unique curve can be found for the shadow tip. The length of the shadow will be unique at any time of the day. Once the path of the shadow is calibrated, the user rotates the compass until the shadow just touches the traced path, orienting the Sun compass.

Rather than calculating the path of the shadow tip using a laborious computation, a person can calibrate it on one day from a fixed location by tracing the shadow tip. On subsequent days a navigator can then travel with the Sun compass and use it for orientation. The curve traced on one day can be used with reasonable reliability for a week or more after the calibration date.

A version of the Sun compass was used by Allied troops in the North African Campaign in World War II. Magnetic compasses on jeeps were unreliable, as the iron on the jeep would deflect the

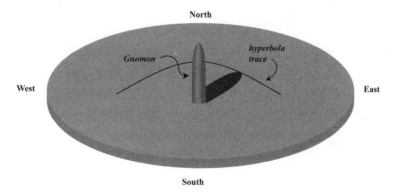

Figure 81 Principle of a Sun compass. This particular example shows the shadow at some time in the afternoon during the summer in the Northern Hemisphere.

compass needle. A Sun compass doesn't suffer from this problem and was rapidly deployed to allow the army to reliably traverse vast sections of the Sahara.

In 1948 C. L. Vebaek, a Danish archaeologist and historian, unearthed a wooden artifact in a Benedictine convent near Uunartoq Fjord in southern Greenland. The artifact bears a resemblance to a Sun compass, with a series of points laid out on a circle. A photograph of this is shown in Figure 82. What remains of the object is roughly three inches high and has seventeen points carved into the circumference and what appears to be a hole in the middle. It is only part of a full circle, and if one completes the circle, it appears to have 32 points in all.

Captain Soren Thirslund, a curator at the Danish Maritime Museum, championed the idea that this artifact was a Sun compass used by the Norse. In addition to the compasslike markings on the

Figure 82 Artifact found in Greenland, believed by many to be the remains of a Norse sun compass. Some of the scratches are interpreted as the paths of shadows cast by a gnomon located in the center (photo credit: Erik Torpegaard, etimage.com, used with permission).

rim, the scratches in the surface are said to be the hyperbolic traces of the shadows cast by a gnomon that extends through the hole in the center.[8]

The interpretation of the Vebaek artifact as a Viking Sun compass has met with some skepticism. The compasslike points, while suggestive, could have been merely decorative or served some other purpose, according to critics of this explanation.

THE SUNSTONE

A device called a *sunstone* is purported to be another item in the tool kit of Viking navigators. I have to first relate some details of skylight to explain the principles of a sunstone.

Many people have an intuitive sense of the time of day based on the character of sunlight: its intensity and color. The blue color of the sky comes from scattering of sunlight from air molecules. Blue light scatters off air molecules more easily than red light.

On the other hand, when the Sun is low in the sky, its light has a red coloration. Seen at low altitudes, sunlight takes a long path through the atmosphere, and most of the blue gets scattered out, leaving mainly red. Photographers often take portraits around sunrise and sunset, as the rosy coloration produces flattering results. This is the case when light scatters once off an air molecule before reaching your eyes. On the other hand, when light scatters many times, as through a cloud, all the color information gets scrambled, and the cloud appears white.

The scattering that creates the blue sky also produces another effect: polarization. Light can vibrate in different directions, but if it vibrates in only one direction, it is said to be *polarized*. When light reflects from a surface, such as from a puddle of water or the roof of a car, only the horizontal vibration survives: the direction parallel to the reflecting surface. This is how polarizing sunglasses work: they have a filter that will block horizontally vibrating light

Polarization of the sky

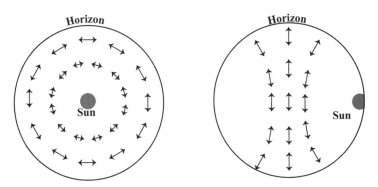

Figure 83 Polarization patterns of the sky, depending on the position of the Sun. The arrows represent the direction of vibration of light from the blue sky.

to the eyes, reducing glare from horizontal surfaces. Light reflecting from vertical surfaces, such as windows, however, will pass through the sunglasses. If you want to see this yourself, take a pair of polarizing sunglasses and rotate them against the glare of light reflecting off a surface and you will see the change in intensity with the rotation angle.

Light also gets polarized when it scatters off molecules in the atmosphere, the same process that creates the blue coloration. When you look at a patch of blue sky, its polarization is determined by the path light takes from the Sun to the patch of sky to your eyes. Depending on the angle between the Sun and the patch of sky you are looking at, the sky will have differing amounts of polarization. In fact, there is a pattern of polarization in the sky that can be mapped out simply with a polarizing filter, just as sunglasses do. Many photographers take advantage of this effect with specific filters. The light from a cloud has all of its polarization scrambled, so the light is not polarized, while the light from the blue sky is. By rotating the filter the photographer can extinguish most of the light from the blue sky, highlighting the cloud.

The pattern of polarization of the sky is shown in Figure 83 for both midday and near sunrise or sunset. The arrows in Figure 83 indicate the direction of the vibration. At an angle of 90 degrees to the Sun, the blue sky is fully polarized, while there are differing states of polarization, depending on the angle the light scatters.

When the Sun is near the horizon, the polarization directly overhead is strongest (right hand side of Figure 83). If you have some way of finding this polarization, you can find the bearing to the Sun. To be precise, the direction of polarization will be perpendicular to the direction toward the Sun in this case.

Some ant species navigate using the polarization of the sky. You may recall the step-counting desert ant, *Cataglyphis,* from chapter 4. The same ant uses the sky polarization to establish direction. If ants can do this, why shouldn't humans be able to use this trick? They could, given the right tools. Unlike ants, our eyes don't directly detect the polarization state of light, but there are naturally occurring crystals that will separate out light into its polarization states. One of these is called calcite, shown in Figure 84.

Figure 84 Light passing through a calcite crystal (Iceland spar) will take different paths depending on its polarization state and orientation of the crystal.

In 1967 Danish archaeologist Thorkild Ramskou suggested that the Vikings used something called a "sunstone" to find the direction of the Sun using the polarization of the sky. Part of this was based on a section of one of the sagas, *Harafns Saga*: "The weather was thick and stormy.... The king looked about and saw no blue sky ... then the king took out the Sunstone and held it up, and then he saw where the Sun beamed from the stone."[9]

Calcite is also known as Iceland spar and is found in large quantities in eastern Iceland. According to a number of accounts, spar crystals were highly prized during the Viking era. Spar has an unusual property of birefringence, meaning that light is bent through the crystal at two different angles, depending on the polarization state of the light. This will give rise to two images of light from an object passing through the spar, as seen in Figure 84. Pure crystals of calcite are *rhombohedral* in structure, meaning that their sides each describe a rhombus. The sides of the crystal are associated with the polarization state, so when the side of the crystal is aligned properly, one of two polarization states is extinguished.

Much has been written about sunstones since Ramskou's original paper. One of the most thorough investigations of the use of the sunstone comes from Leif Karlsen, a sailor and author who reported his work in a book titled *Secrets of the Viking Navigators*. Karlsen puts a small dot on the spar, perhaps from pine pitch, then holds it up to the sky. The light from the blue sky will ordinarily produce two images of the dot when viewed through the spar, but when the crystal is rotated so that its axis is properly oriented, one of the two images will disappear and the long axis of the crystal will point roughly in the direction of the Sun. This works particularly well when the Sun is near the horizon and the spar is pointed near vertical.[10]

The astute reader may be wondering, "If this trick works when the blue sky is visible, what's the point? The Sun should also be visible at the same time." This is a valid concern, but consider the conditions in the North Atlantic at the latitudes where the Norse were sailing: the Sun is typically low in the sky, and fogbanks often obscure everything except a blue patch of sky directly overhead. Under these conditions a sunstone would be ideal.

I've used the technique described by Karlsen and attest that it's a workable way of finding the Sun from a patch of blue sky. Some of my students also tested the sunstone idea as part of a project to see if it could be used as a compass. They tried to find north around the time of sunset using the calcite. As a measure of these students' honesty, they came to me saying that they had failed because their sunstone systematically pointed to the west of due north by quite some amount: 30 degrees. I asked them a few questions and rapidly figured out that they had forgotten to factor in the Sun's setting position, which was 30 degrees south of due west. It was late November in Boston. They had inadvertently not only proven their honesty but also the accuracy of the sunstone, which they ultimately reckoned to be to within about 5 degrees.

GETTING PRECISE WITH THE SUN

What I describe above was good enough for navigators until the fifteenth century. At the start of the Western European era of discovery, the range of voyaging increased to cover the globe and presented more challenges for navigation. Portuguese explorers ventured south along the coast of Africa, crossing the equator and out of sight of Polaris. While Polaris was not visible from the Southern Hemisphere, the Sun was. The use of Sun sightings with instruments of increasing precision pushed navigation into a new realm, where effects at the level of a degree or less became important. Tables of the positions of the Sun, stars, and planets were nothing new for astronomers and astrologers from the time of Ptolemy (AD 200) onward, but this was the first time these were employed by navigators. While the positions of the stars could be taken as relatively fixed, the motion of the Sun in the sky presented more of a challenge.

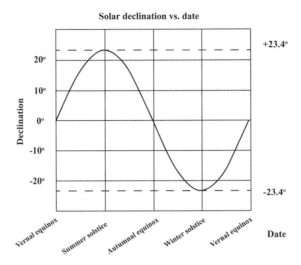

Figure 85 The declination of the Sun over the course of the year, assuming a perfectly circular orbit.

Figure 85 shows the declination of the Sun over the course of a year based on the assumption that the Earth's orbit around the Sun is a perfect circle. The curve describes a mathematical function called a *sinusoid*, which arises from the projection of points around a circle onto a line. You can see that the Sun's declination rises to a maximum at the summer solstice, crosses through zero at the equinoxes and drops to a minimum at the winter solstice.

Some manuals on navigation in emergency situations propose this sinusoid as a way of finding the Sun's declination for different dates. The manuals effectively assume a circular orbit, although this is rarely stated. I naively tested out this approximation to see how well I could do. I constructed a fairly precise device to measure the Sun's altitude from spare parts lying around my basement. I reckoned I could measure the altitude to a precision of about fifteen arc minutes. After making my observations over the course of a day, when I calculated my latitude, I found that I was off by well over a degree, tantamount to an error of

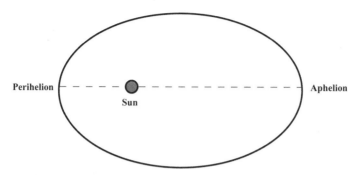

Figure 86   Earth's orbit around the Sun is an ellipse. It moves faster at the point of closest approach (perihelion) than at the farthest distance (aphelion). The figure greatly exaggerates the elongation of the Earth's orbit.

a hundred miles in position, ten times worse than my accuracy! What went wrong?

The Earth's orbit around the Sun isn't a perfect circle but really an ellipse. The Sun isn't centered in the ellipse; it's displaced to one side. Figure 86 illustrates the shape of an elliptical orbit, where the Earth is closest to the Sun at some point in its orbit, called the *perihelion* and furthest at another point, called the *aphelion*. The Earth moves faster near the perihelion and slower near the aphelion. (The figure itself greatly exaggerates the elongation of Earth's orbit.)

Although the deviation from a perfect circle in Figure 86 is vastly exaggerated, the elliptical nature of the orbit is sufficient to create an observable deviation from the sinusoid I showed in Figure 85. Figure 87 shows the difference in declination over the course of a year between the assumption of a perfectly circular orbit and the true orbit. You can see that differences as large as 1.5 degrees can arise at some times of the year, which is large enough to cause a large shift in latitude. From Figure 87 you can also see that there are shifts in the declination at any given date from year to year. Part of this shift is because the calendar is not perfectly synchronized with the Earth's orbit. There are 365.24 days in the

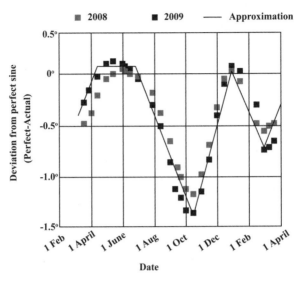

Figure 87 Difference in the Sun's declination from the model of a perfectly circular orbit. The black and gray dots are from different years.

year, which requires a leap year every four years and produces a yearly "creep" in declination on any given date.

If I was able to find the discrepancy between a circular orbit and an elliptical orbit over a weekend with a crude device, you might wonder why the ancients didn't hit on this and continued to use perfect circles to describe planetary motion. Ancient Greek astronomers actually did observe this effect but wanted to preserve the notion of an Earth-centered (geocentric) universe *and* perfect circles. Ptolemy found two constructions to create reasonably precise tables of declination. He had the Sun orbit the Earth (geocentric orbit) with a circular orbit but with the Earth displaced from the center of the circle by enough to ensure that his calculations agreed with observations. This is shown in Figure 88, along with an alternative that involved the use of "an orbit around the orbit," called an *epicycle*. Ptolemy showed that both the epicycles and the displaced circular orbit gave rise to the same declinations. This kind of construction worked reasonably well for fourteen hundred years in the construction of tables for

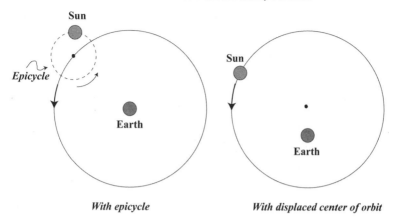

Geocentric orbits in Ptolemy's schema

With epicycle   *With displaced center of orbit*

Figure 88 Two possible solutions to the problem of solar declination with circular orbits around the Earth proposed by Ptolemy. In one case an epicycle is added to the Sun's orbit. In the second case the center of the orbit is displaced from the center of the Earth.

the motion of the Sun and planets in the sky. It was only when Johannes Kepler pushed the precision of calculations using data from Tycho Brahe that the modern model of the solar system emerged.

Time

The rhythm of the day is dictated by light, and time is based on the passage of the Sun. Three points during a day can be easily established: sunrise, midday, and sunset. The subdivisions of time within a day are arbitrary. While the Saxons used eight *tides* as subdivisions, the ancient Roman tradition used twelve hours. The word "noon" comes from the Latin *nona hora*, the ninth hour of the day, or 3:00 p.m. This was a time for church prayer, which shifted to midday around the twelfth century but continued to carry the name "noon" along with it. From that period the word "noon" became synonymous with midday. The overall length of a day shifts with the season, as does the rising and setting position of the Sun, but midday can always be found from the time of the shortest length of a shadow.

Before the widespread use of mechanical clocks, time was often taken to be that cast by a sundial. A town forty miles to your west might have noon four minutes later than you, while a town forty miles to the east might have it four minutes earlier. As long as you're moving from one place to another on foot, or on horseback, this shift is inconsequential. However, when communications and travel became swifter, the idea of a purely local time created challenges. How can you run a railroad and avoid collisions if every town had its own standard?

The concept of time zones arose toward the end of the nineteenth century to solve this problem. Only with the advent of reasonably precise mechanical clocks could time be synchronized across an entire country. Every 15 degrees of longitude equals one hour of shift in the noontime position of the Sun. Accordingly, each time zone is drawn to place the Sun approximately within thirty minutes of the meridian for all observers in the zone. There are notable exceptions to this. Time-zone boundaries shift east and west to avoid large population centers, and some countries choose to be thirty minutes off the standard. In 2007 President Hugo Chavez declared that Venezuela would be in a time zone half an hour different from its neighbors based on the notion that it would improve productivity and the health of its citizens. None of these issues are of much concern for navigation but can be important for local commerce.

Time had importance for navigators much earlier than the creation of time zones. Longitude could be established by comparing the local noon to a standard clock that kept the time of the prime meridian. The invention of a reliable nautical chronometer by John Harrison in the eighteenth century allowed navigators to compare the time of the local meridian passage of the Sun to the time in Greenwich. With 360 degrees in a 24-hour day, this gives a rotational speed of 15 degrees per hour. The conversion of 15 degrees to the hour made the calculation of longitude straightforward. There is an important systematic shift between the position of the Sun and the time kept on a clock: the *equation of time.*

Our clocks and watches are all set to something called *mean solar time.* In principle we expect the Sun to be at the meridian at noon in the center of our time zone. This isn't always the case. With a reliable clock we'd find the following: On some days of the year, the Sun crosses the meridian as much as sixteen minutes earlier than 12:00, and on other days of the year, it crosses as much as fourteen minutes later (Figure 89). Now, you might excuse an appointment for being fourteen minutes late, but for a sailor this represents an error as large as 210 nautical miles.

Two effects cause these shifts. We already encountered the first in the context of declination: the eccentricity of the Earth's orbit. As the Earth speeds up or slows down in its orbit, the position of the Sun in the sky will shift relative to some average circular orbit. The point of closest approach of the Earth to the Sun is early January. In addition to affecting clocks, this effect of speeding up and slowing down requires adjustments in the calendar. The lengths of the months are trimmed so the solstices and equinoxes fall close to the twenty-first days of June, December, September, and March. There are 180 days between September 22 and March 21 and 185 days between March 21 and September 22, a difference of five days.

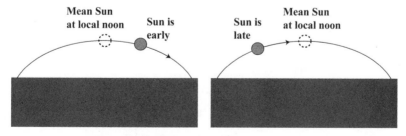

Figure 89 Mean Sun in the sky represents the mean solar time — the basis for our clocks. The true Sun can cross the meridian passage early or late, depending on the time of year.

The second effect is subtler. As you may recall from chapter 7, the Sun appears to move on a path, called the ecliptic, that changes its declination and celestial longitude. The motion through the ecliptic is just a little less than 1 degree per day. Our definition of "mean solar time" implies that the shift of the Sun's celestial longitude is constant throughout the year, but in reality it has an up-and-down motion. Sometimes it advances its celestial longitude faster than the average, and sometimes it advances it more slowly.

Both effects average out to zero by the construction of mean solar time, but that means that the Sun will be faster or slower, depending on the particular date. Figure 90 shows the equation of time for a year. The shape comes from the combination of the two effects above. The word "equation" in the phrase "equation of time" is a bit of a misnomer. Although the effect can be represented as an equation, it has come to mean the difference between mean solar time and the position of the Sun in the sky.

In Figure 90 the negative values mean the Sun is slow with respect to the "mean" Sun, a fictitious object that represents the position of the Sun without the two effects coming into play. The

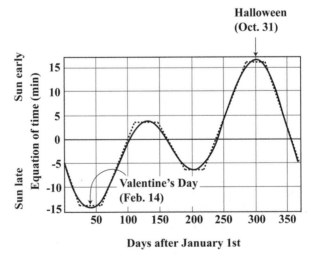

Figure 90 Equation of time and approximation on top of it (dotted, after Burch, *Emergency Navigation*, 194).

dashed line is a crude aid that some people use to memorize the equation of time; it's called a "trapezoidal approximation" and allows a person to bypass difficult trigonometric calculations in an emergency situation.

THE MOON

The passage of the Moon through the sky is more complicated than the passage of the Sun, which is already tricky to follow. The motion of the Moon is an example of what physicists call a "three-body problem." Newton's theory of gravity is very precise, and calculating the paths of two objects is a straightforward process. But when three objects get into the act, the resulting motion can be very difficult to solve, even though we know the forces in principle. This is the case with the Moon, on which both the Sun and the Earth exert competing forces. This situation created a major challenge for some of the best astronomers and mathematicians of the eighteenth century to solve.

A thirteenth-century publication from Norway called *The King's Mirror* reflects the complexity of the Moon's motion: "But these things tradesmen can hardly mark for the sake of their swift course because the Moon takes so large steps either up or down that people can for that reason hardly determine directions from its courses."[11]

The planets and Sun lie in the path of the ecliptic. If you were to look at the solar system from a distance, you would see that the orbits of the planets lie roughly in a single plane, called the *plane of the ecliptic*. This existence of the ecliptic plane hints at a common origin of the planets in the solar system. As a result of this, the motion of the planets in the sky, along with the Sun, are largely restricted to paths along the ecliptic. The Moon's orbit, on the other hand, is inclined to an angle of approximately 5 degrees with respect to the plane of the ecliptic. This has to do partly with a competition of gravitational pull between the Earth and the Sun over the Moon.

As a result of the complicated interactions between the Earth and the Sun, the timing of the Moon's orbit with the Earth's creates some periodicities. One cycle, called the *Metonic cycle*, is nineteen years long and predicts lunar events with good accuracy. Over the course of this cycle, the Moon can show a maximum declination as large as 29 degrees and as small as 18 degrees. When at its highest declination, the Moon is said to be at a *major standstill*; at its smallest declination, it's said to be at a *minor standstill*.

The use of the Moon for precise navigation requires cumbersome calculations that only came into widespread use in the late eighteenth century, when accurate tables could be produced. These had a particular utility for sailors who couldn't afford expensive marine chronometers to find longitude. For a more seat-of-the-pants approach, the Moon is still useful for crude navigation, where a few simple rules can help.

The Moon always has the same face pointing toward the Earth (Figure 91), the result of tidal forces between the Earth and the Moon. The Moon is illuminated by the Sun, so the full Moon appears when it is on the opposite side of the Earth from the Sun.

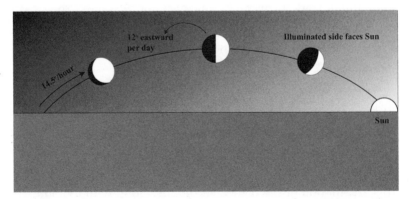

Figure 91 General characteristics of the Moon. The illuminated side of the Moon faces the Sun. The Moon moves across the sky at 14.5 degrees per hour and moves to the east at 12 degrees per day (after Burch, *Emergency Navigation*, 102).

This produces the effect of the setting Sun's being accompanied by a rising full Moon.

A new Moon occurs when the Moon is on the same side of the Earth as the Sun. When the Moon is partial, the illuminated side is always facing the Sun. Like other celestial objects, the Moon will move in an arc across the sky at roughly the same rate as the Sun and stars. The Moon will make a complete orbit around the Earth in 27.3 days. As a result of this rapid orbit, the Moon will rise and set progressively about fifty minutes later each day. It moves eastward against the fixed background of stars by about 12 degrees per day, or half a degree per hour. Its orbital speed and proximity to the Earth makes the Moon the fastest moving celestial object in the sky.

Since the Sun illuminates the Moon, the Moon's terminator will be oriented along an approximate north–south axis. This allows a trick for finding south in the midlatitudes using the edge of the Moon's terminator. A crescent Moon has two "C"-shaped arcs that intersect at two points, called the "horns of the Moon." With an extended arm make an arc that connects the points of the horns of the Moon and swing that arc down to the horizon. The intersection of that arc with the horizon will crudely indicate south.

The phases of the Moon depend on the relative position of the Sun to the Moon, so one can use the Moon in the sky as a crude kind of clock. Here are some examples:

1. Full Moon: The full Moon is on the opposite side of the Earth from the Sun. If the full Moon is at its highest point in the sky (meridian passage), it's roughly midnight local time.
2. Half Moon with illuminated side pointing west: The Sun is 90 degrees ahead of the Moon. If the Moon is at its meridian passage, it's six hours behind the Sun, or roughly sunset. When this Moon sets, it's approximately midnight.

3. Half Moon with illuminated side pointing east: The Sun is 90 degrees behind the Moon. If this half Moon is rising, it's roughly midnight. If this half Moon is at its meridian passage, it's roughly 3:00 a.m.

You can fill in the details for other phases of the Moon. Remember that these rules of thumb are extremely crude, and estimates can vary by a few hours, depending on the time of year, the orientation of the Moon, and your ability to discern its phase — remember that one day is worth 12 degrees of lunar motion. This means that the Moon rises fifty minutes later each day.

The positions of the stars and the Sun in the sky are the basis for *celestial navigation*, the branch of navigation concerned with using objects in the sky to get a position fix. The measurement of the altitude of a heavenly object is critical to celestial navigation and requires both precise instruments and some detailed correction factors to take into account the effect of light bending in the atmosphere and the distance to the horizon. Before jumping into a full-scale treatment of latitude and longitude, I make a brief diversion into these corrections in chapter 9.

9. Where Heaven Meets Earth

· ·

A NUMBER OF creation mythologies speak of a time when Heaven separated from the Earth. The evident human fascination with the interface between the sky and the ground may have its roots in the common experience of distant views. As you stare off into the distance, across the sea or a broad plain, objects appear smaller and smaller. Features get distorted in the far-off haze until your eyes rest on a boundary where the Earth and sky meet: the horizon.

A heavenly body's altitude above the horizon is measured in celestial navigation. Light from an object in outer space must traverse many miles of air before reaching the eyes of an observer on the ground. Along the path the light takes, the growing density of air will bend the light, causing a shift between the observed and the true altitudes. To extract the most precision from celestial observations, the navigator must correct for the effects of light bending in the atmosphere.

A second effect comes into play when the navigator uses the image of the horizon to find an altitude. The altitude of the horizon itself is not always zero, because of the curvature of the Earth's surface. The higher the observer, the lower the horizon appears. Again, to obtain a precise measurement of the altitude of a star or the Sun, the navigator must make a correction for the apparent shift of the horizon. When a position fix to a precision of a few miles is important, both the effect of light bending in the atmosphere and the horizon shift are relevant.

Whenever light passes through the boundary between two transparent materials, it bends, undergoing a process called *refraction*. We're familiar with this when we see the distortion of the image of a spoon halfway immersed in a glass of water. The amount of bending is characterized by a number called the index of refraction, n. The index of refraction is an intrinsic property of

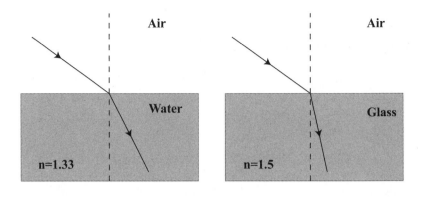

n=index of refraction

Figure 92 Bending of light at the interface between two media with different indices of refraction. The bending toward the normal (dotted line) is larger when the index of refraction is larger.

transparent materials. A pure vacuum, free space, has an index of 1.0. Water has an index of refraction of 1.33, glass 1.5. In Figure 92 you can see that when light passes from air into glass, it bends more than when it passes from air into water. The index of refraction of air at sea level is approximately 1.0004, but this varies with pressure and temperature. The index of refraction from the top of the atmosphere to the surface of the Earth varies continuously with altitude and gets larger as the air gets thicker.

For air the index of refraction depends on density: denser air has a higher index of refraction. Air gets denser the closer you get to sea level, creating a continuous transition from rare to dense media. It's as if there are a large number of tiny boundaries from the upper reaches of the atmosphere to the denser air at the height of the observer. At each one of these boundries the light bends toward the surface of the Earth. Figure 93 shows the effect of this bending. Stars and other objects appear higher in the sky than they really are.

When sighting a heavenly body, a navigator notes the air temperature and pressure. To get a proper value of the "real" altitude of the star above the horizon, he has to apply a correction, based on a model of the density of the atmosphere from the upper reaches

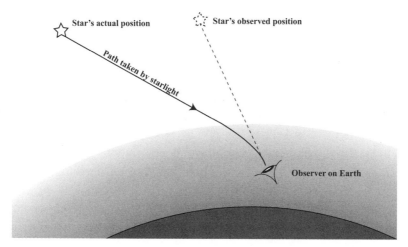

Figure 93 Effect of refraction in the atmosphere raising the observed position of a celestial object. The size of the effect is greatly exaggerated in this figure.

down to sea level. The correction is usually found from a quick calculation. In most cases refraction places the observed altitude higher than the real altitude, as should be apparent in Figure 93.

For average conditions, if a star is observed at an altitude of 60 degrees, the correction is only one arc minute. At lower altitudes the light is entering the atmosphere at increasingly steep angles, and it traverses longer paths through the atmosphere. As a result, the light gets bent more, and the refraction gets larger. A typical correction for a star at the horizon is about thirty-four arc minutes or over half a degree. An error in a star's altitude will directly enter into an error of position. Since one arc minute of latitude is equivalent to a nautical mile, refraction corrections can be substantial and crucial for accurate navigation.

One strange consequence of refraction is the timing of sunrise and sunset. When you see the Sun's image touch the horizon, it has already set. How can this be? The Sun has a diameter of thirty-two arc minutes, and the average refraction at the horizon is about thirty-four arc minutes. It is somewhat remarkable, but it is an accident that they are nearly the same. Since the Earth rotates through 1 degree every four minutes, this means that it takes a little over two minutes for the Earth to rotate by an amount equal to the diameter of the Sun. At the moment you see the lower limb of the Sun touch the horizon, it is already physically below the horizon. At sea level, the true moment of the sunset is roughly one minute before you see the lower part of the solar disk just touch the horizon.

Shifts in refraction are largest at the horizon and will often make the image of the Sun appear to be flattened horizontally. In extreme cases, if the atmosphere has thermal layering, it will appear that a piece of the Sun detaches from the main body, and the light will extinguish with a brief burst of green, called the "green flash."

MIRAGES

The usual corrections used by navigators work well under most conditions, but strange things can happen near the horizon with distant images. The discussion above assumes that the air temperature and pressure drop slowly the higher you go in the atmosphere. Weather conditions can completely change this, however. In the summer, when the hot Sun beats down on a road, mysterious "puddles" seem to appear on the surface of the road. This is sometimes called the hot-road mirage and is illustrated in Figure 94.

Cool air is denser than hot air and has a higher index of refraction. In the summer Sun the surfaces of roads heat up rapidly, creating a layer of hot air trapped just above the surface, with cooler air overlaying it. Light rays approaching the surface will get bent upward from the hot surface. The observer will see both the direct rays — the "real" image — and also rays that look as though they've been reflected, creating a second image. This second image is *inverted*, meaning that top and bottom are reversed in the refracted image. The effect that looks like puddles is created by the reflection of light from the sky and objects above. Another feature of images seen in the distance on a hot day is the shimmer-

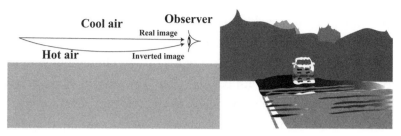

Figure 94 The hot-road mirage.

ing quality of distant images. Hot air rises in waves and will focus and defocus the light rays reaching our eyes.

What do the mirages have to do with navigation? Any celestial object near the horizon will be subject to the same vagaries of refraction that create mirages. This is a nearly insurmountable problem, but sometimes sights at low altitudes are a necessity. The position of the horizon can be distorted by layering of air in the distance. The appearance of landmasses can also be distorted. A continuous island seen in the distance might appear to be broken up into a series of small islets, depending on atmospheric conditions. Careful navigators who take sights on a day with hot air overlaying cold water often find their results surprisingly inaccurate. It's difficult to correct for the conditions that create mirages, but at least some awareness of the phenomenon is helpful.

As an example of thermal effects for distant objects near the horizon, Figure 95 shows three photographs taken at different times of the day from the same location. The photographs are of an abandoned lighthouse on an island eight miles away from the camera, very close to the physical horizon. The most striking difference among the three photographs is the change in resolution of the distant features throughout the day. The photographs were taken with an identical setup.

In the top photograph there is layering of cool air over warm air right at the ocean's surface, creating an inverted image of the lighthouse and parts of the island. The image of the island itself is broken up into a series of smaller islets. In the middle photograph, taken at midday, the inverted image has gone away, but the continuous island still appears as a set of broken islets. The effective horizon is just at, or above, the lower reaches of the island. The last of the three photographs was taken just before sunset. The effective horizon has dropped, allowing the entire island to be seen, and it now appears as a continuous landmass. The lighthouse in the image is fifty feet tall. I can sight the horizon to a precision of one-third of the height of the lighthouse and no better. This implies that the thermal effects would push around my sights by about one and a half arc minutes, which translates into a position uncertainty of one and a half nautical miles.

The shift in the horizon in these photographs illustrates the limits of precision in celestial navigation: one arc minute, or 1/60th of a degree, is about the best one can expect to do. For precision beyond this, one begins to need the kind of equipment surveyors or astronomers employ, which use the local pull of gravity as a reference, rather than the horizon. Unfortunately, this isn't practical on board a ship or on a long journey.

Sometimes atmospheric layering can create very strange conditions. Here is a report of a freak sighting of Milwaukee, seventy-five miles away across Lake Michigan:

> An example occurred on the evening of April 26, 1977, when the residents of Grand Haven in Michigan looked west over Lake Michigan and saw city lights. The city of Milwaukee, seventy-five miles distant, is geometrically well below the horizon and out of sight from Grand Haven. An observation of a blinking red light convinced the viewers that Milwaukee was the city-above-the-lake: an ingenious observer timed the blinking and called someone in Milwaukee to find out what it could be. His measurement agreed with the frequency of the flashing red light beamed eastward from the entrance of the Milwaukee harbor. Weather

Figure 95 Three photographs of a lighthouse on an island eight miles in the distance. The photos were taken in the morning, at midday, and in late afternoon with the same optics.

Service records for that time showed a strong temperature inversion over the lake.[1]

In the late seventeenth and eighteenth centuries, the phenomenon of refraction close to the horizon was called *looming* by sailors.
In addition to the simple hot-road mirage and distorted horizons,
layers of hot air over cold water could cause unusual distortions.
On a kayak trip I took off the coast of Downeast Maine, the water
temperature was in the 50s, but the air temperature was in the
90s. During the course of the trip, there were many strange sightings: oddly elongated structures, magnification of distant trees,
container ships apparently floating in midair, and lighthouses
appearing where there were no signs of any on maps until the
coast of Nova Scotia, some eighty miles away.

Distortions like the ones I saw in Maine are relatively common
in arctic waters, where warm air frequently settles over very cold
water in the summer months. Arctic explorer William Scorseby
wrote of looming in 1820:

> There are several phenomena of the atmosphere depen-
> dent on reflection and refraction deserving of notice.
> Ice-blinks have been already mentioned, when speak-
> ing of the ice. Under certain circumstances all objects
> seen on the horizon seem to be lifted above it a distance
> of two to four, or more, minutes of altitude, or so far
> extended in height above their natural dimensions. Ice,
> land, ships, boats and other objects, when thus enlarged
> and elevated, are said to loom. The lower part of loom-
> ing objects are sometimes connected with the sensible
> horizon by an apparent fibrous or columnar extension
> of their parts, which columns are always perpendicular
> to the horizon; at other times they appear to be quite
> lifted into the air, a void space being seen between them
> and the horizon. This phenomenon is observed most

frequently on, or before, an easterly wind, and is generally considered as indicative as such.[2]

Elaborate images near the horizon sometimes are called *fata
morganas*. This comes from the name used by Italian poets in the
thirteenth century for Morgan le Fay, the fairy half-sister of King
Arthur, who would float in the air and assume strange appear-
ances. Fata morganas can sometimes appear like castles in the
distance. These are often the result of a real and inverted image
merging together, tethered together by a thin column. Having
seen some of these myself, I will attest that they look like a man-
made structure floating in the air.

Chasing a Mirage

Fata morganas in the arctic can be particularly misleading. In 1906
American explorer Robert E. Peary was on an expedition to the
Canadian archipelago west of northern Greenland. A major goal
of the expedition was an attempt to reach the North Pole, but Peary
also did reconnaissance of the area. He followed the northern
coast of Ellesmere Island to the remote and desolate Axel Heiberg
Island (Figure 96). From the two-thousand-foot-high summit of
Ellesmere Island's Cape Colgate, Peary reports: "North stretched
the well-known ragged surface of the polar pack, and northwest it
was with a thrill that my glasses revealed the faint white summits
of a distant land, which my Eskimos claimed to have seen as we
came along from the last camp."[3]

Six days later, from the summit of Cape Thomas Hubbard, on
Axel Heiberg Island, he wrote: "The clear day greatly favoured
my work in taking a round of angles, and with the glasses I could
make out apparently a little more distinctly, the snow-clad summits
of the distant land in the northwest, above the ice horizon."[4]

He named the land Crocker Land in honor of George Crocker,

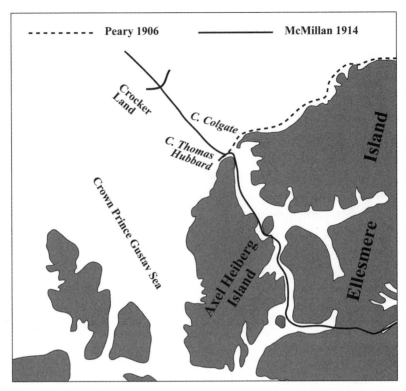

Figure 96 Peary and MacMillan expedition routes near Axel Heiberg and Ellesmere Islands. The supposed position of Crocker Land as reported by Peary is indicated.

a supporter of the Peary Arctic Club, and estimated its position at 83 degrees N, 100 degrees W, more than a hundred miles away from the sighting. He claims that this discovery was one of the major findings of his expedition, along with an attainment of the farthest north, 174 nautical miles from the Pole.

The concept of Crocker Land gained more credence. In 1911 Rollin Arthur Harris of the U.S. Coast and Geodetic Survey wrote a ponderous tome titled *Arctic Tides*. In it he presented an exhaustive compilation and analysis of the tides and currents reported in the Arctic Ocean. From his examination he reached the conclusion that the tidal data could only be explained by the existence of a huge landmass in the unexplored region of the arctic, of which

Crocker Land represented only a small part. In reality the ocean is over thirty-five hundred feet deep where Crocker Land and Harris's huge island were alleged to be located.[5]

In 1913 Donald B. MacMillan, a veteran of Peary's 1906 expedition, organized the U.S. Crocker Land expedition to cross the sea ice from Axel Heiberg Island to find Crocker Land. His party made its way up toward Greenland on the steamship *Diana*, which struck an iceberg and sank. They managed to catch a ride on another vessel up to Etah, in northwest Greenland. There, they constructed a base camp and the next spring set out on a march toward Axel Heiberg Island.

The trek was tortuous, plagued with illness and frostbite. Only four members of the party were able to continue on from Axel Heiberg Island across the sea ice toward Crocker Land: MacMillan, navy ensign Fitzhugh Green, and two Inuit veterans of previous expeditions, Pee-a-wah-to, and E-took-a-shoo. They ventured a hundred miles onto the sea ice to the northwest. One day the white men thought they spied Crocker Land, but their Inuit companions saw it as a mirage:

> April 21st was a beautiful day; all mist was gone and the clear blue of the sky extended down to the very horizon. Green was no sooner out of the igloo than he came running back, calling in through the door, "We have it!" Following Green, we ran to the top of the highest mount. There could be no doubt about it. Great heavens! What a land! Hills, valleys, snow-capped peaks extending through at least one hundred and twenty degrees of the horizon. I turned to Pee-a-wah-to anxiously and asked him toward which point we had better lay our course. After critically examining the supposed landfall for a few minutes, he astounded me by replying that he thought it was poo-jok [mist]. E-took-a-shoo offered no encouragement, saying, "Perhaps it is." Green was

still convinced that it must be land. At any rate, it was worth watching. As we proceeded the landscape gradually changed its appearance and varied in extent with the swinging around of the sun; finally at night it disappeared altogether. As we drank our hot tea and gnawed the pemmican, we did a good deal of thinking. Could Peary with all his experience have been mistaken? Was this mirage, which had deceived us, the very thing which had deceived him eight years before? If he did see Croker Land, then it was considerably more than 120 miles away, for we were now at least 100 miles from shore with nothing in sight.[6]

Soon after, MacMillan became certain that he had been fooled by a fata morgana: "We were convinced that we were in pursuit of a will-o'-the-wisp, ever receding, ever changing, ever beckoning." The sea ice began to melt underneath them, and they beat a hasty retreat back to Axel Heiberg Island. From the same point at which Peary stood in 1906, they again saw the apparition of Croker Land in the distance. MacMillan wasn't out of the figurative woods, however, as they still had a thousand-mile trek back to their base camp.

Looming

Over time the term *looming* has acquired different meanings from the original sense of a mirage. In the common usage it is an event or object that seems larger than it really is. In nautical terms it has also come to mean a glow in the sky associated with bright lights at night. The lights themselves are over the horizon, but their glow can be seen hovering just on the horizon. The modern usage seems to have developed at the same time as the invention and use of the electric lights in lighthouses, circa 1880. Previously,

gas-powered lighthouses didn't produce enough luminous intensity to create a *loom*.

Often at night, lights from a city or lighthouse that would not
be directly visible over the horizon could be seen as a dull glow
in the sky (Figure 97). This is caused by light from a source on
or near the surface being scattered by particles higher up in the
atmosphere. From my house on the south shore of Cape Cod, I
cannot directly see Nantucket, which is twenty miles away and is
relatively low lying. On the other hand, on a clear night I can see
the loom of the lighthouse on the northeast corner of the island
and the loom of the lights of the town of Nantucket proper.

While the distance estimates from the nighttime loom of a
city or lighthouse may not be accurate, the light will give a clear
direction to the object and can be used to navigate. This not only
works at sea but also can work on land, where a large town or
city may not be visible in the distance because of intervening hills
or ridges, but the glow will be distinct in the sky, which can
be helpful in direction finding. Since the light from the loom
is passing through a long distance at a low angle in the sky, the
color appears to tend toward the red-orange. This is likely due to
two effects: (1) the distinctive color of bright sodium-vapor street
lamps and (2) the same scattering effect that creates reddish

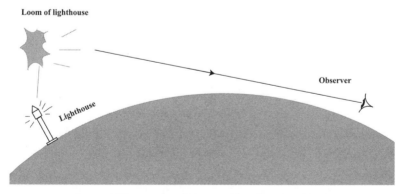

Figure 97 A loom or glow in the sky created by a light source that is over the
horizon.

sunsets. I've spotted the reddish loom of Boston a hundred miles away from a distant mountain in Vermont. For navigators on shipboard the loom of a city over the horizon is often the first sign of landfall. Large cities throw off enough loom to be seen a hundred miles away.

Even during the day, there is a subtle looming effect that can be used for navigation. Sunlight will reflect off the ocean and land differently and create a slightly different coloration in the sky over both. In principle this phenomenon can be used to detect the loom of distant islands that would ordinarily be hidden over the horizon. In his book *We the Navigators*, author David Lewis recounts an interview with a navigator from the Gilbert Islands, Abera Beniata, who was taught the art of navigation by his grandfather. He calls the phenomenon of looming *te ḳimeata*.[7]

Albera says that the loom of land is visible both at night and during the day. He notes that, "when there is no cloud at all, look carefully round the horizon and you may see a brightness coming up over the horizon at one point. This is called *te ḳimeata* and is not marked at all, it is just a little different from the rest of the horizon on close inspection. You can see this appearance in any direction, but most easily when the sun is high around midday. *Te ḳimeata* may be detected up to about 30 miles or a little more." He says that islands that can only be directly visible from ten miles away can be seen by their loom from three or four times this distance.[8]

The Distance to the Horizon

For most of us the horizon is an imaginary place that can never be reached, like the end of the rainbow, or true love. As you try to approach it, it just moves away from you. For the navigator the distance to the horizon has meaning. The horizon, physically, is a boundary created by the curvature of the Earth and determines what you can see in the distance. If your eye is right at ground level, the nearest obstruction defines your horizon. The higher you go, the farther you can see.

We can go back in time and try to view the world as the ancients did. At first blush the curvature of the Earth is difficult to perceive. It would appear that we live on a huge plane that could be infinite or finite in extent. Early Greek, Babylonian, Hindu, and Chinese concepts of the world were that the Earth was a plane of finite extent. If you ventured too far to the horizon, you would fall off the edge. The Chinese clung to the notion of a flat Earth until the seventeenth century, when Jesuit missionaries introduced to them the spherical nature of the globe.

A flat Earth makes some sense as the highest mountains (at eight thousand meters) have a height a tiny fraction of the Earth's radius (6.3 million meters). Only around the time of Pythagoras around 500 BC did the concept of a spherical Earth begin to take hold. By the Middle Ages, although the circumference of the Earth wasn't well established, the concept was widely held. Still, some diehards persist to this day in the Flat Earth Society.

The most direct evidence of a curved Earth to an early navigator is the way objects seem to disappear or appear as you sail away or toward them. Figure 98 shows a sailor approaching a mountain from a distance. When he's far away, the mountain can't be seen. As he gets closer, the top of the mountain can be seen, but the lower elevations are still below the horizon. As he gets closer still, the base of the mountain becomes visible.

If the sailor is sailing away from land, the process happens in reverse. At first the entire mountain is visible, then the bottom of the mountain disappears, but the top is visible; finally, the top disappears altogether. We don't know who first sailed out of sight of land. It could have been a storm-tossed sailor who was trying to follow a coastline but got blown out to sea. It could have been a fisherman who decided that more fish could be found if he sailed farther away. In either case they could see this effect of lower elevations first disappearing, then the higher elevations vanishing, and

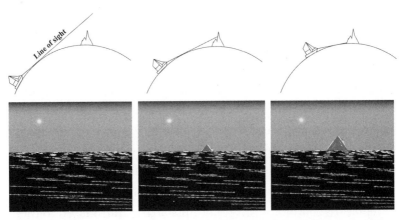

Figure 98 View of a mountain on an island as seen from a ship approaching land over some distance. The effect of the curvature of the Earth creates a situation in which the top of the mountain is visible first, followed by the lower elevations. The height of the mountain and of the ship are exaggerated in this figure relative to the radius of the Earth.

the process happening in reverse as they approached land. A far-sailing fisherman who lived in a society where you were taught that the Earth was flat might get very confused in these circumstances.

The Great Flood

In a large number of early cultures, there are mythological descriptions of a great flood. These include the Gilgamesh epic of Babylon, the Jewish story of Noah, and Chinese and Hindu myths. Early American cultures and Polynesian cultures all had versions of a flood mythology. In many of these legends, a sailor survives the flood and sees land gradually covered by water until it disappears. Eventually, the waters recede, and land is seen, with the highest elevations appearing first.

Many conjectures for the flood mythology have been put forward, but one plausible explanation that has not been advanced is the phenomenon of land disappearing beyond the horizon as the boat departs and on its return the highest elevations appearing first as the ship approaches land. The ubiquity of the deluge myth could possibly be reconciled as the phenomenon of objects disappearing over the horizon for cultures believing in a flat Earth. How else could one account for such a phenomenon if the Earth were flat?

There are many plausible reasons for long-distance voyaging. Fishing and trade are certainly two of the most important, as economic factors can be strong motivators. Long-distance migration over the ocean is less common but is responsible for the Pacific Islands peoples' hegemony. Growing populations with limited resources can be important factors motivating these voyages.

Imagine a patriarch living with his family circa 7000 BC. He lives in a fertile coastal region but is hemmed in by mountains. He makes his living from fishing and farming and is highly regarded in his society. The priests hold to the concept of a flat Earth, but the patriarch is aware of distant fertile lands beyond the horizon. He visited these on long-distance fishing expeditions trying to chase migrating fish and escape the overfished stocks close to his native land. The population has slowly been rising, and arable farmland is becoming scarce. Fights break out among clans, and blood feuds take their toll. Knowing of the uninhabited fertile land over the horizon, the patriarch sets out to save his family by sailing there.

He supervises his family in the construction of a large, scaled-up version of his fishing vessel, capable of transporting them and the possessions they need to make a new living in the far-off land. When the vessel is complete, it's only a question of timing when to leave without arousing suspicion. He waits for a storm, then moves his livestock, plants, water, food, and family onto his vessel and sets off. While the scene is familiar to him, his passengers witness an amazing sight as they depart from land. First the low-lying farmlands disappear under water, followed by the upland forests, and finally, the mountaintops vanish.

For those raised with the notion of a flat Earth, there is only one inescapable conclusion: the Earth has flooded. Everywhere they look they are surrounded by water. Knowing that there's no possibility of return to the precarious situation in their homeland, the patriarch allows this idea to take hold as if to emphasize the impossibility of turning back. After sailing for several weeks, the patriarch knows that he should be close to land and releases the pigeons he has brought along. Although many now believe that pigeons navigate by "seeing" the Earth's magnetic field, ancient sailors knew of their utility as shore-finding birds. Finally, one departs in a direction indicating land.

He alters course, and as he approaches, what do his passengers see? At first, only the tops of mountains are visible, then the lowlands. The floodwaters are receding! They're saved. They make landfall. The patriarch goes to his grave with his secret, and his story becomes one of divine intervention.

The mythology of the great flood is quite common, yet there is no geological evidence for a global flood in any historical time. It's plausible to imagine that the scenario described above happened a number of times in different settings as population pressures forced a long ocean migration, and the witnesses who believed in a flat Earth carried with them the story of a great flood.

The above conjecture on the ubiquity of great flood stories is not without flaws. Landlocked Native American tribes such as the Hopi have a story of a great flood, and presumably, none of the Hopi sailed out of sight of land. Nonetheless, the stories may have originated elsewhere and were preserved in their traditions.

Visible Range

Instead of taking the viewpoint of a sailor on the surface of the ocean, we can adopt the point of view of a person standing high on a mountain, looking out over the ocean and imagining what he sees. Right at the surface of the Earth, the horizon is very close. As you stand up, it may appear to be a couple of miles away. As you gain altitude, it will be farther and farther away. This again

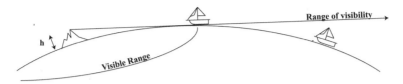

Figure 99 Relationship of visible range to height of observer.

is the effect of the curvature of the Earth — as you gain height, the horizon recedes, in a way that is directly related to both your height and the curvature of the Earth. Figure 99 shows how the effect of the Earth's curvature gives the effect of a visible range that depends on the observer's height.

For an observer precisely at the surface of the Earth, the horizon is right there, zero distance away, but if the observer stands up, the horizon recedes to two miles away. At the height of twenty-five feet, the horizon is roughly eight miles away. This relationship is shown in Figure 100. The relation between height and the visible range to a feature, like a mountain, is the same as the distance to the horizon. A hill one hundred feet high will be visible over the horizon up to a distance of ten miles. Islands with tall mountains, such as the big island of Hawaii can be visible from up to one hundred miles away, while low-lying islands may only be visible from less than ten miles.

This relationship was hugely important in sighting land or other vessels. Lookouts were regularly perched high up on top of

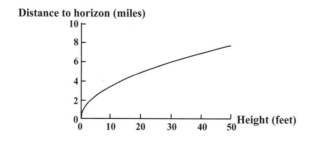

Figure 100 Relationship between the distance to the horizon and the height of the observer.

the tallest mast in a crow's nest to get views over the horizon that would be invisible to sailors on deck. A lookout might be able to see land or vessels ten or more miles distant, while the rest of the crew might only see things five miles away. The job of lookout was one of the regular duties on shipboard. In times of war the lookout also had the duty of sighting enemy naval vessels in the distance and communicating this to officers on deck.

Another major concern on shipboard was landfall. If land was sighted in the distance, there was always a possibility that shallow reefs could tear out the hull of a vessel if the captain wasn't careful. If the lookout sighted land during the night watch, the captain would order the boat to stand off; that is to say, turn away from the sighted land and stay away until sunrise, when a safe approach could be made.

Dip Angle

A related effect is called the *dip angle*, which can be important for precise navigation. On shipboard most navigators use a sighting of the horizon to establish the altitude of a star or the Sun above horizontal. The earliest nautical instruments to measure altitude used gravity to establish horizontal. One example of these early instruments was a *quadrant*. A weight attached to the end of a string hung down and used gravity as a reference while a sighting was made with an arc showing the angle from the hanging weight. As a ship sways under the influence of waves, the string and weight would dance around, making it difficult to find the altitude of a star with any precision.

Columbus was one of the earliest sailors to try out celestial navigation, but he found it next to impossible to make sightings of Polaris using a quadrant while at sea. The string moved around far too much to get any useful measurement. Only when becalmed or safe in a quiet harbor could he use the quadrant.

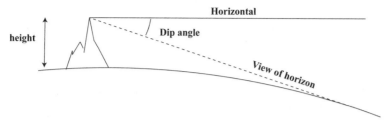

Figure 101 The view of the horizon dips below horizontal, depending on the height of the observer.

On the other hand, the horizon provids a reliable horizontal reference that could be used on a ship being tossed by the waves, and the altitude of a star could be measured against that. This was the principle behind such nautical instruments as the sextant. Although these instruments made the job of finding altitudes of stars easier, the curvature of the Earth created a problem in the interpretation of the horizon: it wasn't precisely horizontal.

As a person gains altitude, not only does the horizon get farther and farther away, but it gets lower and lower with respect to true horizontal. This lowering of the horizon creates something called the *dip angle,* which is illustrated in Figure 101. The dip angle depends on the height of the observer and the curvature of the Earth. The navigator is interested in the altitude of the star above horizontal, but if he views the altitude of the star from his visible horizon, he overestimates its height by some amount and must correct for it.

Figure 102 gives some idea of what the effect might appear to the observer; on the left you see the horizon lying at horizontal for an observer at zero height and on the right from the point of view of an observer at a substantial height. Typical corrections for dip angle range from about one to five arc minutes, depending on the height of the observer. Like the correction of atmospheric refraction, these become important when the navigator is trying to milk the last few arc minutes of precision from his observations.

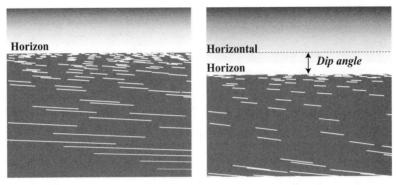

Figure 102 The appearance of the dip angle to the observer at different heights. The vertical scale is exaggerated for clarity.

On shipboard a navigator will note the altitude of the stars, the time he made his sighting, his physical height above the ocean, the temperature, and the barometric pressure in his logbook, then goes off to work on a process called *sight reduction*, where he turns the sightings into a latitude and a longitude. The first step is the work of correcting for both dip and refraction to get a true altitude for a star, a planet, or the Sun.

At precisions of one arc minute, refractive effects start to appear everywhere. It's as if you have discovered that you're living in a carnival funhouse with warped mirrors and optical illusions that distort your reality. Figure 103 shows how refraction will distort the dip angle correction. If a navigator makes a measurement from a relatively high vantage point, he has to not only correct for dip angle but also make an educated guess on the effect of refraction on dip.

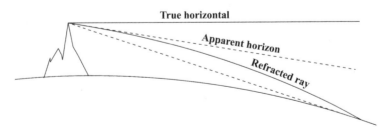

Figure 103 Effect of refraction on the image of the horizon.

The Modern Legend of al-Bīrūnī

Navigators frequently have to convert angles such as latitude differences into distances. The measurements of angles are meaningless unless one important number is known: the radius of the Earth. This is the single most important conversion factor that brings the measurements of the heavens onto Earth. The radius has traditionally been found by inverting the process of celestial navigation. Rather than finding the differences in positions from measurements of stars or the Sun, a known distance is established between two locations and the differences in the angles of the Sun or stars are measured.

Eratosthenes (c. 276–195 BC) of Alexandria is reputed to be the first to undertake this measurement. He knew that the Sun cast no shadow at noon on the summer solstice in Aswan, Egypt, and cast a shadow of 7{1/5} degrees at Alexandria, Egypt, at the same time. Knowing that the distance from Alexandria to Aswan was 5,000 stade, he established the first measurement of the radius of the Earth by using the 5,000 stade = 7{1/5} degrees as a conversion factor.

Most subsequent measures of the Earth's radius used variations on Eratosthenes's original measurement. Caliph al-Ma'mun was a powerful ruler of Mesopotamia. In 830 AD he commissioned a crew of astronomers to repeat Eratosthenes's measurement over a long baseline.

One of the most famous mathematicians during the golden age of Islamic science was Abū Rayhān Muhammad ibn Ahmad Bīrūnī (c. AD 973–1048). He had a brilliant career and made numerous important contributions to geodesy and mathematics. He wanted to reproduce al-Ma'mun's measurement with greater precision but was unable to raise the resources to carry out the measurement of a long baseline and make astronomical observations on either end.

While traveling in the Punjab region (which now straddles the India–Pakistan border), al-Bīrūnī was detained at a fort and came up with a novel technique that used the dip angle as a way of measuring the Earth's radius. His technique was designed to cut out the laborious step of using human measurements to create a long baseline. He first measured the height of a nearby mountain by triangulation from two positions, then found the dip angle to the horizon from the top of the mountain. The horizon was established by long view across the dusty plains of the Punjab. We don't know precisely how he measured the dip angle, but I imagine it could have involved some use of water to establish a true horizontal and rulers to measure the dip. By measuring both the height of the mountain and the dip angle, he could compute the radius of the Earth.

In his writings we only know the reported dip angle and the height of the mountain in units for which we don't know the precise modern equivalent. Historically, this measurement had no effect on navigational practices and was largely unknown. Al-Ma'mun's value for the Earth's radius was employed for some time until Western Europeans subsequently established more precise values.

Al-Bīrūnī's measurement, while clever, had at least one fatal flaw: he neglected to take into account the effect of refraction on his measurement of the dip angle. We cannot fault him, as the science of refraction wasn't understood until the seventeenth and eighteenth centuries. A typical correction for refraction is about one-sixth of the dip angle, but can vary substantially, depending on atmospheric conditions.

Why am I bothering to tell you this? There has been a modern mythology that has grown around al-Bīrūnī, purporting that he made the most precise measurement of the Earth's radius well before modern times. In 1973 an engineer from Pakistan, Saiyid Samad Husain Rizvi, published an article reporting the discovery of a long lost book by al-Bīrūnī, the *Ghurrat-Uz-Zijat*, and brought the measurement to the public eye.[9]

If you assume some standard for the height of the mountain that al-Bīrūnī reported, then take his dip angle of thirty-four arc minutes *without* any refraction correction, you derive a value of 6,339.9 km for the radius of the Earth, which can be compared to the present-day value of 6,356.7 km. This is a difference of only 16.8 kilometers! With a plausible standard atmospheric correction, he *might* have derived a value that was precise to within about 20 percent. This has not deterred the numerous supporters of this measurement, who have uncritically accepted it as highly precise and evidence of his genius.

Although some historians noted the problem with al-Bīrūnī's understandable neglect of refraction, the remarkable agreement of his value with the true value contributed to modern legend building. This appears in websites, in print, and in documentaries on Muslim science. In a recent BBC documentary, *Science and Islam*, which first aired on January 12, 2009, the narrator describes the previous attempts at the measurement of the circumference of the Earth and comments, "A more reliable and sophisticated method for estimating the Earth's size was needed, and two centuries after al-Ma'mun did, it came."

The narrator goes on to describe al-Bīrūnī's dip-angle method, even using a large astrolabe as a dramatic prop, and illustrates the technique, giving the dip-angle formula. He concludes, "With this formula, Biruni's able to arrive at a value for the circumference of the Earth that's within 200 miles of the exact value we know it to be today, about 25,000 miles. That's to within an accuracy of less than one percent; a remarkable achievement for someone a thousand years ago."[10]

In the MacTutor online archives of mathematics, John O'Connor and Edmund Robertson echo this: "Important contributions to geodesy and geography were also made by Biruni. He introduced techniques to measure the earth and distances on it using triangulation. He found the radius of the earth to be 6339.6 km, a value not obtained in the West until the 16th century."[11]

In reality it is quite possible that atmospheric layering could have given al-Bīrūnī a flat or even concave Earth in the same way the layering can give rise to strange mirages. A precision within 20 percent is perhaps plausible, but without any knowledge of the refractive effects, it's impossible to assign any accuracy to this; it's more like winning a lottery.

THE JUMP-UP TRICK

Related to al-Bīrūnī's measurement is an entertaining trick that shows the Earth is curved — in case there are those of you who may still be doubters. This requires you to have a view of the sunset over the ocean. If you are lying on a beach near sunset, you can put your head as close to the sea as possible. Watch the Sun setting over the western sea. While lying down, at the moment the very top of the Sun winks out, start counting and jump up. Your increased height will allow you a view of the upper tip of the Sun, which will then wink out a second time.

If you time the difference and know your height, you can get a rough calculation of the radius of the Earth. I tried this while on vacation in Jamaica and found a radius of twelve thousand kilometers, roughly twice the accepted value. It's not terribly accurate, but it is a fun trick.

10. Latitude and Longitude

· ·

THE NORSE AND Polynesians had a vague sense of latitude. For the Polynesians the concept of a "star overhead" marked the zenith location of an island. It was imprecise but associated a star with a place on Earth. Likewise, the Vinland Sagas show how the Norse identified the latitude of their camp in North America by the path of the Sun. Nautical charts in the fourteenth and fifteenth centuries depicted geographic features faithfully but did not portray latitude and longitude. The use of these coordinates arose in navigation when celestial observations were associated with places on Earth and captured on maps. The process of bringing celestial observations down to Earth evolved over many centuries and is still evolving. Much of the discussion below follows the history of the growth of celestial navigation from the twelfth century onward. A timeline of some major milestones is shown in appendix 2.

During the early Middle Ages, Islamic scientists carried the Greek concept of latitude and longitude forward and made huge advances in mathematics. Many of these developments were in algebra and trigonometry, enabling calculations that related the position of an observer to celestial objects. Although much of the interest of Islamic scientists was related to geodesy and astronomy, the general public had more pragmatic uses for latitude and longitude: astrology and prayer.

In our era you might consult astrological forecasts in newspapers and online. These are usually based on Sun signs, meaning the position of the Sun in the zodiac on the day you were born. There is a more sophisticated form of astrology called natal (or birth) astrology that describes your personality based on the positions of the planets, Sun, Moon, and zodiacal signs relative to their positions in the sky at the time of birth.

In natal astrology the sky is partitioned into a system of twelve houses that wraps around the sky along the ecliptic (Figure 104). Houses have a symbolic content, representing aspects of a person's life, such as his or her wealth, children, or spouse. The eastern horizon is the start of the houses, which are numbered one through twelve and advance counterclockwise around the sky. At some place and moment in time, the zodiacal signs and planets will be aligned with certain houses, and as the Earth rotates, the alignment will shift throughout the day and night. For example, the Moon might be in the eighth house at 4:00 p.m. local time but two hours later will be in the seventh house. There are a number of different systems for constructing houses. The house system shown in Figure 104 is a simple illustration based on my partition of the ecliptic into equal angular swaths. Other house systems can be more complicated. In Figure 104 the characteristic name of each house is listed, based on common interpretations. In addition to the eastern and western horizons, the nadir and the meridian point of the ecliptic, called *midheaven,* are fixed points in the systems.

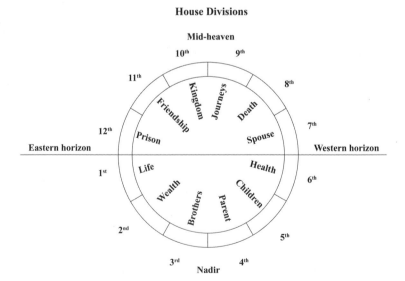

Figure 104 Division of houses for natal astrology. The houses are projected onto the ecliptic, begin on the eastern horizon, and wrap around counterclockwise.

To extract the symbolic meaning of a natal horoscope, the astrologer needs the position of celestial objects in the houses relative to the subject at the moment of birth. Since the houses are specific to the position and time of the individual's birth, the latitude and longitude of the birthplace are crucial pieces of information, in addition to the time of birth and location of celestial objects in the sky. Tables of latitudes and longitudes were important to any self-respecting astrologer. Ptolemy wrote a treatise on natal astrology called *Tetrabiblos* (four books) that codified many of the earlier practices and added some of his own concepts.

The other widespread use of latitude and longitude in the Islamic world is the direction to face Mecca. The faithful pray five times a day toward Mecca, and the direction they face is called the *Qibla*. The usual convention for the Qibla is a great circle route: the shortest distance on the surface of the Earth to Mecca. Mecca is, in effect, a religious North Pole, where all the *Qibla lines* converge on one place, as the lines of longitude converge at the North Pole. Tables of latitudes and longitudes of cities are needed to find the proper direction in which to face Mecca.

The Toledo Tables

While there were many tables of latitude and longitude in the medieval Arabic world, there is no evidence that they were used for maps or navigation. Can we learn something from these tables? One surviving table from the twelfth century, the Toledo Tables (appendix 3), compiled by the Arab scientist al-Zarqali, presents a unique window into the geography of that era.[1]

In the time of al-Zarqali, latitudes were often found from the length of a shadow cast by a stick on the equinox. As I described in chapter 8, the tip of a shadow on the equinox traces a straight line for everyone on the planet (Figure 105), while on other days of the year the path is curved, and the curvature will depend on latitude.

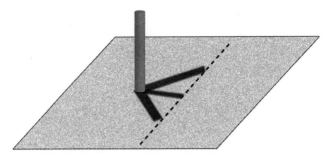

Figure 105 The path traced by the tip of a shadow on the equinox is a straight line. The shortest length of the shadow can be combined with the length of the stick to find your latitude.

The shortest length of the stick's shadow can be used to find latitude, as illustrated in Figure 106. All you need is a string to measure the length of the shadow and the height of the stick. The ratio of the length of the shortest shadow to the length of the stick gives a number that can be converted into an angle. This ratio is called a *tangent*. For a triangle containing one 90-degree angle (a right triangle), the tangent of an angle is the ratio of the length of the side opposite the angle to the length of side adjacent to the angle. As you might see in Figure 106, the tangent of the latitude is the ratio of the shadow length to the stick height. You then look up the tangent in a table and convert the ratio into an angle.

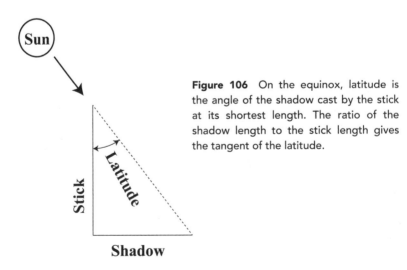

Figure 106 On the equinox, latitude is the angle of the shadow cast by the stick at its shortest length. The ratio of the shadow length to the stick length gives the tangent of the latitude.

Arabs first developed tables of tangents, appropriately called shadow tables, in roughly AD 860. Using a string and a stick, with the tables you can measure your latitude to a precision of about 1 degree. I've tried this myself on a beach in Jamaica and, lacking shadow tables, had to find them myself like al-Zarqali might have done. In the Middle Ages most astrologers probably would have just purchased tables from someone like al-Zarqali or copied them from a friend. Instrument makers often produced geometric constructions on the back of astrolabes to find tangents, eliminating the need for compiled tables.

Longitude is more difficult to determine than latitude. The Earth's rotation presents the same sky over the course of the day and night to any observer at some latitude; there is nothing in the sky to distinguish longitudes. More often than not, longitude was assessed from dead reckoning. Travelers from one city to another would report the length of their journeys, often in days. The distance has to be converted into an angle, requiring the radius of the Earth to make the conversion. The earliest measurement of Earth's radius, attributed to Eratosthenes, and the subsequent measurement by Caliph al-Ma'mun in AD 830 established the most widely used values to convert distance to longitude. The Toledo Tables report longitudes that span from 6 degrees to 125 degrees, the extent of the world known to al-Zarqali.

The measurement of the Earth's radius by Eratosthenes was based on the angles cast by shadows on the ground during the summer solstice (Figure 107). The city of Aswan (called "Assuen" in the Toledo Tables) is located at the Tropic of Cancer. A vertical stick in Aswan casts no shadow at noon on the solstice. On the other hand, a vertical stick on the solstice casts a 7{1/5} degree shadow in Alexandria, Egypt. The distance between Aswan and Alexandria was reckoned to be 5,000 stade in Eratosthenes time. This gives a conversion of 694 stade per degree. In the modern era sixty nautical miles is 1 degree of latitude.

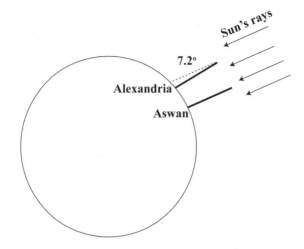

Figure 107 Eratosthenes's measurement of the radius of the Earth. He compared the angles cast by shadows on the summer solstice at Aswan and Alexandria, Egypt. By knowing this angular difference and the distance between Aswan and Alexandria, he calculated the radius of the Earth and developed a conversion factor from degrees to distance (miles, leagues, stadia, farsakhs).

A table of latitudes and longitudes requires a prime meridian as a reference for the "zero" of longitude. Many tables in the Middle Ages used different prime meridians. Ptolemy used islands referred to as the "Fortunate," "Eternal," or "Blessed" Isles for his choice of prime meridian. This was consistent with the view of an occupied half of the world (or *oecumene*) that stretched to the east from this westernmost parcel of land. An encircling ocean surrounds the oecumene. The name "Fortunate Isles" comes from Greek and Celtic mythologies describing a precursor to the Elysian Fields where heroes would be welcomed after death into an earthly paradise. Many uninhabited islands lay to the west, off the coast of North Africa and Europe, and could have been visited by sailors who left no written record. The Canaries, the Madeira Islands, the Azores, and the Cape Verde islands are all candidates. It is possible that mythology and a chance encounter of a storm-tossed sailor with a beautiful uninhabited archipelago combined to create a physical location for the Fortunate Islands.

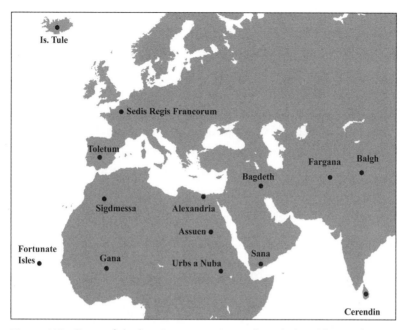

Figure 108 Some of the locations appearing in the Toledo Tables in what are likely their modern locations.

Figure 108 shows some notable entries appearing in the Toledo Tables. Balgh is likely the city Bactria, along the Silk Road, as Fargana is Fergana in Uzbekistan. Cerendin was a name given in that era to what is now Sri Lanka. Is. Tule, at 58 degrees N in the tables, could be the Thule of Pytheas, perhaps Iceland. Sana is the modern capital of Yemen, Sanaa.

Three locations mentioned in the tables deserve some discussion, as they correspond to places that appeared some time after Ptolemy's death but have long since vanished from the face of the Earth; only their ruins remain:

> **Gana:** This is not the country Ghana, but the Empire of Ghana, roughly dating from AD 850 to AD 1100. The capital of the empire was likely Kumbi-Saleh, an archaeological site in southern Mauritania. The Empire of Ghana was a major exporter of gold northward.

Caravans laden with salt and gold from the productive mines of West Africa would leave Ghana on their trek north across the Sahara. After the disintegration of the Empire of Ghana, the Mali Empire took over much of the lucrative gold trade.

Sigdmessa: This is Sijilmasa, a major city for northbound caravans crossing the Sahara. It was part of the Almoravid Empire but was torn apart in civil war and largely deserted by the sixteenth century AD. The remains of the city are found in southeast Morocco.

Urbs a Nuba, or "City of Nuba": The Toledo Tables reference the Christian kingdom of Nubia that spanned from roughly the eighth to the fourteenth centuries, resisting Arab capture. Remains of Nubian Christian culture, including churches, are found scattered in many sites along the Nile in what is now Sudan.

In addition to place names, the Toledo Tables gives some clues about the reckoning of latitude and longitude in the eleventh century. I analyzed the tables to get some sense of the accuracy of the longitude and latitude determinations. In addition, I wanted to find the location of the prime meridian, and whether the latitudes reported were consistent with the equator as the "zero" for these measurements.

For the analysis I chose a subset of locations from the tables to improve the accuracy. Islands were excluded from consideration because they are too large to establish with any precision. Places like Gana were also excluded, as the latitudes appeared to come from dead reckoning and the location of the empire city of Gana is not agreed upon, although Kumbi-Saleh is considered a leading candidate. This left me with thirty entries that I could compare to modern latitudes and longitudes. Such entries as Bagdeth (Baghdad), Toletum (Toledo), and al Medina (Medina) are examples of cities that have good correspondences between medieval and present-day locations.

Comparing the latitudes in the table with the modern values, I found that, on average, there was a spread of about 1.4 degrees but centered on the modern values. The spread of 1.4 degrees indicates that measurements such as the shadow-stick method were likely used to find latitude in that era, as this is a typical accuracy. By averaging over thirty entries, I found the equator to be consistent as the "zero" of latitude to within 0.25 degrees.

The first question for longitude is the location of the prime meridian. Although the values of longitude in these tables are less precise than the values of latitude, by taking an average over the thirty entries, I can improve the statistical power and compare the prime meridian used in the Toledo Tables to the Greenwich Meridian. For the Toledo Tables I found that its prime meridian lay 23 degrees west of the Greenwich Meridian, with an accuracy of 1 degree. Of all the possible candidates for the prime meridian, this is an excellent match for the Cape Verde islands at 23 degrees W and a poor match for other possibilities. The Azores are much farther west, at about 39 degrees W. The Canary and Madeira Islands are too far east, at roughly 16 degrees W.

Although most history books say the Portuguese discovered the Cape Verde islands around 1460, that is really the start of their continuous human habitation. There are hints in older stories of fishermen from Senegal visiting the Cape Verdean island of Sal to get salt, of an expedition from the Kingdom of Mali to the Cape Verde islands, and there are possibly other encounters. In chapter 9, on the distance to the horizon, I mentioned the Persian polymath, al-Bīrūnī. In addition to making numerous advancements in astronomy and mathematics, he documented the state of geographic knowledge in the eleventh century. Al-Bīrūnī commented on confusion arising from different choices of the prime meridian in various latitude-longitude tables:[2]

Because the practice of this science has been derived from the ideas of the Greeks, who reckoned the oecumene from its extremity that is nearest to them, which is the western extremity, consequently the longitude of a town is calculated on the basis of its distance from the west. There is disagreement among the astronomers, however, regarding this extremity. Some of them calculate longitude from the coast of the Western Ocean, which is the Encircling Sea, while others calculate it from the islands that lie far out in the Western Ocean, some 200 farsakhs from the coast. These islands are known as the Fortunate Islands [Jaza'ir as-Sa'adat] and the Eternal Islands [al-Jaza'ir al Khalidat], and they are off the coast of the country of the Maghrib. For this reason there may be found in the books two sets of longitude, with a difference of ten degrees, for the same town. Intelligence and skill are required in order to distinguish one from the other.

"Maghrib" is the general term for the Berbers, inhabitants of northwest Africa in that era. Al-Bīrūnī places the Fortunate Islands two hundred farsakhs off the coast of the country of the Maghrib, which is six hundred miles west of the coast of Africa and corresponds well with the Cape Verde islands. However, it is a poor match for the Canaries, Madeira Islands, or the Azores. The separation of 10 degrees longitude between the westernmost part of continental Africa and the Fortunate Isles is also a good match for the Cape Verde islands. From al-Bīrūnī's commentary and the Toledo Tables, it appears that many Arab geographers in this period took the Cape Verde islands as the westernmost known land for their prime meridian.

THE DEVELOPMENT OF LATITUDE MEASUREMENTS

The art of navigation has many components to it, but when someone mentions the word "navigation," it often brings to mind the specialty of celestial navigation as developed by Western Europeans. This variety of celestial navigation has its roots in the same measurements used to construct the Toledo Tables. There is an additional element of combining dead reckoning with celestial navigation to plot a position on a chart. The evolution of celestial navigation took centuries to mature.

The first revolution in navigation began with the use of compasses in conjunction with the earliest nautical charts of the thirteenth century, called portolan charts (described in chapter 6). Only toward the end of the fifteenth century did Western European sailors begin to experiment with celestial navigation. Dead reckoning was never abandoned, but because of the accumulation of uncertainties over time as a voyage progressed, periodic updates with celestial observations improved the accuracy of navigation tremendously. A full realization of the power of celestial navigation didn't emerge until the latter half of the eighteenth century. Taking the techniques of the Arab astrologers as a starting point, below I trace the evolution of the art of position finding using stars and the Sun.

One of the principle tools for celestial navigation is an instrument that will give an accurate measurement of the altitude of a body in the sky. The Arabs developed a device called a *quadrant* for astronomy and astrology. It measures one-quarter of a circle, hence its name. Figure 109 illustrates how it is used to find the altitude of a star. You sight the star along one edge of the quadrant. A weight at the end of a string (plumb bob) hangs down, using gravity as a horizontal reference. A scale marked in degrees is etched onto the side of the quadrant. As you sight the star, you allow the weight to hang freely, then use the position of the string along the scale to measure the altitude of the star.

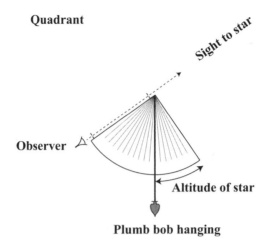

Figure 109 Principle of a quadrant. The observer sights the stars along one edge through two holes, and the altitude of the star is given by a weight (plumb bob) hanging at the end of a string.

There was money to be made in astrology during the height of the Arab Caliphates (and still is). The manufacture of quadrants and tables of celestial positions, along with tables of latitude and longitude, was a reputable occupation for learned scholars during that era. For the Sun a table of the declinations (celestial latitude) for any day of the year could be used to convert the observation of the solar altitude into your latitude with simple addition and subtraction. You observe the height of the Sun many times around the local noon. The highest point is the altitude at meridian passage. This altitude, combined with the declination of the Sun on that day gives the latitude from a formula; for example, the equation[3]

$$\text{Latitude} = (90° - \text{noon altitude} + \text{declination})$$

gives latitudes for observers north of the Tropic of Cancer.

By the fifteenth century Portuguese explorers were slowly making their way south along the western coast of Africa. By cutting out the need for trans-Saharan caravans, the Portuguese

could directly access the gold and slave trade. Ships called caravels using triangular (lateen) sails plied the West African waters.
Although navigation was done principally through dead reckoning,
celestial sights began to be used to establish the latitude of major
landmarks. Christopher Columbus and his brother Bartholomew
lived in Lisbon from 1477 to 1485, making a partial living as
mapmakers. Columbus paid close attention to the Portuguese
navigational techniques and recorded some of the earliest attempts
to ascertain the latitudes along the coast of Africa.

According to Columbus, in 1485 King D. João II sent his physi-
cian and astrologer, José Vizinho, on a voyage along the coast of
Africa to determine the latitudes of outposts using a quadrant. As
related by Columbus, José used Sun sights to establish the latitude
of an island called Los Ydolos near Sierre Leone at 1° 5′ N.[4] Sierra
Leone is at roughly 8 degrees N latitude. This is actually a fairly
respectable measurement, given some of the problems of taking
altitudes on shipboard with a quadrant.

D. João II had his eyes on riches beyond gold and slaves in West
Africa; Arab traders at that time had a monopoly on the spices
and silk coming into Europe from Asia. By establishing a route
around the southern extremity of Africa, the Portuguese could
again cut out the middleman and make a handsome profit. D.
João dispatched Bartholomeu Dias to sail as far as he could to the
southernmost extremity of Africa. Dias carried with him a mariner's astrolabe.

Figure 110 shows the principles of a mariner's astrolabe. Like
the quadrant, it uses gravity to establish the horizontal, by hanging
freely from a ring. The navigator sights the Sun or a star using a
pair of aiming slots on a shaft that freely rotates. The Sun's altitude
can be read off a scale of degrees engraved on the side of the astrolabe. There are four holes cut out of the side to reduce the effect of
the wind's buffeting it about on shipboard. Both the quadrant and
the mariner's astrolabes were adapted from instruments used by
astrologers but were vastly simplified. While the mariners' versions

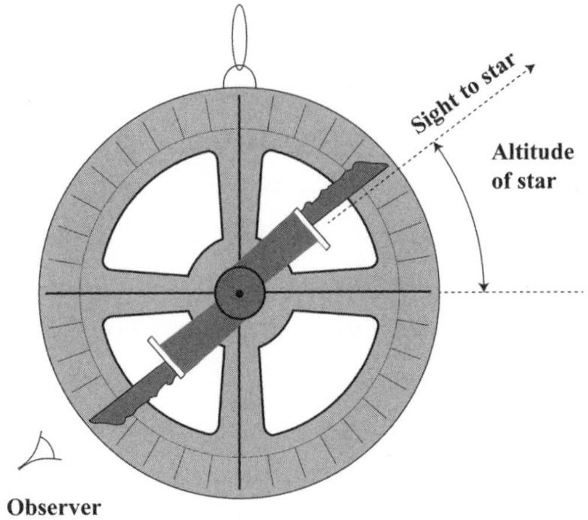

Mariner's astrolabe

Figure 110 A depiction of a mariner's astrolabe. This was adapted from the astrolabes used by astronomers and astrologers and was simplified for shipboard use.

of the quadrant and astrolabe only had angles for finding the altitudes of stars, astrologers' instruments were more sophisticated.

On his trip to the southern extremity of Africa, Bartholomeu Dias carried a table of solar declinations with him. When he reached the Cape of Good Hope, he used the Sun to find his latitude at 45 degrees south of the equator.[5] The Cape of Good Hope is actually located at 33 degrees S. This measurement is typical of early attempts at position fixing with celestial observations. Dead reckoning continued to be far more reliable than celestial fixes for quite some time. Nonetheless, the establishment of latitudes and longitudes of newly found lands slowly emerged.

Christopher Columbus adopted some of the Portuguese methods of determining latitudes, bringing a quadrant and an astrolabe on his first voyage to the Americas. On November 2, 1492, he tried to find the latitude of a sheltered bay on the north coast of Cuba by sighting the altitude of Polaris with his quadrant. The north coast of Cuba is around 21 degrees N latitude, but Columbus measured it as 42 degrees, the latitude of Boston.[6] It is likely that he took a sighting of a different star than Polaris by mistake. In reporting a latitude for this point, he chose 28 degrees N, based on dead reckoning from the Canary Islands, his point of departure.

On his return voyage from the West Indies in 1493, Columbus sailed a northerly course to catch prevailing winds from the west. After eighteen days at sea in the mid-Atlantic, he attempted a sight of Polaris with quadrant and astrolabe. Unfortunately, the ship was pitching in the waves, bouncing the astrolabe and quadrant around so much that it was impossible to get an accurate altitude. He eyeballed the height of Polaris, declaring it to be roughly the same altitude as that at Cape St. Vincent, Portugal. At the southwestern extremity of Portugal, Cape St. Vincent is a natural point of departure for many sailors venturing into the Atlantic. At 37 degrees N, he would have been far enough north to sail due east to reach Europe. This rough and ready estimate of the altitude of Polaris by eye was common.

On his fourth and last voyage to the New World, Columbus was marooned in Saint Ann's Bay in Jamaica. Some of his party went to find help from the Spanish colony at Hispaniola, across the Jamaica Passage. During his stay Columbus again measured the altitude of Polaris and obtained 18 degrees N latitude. Saint Ann's Bay is actually at 18° 30′ N. Although this result may overstate the precision of the measurement, Columbus definitely improved in finding latitude from the stars over the course of his voyages.

The techniques used by the Portuguese and Columbus were most useful in determining the latitudes of landfalls and at a modest level could supplement dead reckoning in midocean. For most sailors, however, the mathematics required for celestial navigation and the unreliability of quadrants and astrolabes at sea were large impediments. This refusal to adopt the techniques of astrology and astronomy was incomprehensible to many land-bound scholars. The Portuguese mathematician Pero Nunes wrote: "Why do we put up with these pilots, with their bad language and barbarous manners; they know neither Sun, Moon nor stars, nor their courses, movements or declinations; or how they rise, how they set and to what part of the horizon they are inclined; neither latitude nor longitude of the places on the globe, nor astrolabes, quadrants, cross staffs or watches, nor years common or bissextile, equinoxes or solstices?"[7] Nunes's comments are typical of the chasm between the world of supposedly learned scientists on one hand and navigators on the other. Neither could understand the other's world.

The Mercator Projection

Mapmakers were a bridge between the provinces of scientists and sailors. Their stock in trade from the fourteenth century onward was the production of high-quality maps. To be competitive they had to portray the most up-to-date and accurate representations of known and newly discovered lands. Oftentimes, they relied on the reports of mariners returning from distant voyages to fill in blank spaces. The imagination of mapmakers and tall tales of returning sea captains created some unusual imaginary lands. For example, in the sixteenth century a map of the regions near the North Pole included a huge magnetic mountain and a land inhabited by pygmies.

In chapter 6 we saw the rise in the thirteenth century of the portolan charts that had accurate depictions of scale and sets of rhumb lines crisscrossing the map. The charts did not display latitude and longitude, however. As the sixteenth century progressed, an increasing amount of geographic information from discoveries began to flood Europe in the form of reports and estimates of latitudes and longitudes of major landmarks.

For the mapmaker the challenge of capturing a round Earth on a rectangular map is a major issue. It is impossible to represent the globe on a flat, two-dimensional surface without introducing distortions. The question is, what distortions do you choose?

The Spanish cartographer Diogo Ribeiro, using tables and reports from explorers, produced the first known world map with latitudes and longitudes in 1527. This is some four hundred years after the creation of the Toledo Tables. Diogo chose to use a grid of lines where latitude and longitude were equally spaced, much like a simple sheet of graph paper. This representation was advocated by Ptolemy and is what a sensible person might choose as a starting point.

This mapping, called a plane chart, or *equirectangular* map, is a perfectly viable way to represent the world. It has one failing: headings of a constant compass direction (rhumb lines) are distorted. This may not be so obvious to the casual observer. The issue is the following: near the equator the distance between two lines of longitude and two lines of latitude is approximately equal. If you travel away from the equator, lines of longitude get closer together. Near the equator 1 degree of longitude is sixty nautical miles, but near Iceland 1 degree of longitude is twenty-six miles. If you took Ribeiro's map and tried to find the proper heading by connecting two points on the map, you would find yourself in trouble; a compass heading extended from one location would become distorted when extrapolated over a large distance. A map was needed on which rhumb lines (lines of constant heading) could be extended through all latitudes.

In principle the problem was solved in 1569, when Flemish mapmaker Gerardus Mercator created a world map that preserved rhumb lines everywhere on its surface. Figure 111 illustrates how Mercator's map (also called a "Mercator projection") works. The globe is projected onto a cylinder, which is then unwound to produce a flat map. It is impossible to get the entire Earth on a Mercator projection, as the representation of the surface of the Earth becomes infinitely long near the poles. On a Mercator projection the separation between lines of latitude increases closer to the poles, while on an equirectangular projection the spacing remains constant.

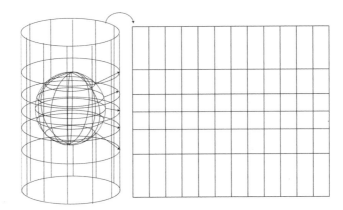

Figure 111   Principle behind a Mercator projection. Features on the globe are projected onto the surface of a cylinder, which is then "unrolled" to form a map that preserves angles everywhere.

Although it wasn't widely appreciated at the time, Mercator's invention allowed an improvement in navigation practices. Because his projection preserved rhumb lines, a navigator could set a fixed heading based on a compass direction and would be assured that he could maintain that angle across the chart. In addition, it was now easy for a navigator to make a reliable plot of dead reckoning. Even though distances were distorted depending on the latitude of the navigator, a simple scale could be used to compensate for this. Finally, any celestial measurement giving latitude and longitude could be directly plotted and compared to dead reckoning.

A writer who recognized the utility of the Mercator projection for navigation was mathematician and cartographer Edward Wright. In 1599 Wright published *Certaine Errors in Navigation*. This was a treatise that included the use of the Mercator projection to improve dead reckoning at different latitudes. Wright's work wasn't widely appreciated immediately. To work dead reckoning navigators often used the equirectangular projection or plane chart. The reliance on the plane chart created significant errors in dead reckoning when used over long distances. Wright showed how to make the appropriate corrections.

The Cross Staff

As the ability to use dead reckoning improved with better charts, celestial navigation began to come into its own. Using dead reckoning, a navigator could estimate latitude from his last fix. He could then directly compare this estimate to his latitude as determined from a celestial observation. The quadrant and astrolabe, which both used gravity to find horizontal, were notoriously unreliable on a ship in heavy seas. A more reliable device was the cross staff (Figure 112), a long stick and a crosspiece that slid back and forth. A navigator could sight the lower end of the crosspiece on the horizon and shift it back and forth until the upper end of the rod coincided with the position of the Sun or a star. The stick itself had a scale on it that allowed the navigator to find altitudes. Cross staves could achieve a fairly high degree of precision if built properly and used by an experienced navigator.

The cross staff was first used in surveying and astronomy, as angular measurements were important for both. It became common on shipboard in the latter half of the seventeenth century. A good practice in navigation is to make a best use of dead reckoning (then called *judgment* in many logbooks of the era) and when a good altitude sight was available correct the dead reckoning with the new latitude.

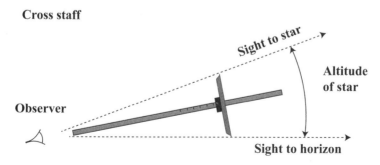

Cross staff

Sight to star

Altitude of star

Observer

Sight to horizon

Figure 112 Principle of a cross-staff. A long stick has a crosspiece that can slide back and forth, allowing the navigator to simultaneously sight the horizon and altitude of a star or the Sun.

Ships' logs often give insight into navigational practices. The log from an attempt to find a northeast passage to Asia along the coast of Siberia from a Captain John Wood provides insight into the navigation of the late seventeenth century. Wood used a cross staff. Here is one of his logbook entries from 1676 while he was en route to Siberia:[8] "Saturday, Jun 17. From the 16. noon to this day noon, a fresh gale at west-north-west and west with rain and cloudy weather. Course per compass north-east, distance sailed by the log 127 miles, difference of Lat. 90 Miles, departure east 90 miles, Lat per judgment 69° 48′. Meridian distance 303 miles, but by a good observation at noon Lat. 69° 53′. Difference of latitude between the dead reckoning and observation is 9 miles, which imputed to be westerly variation, which is found by an azimuth of 7°. Meridian distance corrected is 300 miles; fair weather."

Logbook entries are often telegraphic in nature but can be decoded. It was common practice to make entries based on the period from noon on one day until noon on the next. This would be the time when the altitude of the Sun at its meridian passage could be measured, weather permitting. In the above entry Wood first notes the weather over the course of twenty-four hours. He then records the compass heading ("north-east") and the distance sailed by use of a log line. Note that the distance sailed (127 miles) implies an average speed of five knots.

Wood then noted his estimate of the change in latitude by taking projection of his dead reckoning distance onto a north–south axis (ninety miles) and converting it to a latitude difference. By adding this to his previous latitude measurement (from the previous entry, not shown), he derived a latitude of 69° 48′ from "judgment."

The longitude was not reported as an angle but rather a distance from the prime meridian at Greenwich in miles. The dead reckoning value for the distance run east is ninety miles. In this case he estimates that he was east of the prime meridian by 303 by adding his previous measurement.

Wood records a latitude from the observed meridian passage of the Sun at 69° 53'. In comparing this to his dead reckoning estimate of his latitude of 69° 48', he attributed his discrepancy to magnetic variation, which he put at 7 degrees west and corrected his estimate of easterly progress from 303 to 300 miles from the prime meridian.

Although he doesn't report a longitude, it's straightforward to take Wood's three-hundred-mile estimate from the Greenwich meridian and turn it into 14° 32' E. His position estimate places him roughly one hundred miles east of the northwest coastline of Norway.

The Sextant

The cross staff suffered from a number of drawbacks. It did not do a terribly good job for sights at high altitudes, as it was difficult to view one end of the crosspiece at the horizon and the other at the star or Sun. Direct observations of the Sun were also painful for the eyes. A number of variations on the cross staff were employed to get around these problems, but more precision was achieved with the development of high-quality optical systems for taking sights.

Thomas Godfrey of Philadelphia and John Hadley of London are both credited with the independent creation of the octant, circa 1730. The octant spans one-eighth of the arc of a circle and as originally devised had two mirrors that allowed for a simultaneous sighting of a celestial object and the horizon. The octant has an angular range of 90 degrees. By the late eighteenth century, further refinements were employed in a device known as a sextant.

The sextant (Figure 113) consists of an arc of one-sixth of a circle, hence its name. By using the optics of reflections off a sighting mirror, the sextant can be used to measure any altitude reliably. The sighting mirror can be swiveled back and forth along a

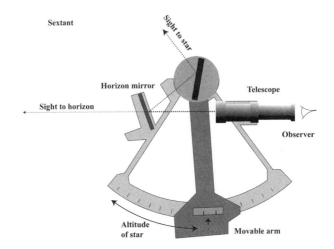

Figure 113 Operation of a sextant. The observer looks through a telescope and simultaneously views the horizon and a star or the Sun using a half-silvered horizon mirror. The navigator adjusts the altitude using a movable arm, then reads off the altitude from a graded scale.

movable arm. The image of the celestial object is reflected onto a horizon mirror, which is half-silvered. The observer will then see the image of both the celestial object and the horizon through the telescope. A set of filters can be placed over either the image of the horizon or the celestial object to make the observations easier (e.g., blocking out most of the solar intensity). By adjusting the movable arm, sometimes using a fine screw mechanism, the navigator lines up the celestial object's image with the horizon, then takes a reading. Under ideal conditions, and with some practice, a good navigator can obtain an altitude that is precise to within one or two arc minutes. The precision is often limited by the ability to observe the horizon itself.

With the precision of a sextant, the effects discussed in chapter 9 become important: dip angle and refraction. Recall that the dip angle results from the height above sea level of the observer to the horizon and increases the observed altitude of celestial objects. Likewise, refraction also increases the observed altitude. When an object is first sighted, the navigator must correct for both effects to obtain the true altitude. These corrections can be as large as half a degree, particularly in the case of sightings near the horizon. With these corrections and the precision of a modern sextant, altitudes of celestial objects can reliably give latitudes to within one or two nautical miles. While the octant has a range of 90 degrees, the sextant has a range of 120 degrees, which allows the user to find angles between two widely separated celestial objects, such as the Sun and Moon, which is often important in a technique to find longitude called the *lunar method*.

THE LONGITUDE PROBLEM

Latitude has natural reference points against which the altitudes of the stars and Sun can be determined: the equator and the poles. Longitude is a thornier measurement than latitude. The Earth is effectively a large merry-go-round. Everyone at some latitude will see the same objects in the sky rising in the east, climbing past their meridian and setting in the west, regardless of their longitude. There is no obvious distinction between how far east or how far west someone is. There is no "absolute" west in Ptolemy's sense of the Fortunate Isles being the farthest west; it is arbitrary.

The longitudes reported in the Toledo Tables were based on distances traveled between cities, and referenced to a prime meridian, which appears to be the meridian of the Cape Verde islands. Through the end of the eighteenth century, dead reckoning was the most common way of establishing longitude.

Figure 114 shows the challenge of longitude for an observer. The view is looking down at the North Pole of the Earth, with the parallel Sun's rays reaching the surface. If you had a perfect clock, synchronized to Greenwich Mean Time (GMT), you would see the Sun directly at its meridian passage at 12:00 GMT over 0 degrees longitude: the prime meridian. The Earth rotates at 15 degrees per hour. Two hours after noon at 14:00 (2:00 p.m.), the

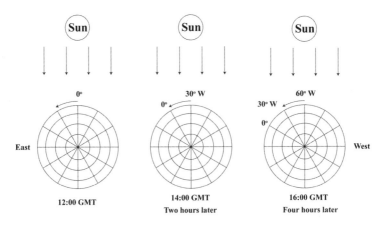

Figure 114 Longitude by meridian passage of the Sun.

Sun will be at its meridian passage at a longitude of 30 degrees W. Four hours later the Sun will be at its meridian passage at a longitude of 60 degrees W. With a set of clocks all synchronized to the same time, longitude differences could be established by the timing of the meridian passage of the Sun. The use of precise clocks to find longitude didn't exist until the eighteenth century.

In the Middle Ages scholars understood the concept of measuring longitude using celestial events. Imagine that everyone on the planet witnessed a simultaneous global event, such as a lunar eclipse. People at different longitudes could compare the relative positions of objects in the sky at the time of the event and get relative longitudes by comparing notes. An eclipse functions as a "global clock" that allows observers across the Earth to establish the timing of the eclipse relative to sunset. This could be measured with hourglasses, which had sufficient precision to measure relatively short time intervals. Although this technique was known in principle, it was rarely used in practice.

On his fourth and final voyage to the New World, Columbus tried to find his longitude using the timing of sunset to a lunar eclipse. He had tables of celestial events from Regiomontanus of Nuremburg, a famed astronomer and astrologer. Sailing along the coast of Central America, Columbus's ship became infested with worms and took on water faster than it could be bailed. He became marooned in what is now Saint Ann's Bay on the north coast of Jamaica, and a local native tribe fed him and his crew in trade for goods and out of kindness. Columbus dispatched a small crew of men in a dugout canoe to try to raise a rescue party from Hispaniola by crossing the Jamaica Passage. Some of his crew mistreated the natives to the point where they balked at supplying more food.

Columbus knew from Regiomontanus's tables that a lunar eclipse was going to commence and used this piece of information for two purposes: determining his longitude and frightening the natives into bringing him food. He threatened to make the Moon disappear unless they agreed to resupply him. Columbus used a half-hour glass called an *ampolleta* to measure the time between sunset and the eclipse. When the Moon was blotted out from the sky, the frightened natives begged him to make the Moon return. He said a solemn prayer, and the Moon returned. Cowed, the natives relented, bringing provisions to Columbus and his crew.

From his measurements with the ampolleta, Columbus concluded that the Sun set seven hours and fifteen minutes later in Saint Ann's Bay than in Cadiz, Spain. This would put him at 110 degrees W longitude, compared to Saint Ann's Bay's modern value of 77 degrees W. Although this is a substantial mismatch of 33 degrees, it represents one of the early attempts to measure longitude using a lunar eclipse.[9]

The lack of a precise longitude was dangerous. Many voyages across the Atlantic were based on the strategy of "running down latitude," whereby the captain would sail due east or due west along a parallel of latitude, then make landfall. With coastlines running mostly north to south in the Americas, an imprecise estimate of longitude could end in disaster with a ship striking a reef in the middle of the night. With the rise of major naval powers in Western Europe, the number of vessels plying the Atlantic increased tremendously, as did the number of shipwrecks. This was a strong enough incentive to solve the longitude problem.

Investigators pursued two main paths to a robust determination of longitude. The first involved the motion of the Moon and was based on its rapid motion against a fixed background of stars: half a degree per hour. A navigator in the open ocean didn't have the luxury of waiting for an eclipse to find longitude, as Columbus did. In principle, if a navigator had a table of the position of the Moon in the sky over the course of time, he could look at the angular distance between the Moon and the Sun or a bright star and use the tables to figure his longitude.

This *lunar method* suffered from multiple problems. In the sixteenth century there was no adequate theory to calculate the orbit of the Moon. Even with Newton's discovery of the laws of gravity, the calculation of the Moon's orbit around the Earth was a mathematical nightmare and took until the mid-eighteenth century to be adequately solved. Another difficulty with the lunar method was and is the visibility of the Moon, which is limited, depending on the time of the month.

Galileo proposed another method. The moons of Jupiter orbit the giant planet fairly quickly, and they are numerous enough that the timing of their disappearances behind the Jovian disk could give information on longitude; again, given the proper tables. If the lunar method had its drawbacks, this one was even tougher: a telescope would have to be carried on board every vessel, along with a table of the motions of the Jovian moons, which would have been difficult to produce with any precision.

The other path to finding longitude involved the use of a mechanical clock that could be set to a global time reference. By comparing the clock time to the time of the local meridian passage of the Sun, longitude could be calculated. Gemma Frisius, who developed triangulation, first suggested the idea of finding longitude with a clock in 1530. At the time there was no reliable clock that could keep time adequately. In the mid-seventeenth century, the Dutch physicist Christiaan Huygens developed a clock based on a swinging pendulum, familiar to many as a "grandfather clock."

Like Frisius, Huygens believed that the determination of longitude could be solved with a clock, but his invention was not suitable for use on shipboard. When a boat is pitching around in waves and subjected to large variations in temperature, the mechanical mechanisms of pendulum clocks are thrown off, rendering them useless.

THE LONGITUDE ACT

The British Royal Navy lost many vessels because of poor navigation, but the most infamous loss was the Scilly Islands naval disaster. In 1707 Great Britain dispatched a large fleet to Toulon, France, under the command of Admiral Sir Cloudesley Shovell to besiege the city during the War of the Spanish Succession. The fleet returned to England by way of the Straits of Gibraltar, then north. On this leg of the voyage, the fleet encountered a long stretch of bad weather, blowing the fleet off course, and making it difficult to get latitude sightings. On turning east toward what Shovell thought was the English Channel, they discovered that their position was well off their reckoning. Some of the fleet ran aground and foundered near the Scilly Islands off the coast of Cornwall. Thousands of crewmen were lost.

Bodies and wreckage washed up on the coast of Cornwall for months afterwards. It was a national scandal. Partly motivated by the disaster, in 1714 the British Parliament passed the Longitude Act, establishing a Longitude Board and a prize for the inventor of a practical method of the determination of longitude. Here, "practical" is the key adjective, as Galileo, Frisius, Huygens, and others had proposed schemes that, although plausible, weren't suited to a sailing vessel.

In response to the Longitude Act, many methods were proposed, including one highly impractical one in which a string of vessels moored far out at sea would all fire off rockets at the same time. The path of Gemma Frisius and Christiaan Huygens was pursued by clockmaker John Harrison, who labored for a substantial fraction of his life on a *practical* naval chronometer. To solve the problems of a pitching vessel and temperature shifts, he created elaborate compensation mechanisms and nearly friction-free bearings that would allow a clock to keep time to a precision of a few seconds over a trip the length of a transatlantic voyage. In a sea trial from England to Jamaica in 1761, a clock built by Harrison demonstrated an accuracy of better than within ten seconds, which corresponds to less than two nautical miles in position after a voyage of over sixty days.

The parallel path of the lunar method was followed by a number of physicists and astronomers. This task required the calculation of the Moon's orbit to a high precision, which was quite difficult.

LONGITUDE AND EQUAL ALTITUDES

The *equal altitude method* is a conceptually simple way to find longitude with a chronometer and a sextant. Assume you set your chronometer to Greenwich Mean Time (GMT), which corresponds to the Sun's being directly over the Greenwich meridian (0 degrees longitude) at noon, on average. In the morning you observe the Sun as it arcs higher and higher in the sky. It finally reaches its highest altitude and begins to drop. As the Sun is rising higher, then sinking lower, you record a series of altitudes and note the time of each. If you plot the curve of the rising and setting altitudes of the Sun, you can find the midpoint of a line connecting one altitude during the rising and setting periods. This midpoint of time corresponds to the moment the Sun crosses your meridian. You can extract a longitude by converting the time between the Greenwich meridian passage and your local meridian passage into degrees.

In an attempt to illustrate the equal altitude method, I created my own challenge. I pretended I was marooned on a desert island

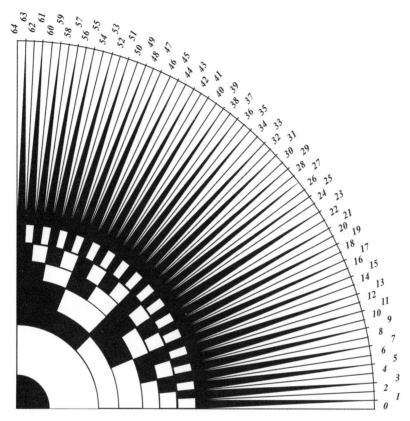

Figure 115   Graded angle scale for a homemade quadrant made by the author.

with only my wristwatch to act as a chronometer and some spare parts. I made a quadrant out of materials I found in my basement: a wooden board, a metal tube, fishing line, a rock (for the plumb bob), and a graded angle scale I made from scratch (Figure 115). The angular scale for the quadrant was based on a method taught in high school geometry classes: I successively halved angles until I subdivided the quadrant into sixty-four equal angular divisions. I then drew elongated triangles to subdivide each division, as seen in the figure. The triangles allowed me to subdivide the angular divisions more finely. I used the shadow of the Sun cast through a tube attached to the quadrant to make sightings over the course of the day.

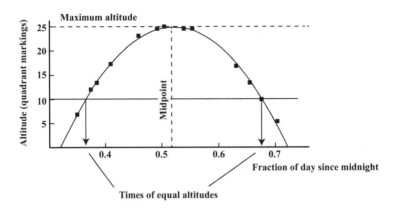

Figure 116 The maximum altitude can be converted into the latitude of the observer. The midpoint of two positions of equal altitude provides the longitude.

Figure 116 shows the altitudes of the Sun I measured on October 27, 2008. The time is reported as a fraction of the day after local midnight recorded from my wristwatch. The altitudes are reported as the readings on the quadrant markings. You can see the lowest altitude measurements were around five quadrant divisions, and the highest were around twenty-five. At close to local noon, the height of the Sun doesn't change rapidly, so those measurements could be averaged to find latitude but were not so helpful for finding longitude. Converting the quadrant markings into degrees, the maximum altitude of the Sun was 36.1 degrees. I calculated the solar declination for that day using the approximation shown in Figure 87 in chapter 8. With the computed declination, I found my latitude to be 41.9 degrees N, with an estimated uncertainty of +/– 0.5 degrees. My actual latitude was 42.4 degrees N, consistent with the accuracy of my quadrant.

The equal altitude method is also shown in Figure 116. While the altitude of the Sun changes little around the time of meridian passage, the altitude changes quite rapidly just after rising and just before setting. The two times when the Sun passes an altitude while rising and setting can be averaged to give the time of meridian passage. One traces a curve through the altitudes with a straightedge or even freehand; this gives an interpolation of sightings and is relatively precise if enough care is exercised. A horizontal line is then extended at a constant altitude through the curve of altitudes.

In this case I chose a constant altitude of ten quadrant divisions, which corresponds to an altitude of 14 degrees. The times of the two equal altitudes can be found from the intersection of the curve for the Sun with the altitude (ten divisions). The average of those two times gives a mid-point between rising and setting, which is the time of meridian passage. After finding the midpoint of my curve and correcting for the equation of time, I found a longitude of 71.5 degrees W. The actual value was 71.2 degrees. A large amount of this precision is due to the wristwatch, which is driven by a quartz-crystal oscillator.

To summarize my "desert island" experiment with solar altitudes: I found my position to a precision of thirty nautical miles along a north–south axis and fifteen nautical miles along an east–west axis. A navigator in the nineteenth century would have used a sextant with much better precision than my crude quadrant. In addition, he would have carried a more accurate table of solar declination to achieve a precision of perhaps ten miles or better.

This equal-altitude method is probably the simplest to understand conceptually, but it suffers one drawback: if the measurements are made on a moving ship, the measurements are skewed over the course of a day. In principle this can be compensated for, but this is an additional step and subject to some uncertainty.

Lunars

The lunar method, often called *lunars*, did not die immediately with the invention of the chronometer. In the late eighteenth and early nineteenth centuries, accurate chronometers were expensive devices, and many owners of merchant vessels couldn't afford them. Over the years the attitude of mathematicians toward mariners had softened, and some actively strove to create workable systems of navigation that could be readily employed on shipboard.

Lunars can be taken during times of the month when the Moon is in a favorable position to measure its angle to the Sun or a bright star. Typically, a sextant is used to measure the angle between the Sun and the Moon. Additionally, the altitudes of both bodies are measured to make a correction for an effect called *parallax*. This effect arises from the proximity of the Moon to Earth.

Nathaniel Bowditch (1773–1838) was a brilliant self-taught mathematician from Salem, Massachusetts. He had one foot in the maritime tradition of Salem and the other foot in the tradition of such scholars as Isaac Newton, whose work he read in Latin. The owners of whalers and merchant vessels in Salem could not afford marine chronometers. Bowditch published a simplified scheme for using lunars to calculate longitude and laid out this technique in his book *The American Practical Navigator*, guided by the maxim "to put down in the book nothing I can't teach a crew." To this day navigational treatises published in this tradition are called "Bowditches."

PRECISE NAVIGATION

In July 1843 sea captain Thomas Sumner published *A New and Accurate Method of Finding a Ship's Position at Sea by Projection on Mercator's Chart*. Frenchman Marcq St. Hilaire further refined Sumner's work in a technique called the *intercept method,* which is still commonly in use. Equipped with an accurate sextant, a marine chronometer, tables, and a Mercator chart, a navigator could now find his position to an accuracy of a few nautical miles with the intercept method. This method is based on the concept that, over a short distance, the altitude of a star at a moment in time gives a line of position for the observer. With two lines of position, the navigator can determine the ship's position from the intersection of both. By the twentieth century this became the standard for celestial navigation.

A navigator on board ship prepares for the magical time between the world of day and night when both the horizon and the brighter stars would be visible. This period, called *nautical twilight,* lasts about half an hour. In advance of this prime time, the navigator readies himself by estimating his position by dead reckoning. He also knows roughly which stars appear in the sky just after sunset. Once the stars begin to come out, he takes an altitude of a star above the horizon and observes the time on the chronometer, then quickly takes another altitude above the horizon on a different star and again takes note of the time on the chronometer. Before sunrise the same procedure would be repeated, but the navigator only waits in the dusk with confidence for a sufficient glow in the sky to illuminate the horizon.

The altitudes of the stars have to be corrected for refraction and dip angle. Then, using a Mercator projection, a navigator, with an estimate of his position from dead reckoning, can draw the lines of position for each star on the chart. He knows that he must lie *somewhere* along each line of position, so that his true position must be at the intersection of these two lines of position, which he traces on his chart. Latitude and longitude can also be directly calculated from the lines of position.

This brings me to the end of the story of dead reckoning and position finding by the stars and Sun. Navigation, however, is much more than this. The winds, weather, waves, tides, currents, and even the boats themselves are critical factors in making a safe passage and have to be carefully factored into any successful voyage. Far from being secondary to navigation, the challenges presented by nature were dutifully noted in logbooks and were just as important as, if not more than, celestial navigation. The second half of the book deals with these issues.

11. Red Sky at Night

· ·

AT A NOT-TOO-DISTANT time in the past, weather dictated travel. In the early era of sail, mariners might have to wait weeks in port for a favorable wind to carry them to their destination. The ability to anticipate a storm was a critical skill for any competent navigator. Winds are used in many cultures as a compass and given their names as cardinal directions. In our era people rarely notice the signs of the weather and rely instead on forecasts published in newspapers, appearing online, or from a cell-phone application.

The development of instruments such as the barometer and the thermometer in the seventeenth century allowed numbers to be put to weather conditions, and methods for weather forecasting evolved. Currently, networks of reporting stations and satellites provide input to computer models for predictions. But the processes driving weather systems are complex, even described as "chaotic," making long-range forecasts with any precision difficult. Likewise, predictions for intense weather on very short time and distance scales are elusive.

Before the wide availability of the forecasts we've come to rely on, travelers had to either cast their lot with fate or rely on their ability to read the signs in the clouds and winds to predict the weather for themselves. Many seasoned sailors could predict the weather with confidence. The seeming magic in statements like "Tomorrow, it's a-gonna blow for sure" from an old salt had its basis in signs read from the wind and clouds. It is a skill that anyone can learn through observation and patience.

The easiest forecast is *persistence*: the weather tomorrow will be the same as the weather today. This method is reliable but can be improved upon by the observation of cloud formations and wind direction. Part of this chapter is devoted to the rudiments of amateur weather forecasting.

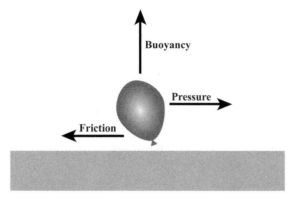

Figure 117 Forces on a parcel of air. Buoyancy will lift lighter parcels of air, or denser parcels will fall due to gravity. Differences in pressure from one region to another will push the parcel along, while friction along the ground will rob it of energy.

You can think of the atmosphere as consisting of millions of small packages of air. Imagine we take a small parcel of air and encase it in a balloon. It might rise or fall or be transported by the wind. Most of the weather on the Earth can be explained in terms of the forces at work on these small parcels. Three major forces are pressure, buoyancy, and friction. They are shown in Figure 117 acting on a balloon. Buoyancy will lift parcels of air lighter than the surroundings, while parcels of air heavier than the surroundings will fall. If pressure is higher on one side of an air parcel than another, it will move in the direction of lower pressure. Air that moves past the ground will also experience friction, slowing it down.

The most important factor at play in the atmosphere results from pressure differences between one area and another. Air flows from high-pressure regions to low-pressure regions, creating wind. On weather maps lines of equal pressure are called *isobars* (Figure 118). When the isobars are close together, they represent a rapid change in pressure, causing a stronger wind to blow. The amount of pressure change over a distance is sometimes called a *pressure gradient*. When the isobars are farther apart, the gradient is small, and winds are light. Large gradients between high- and low-pressure zones can be dramatic, with high winds and storms.

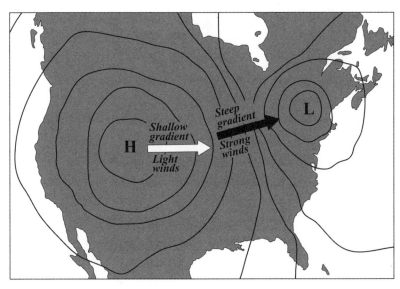

Figure 118 Lines of equal pressure are called isobars. Wind flows from regions of high pressure to low pressure. When the isobars are closer together, the winds will be stronger.

Friction of air moving against the ground is also a significant force. As wind travels over a surface, much energy is lost in the form of displacement of objects, such as the rustling of tree leaves or waves generated on the ocean's surface. A storm over the ocean with few impediments might generate thirty-mile-per-hour winds, but a few miles inland, after crossing many obstructions, the wind speed may drop to ten miles per hour. When hurricanes make landfall, most of their destructive power is lost within a hundred miles of moving inland. Likewise, as a result of friction, wind speeds near the ground can be quite low, but they increase with altitude. The higher wind velocity at altitude is responsible for the distinctive shape of many cloud formations.

The buoyancy of air is directly related to its density — how much of it is squeezed into a volume. When air is heated, the molecules move around more rapidly and the parcel of air expands. Cold air is denser than hot air, so a parcel of hot air will rise, a principle used in hot air balloons. Air pressure falls rapidly with increasing altitude. The weight of a column of air above you is responsible for pressure: fifteen pounds per square inch at sea level. As you move higher, the weight of the column of air above you becomes lighter and pressure diminishes.

Air pressure at the surface of the Earth is very nearly constant. Small pressure variations of about one part in ten thousand drive the winds on the Earth's surface. This constant pressure is maintained by the expansion and contraction of the columns of air reaching into the sky. Although the pressure may stay roughly the same during the summer and winter, this "breathing" of the Earth's atmosphere with the seasons creates a situation in which air at sea level will be denser in the winter than in the summer.

The drop in pressure with altitude is much more dramatic than changes along the surface. At the summit of Mount Everest, 8,848 meters, the air pressure is one-third that at sea level. The rapid loss in pressure with altitude creates a kind of refrigeration system that cools off the atmosphere with altitude. When a parcel of air rises, it expands against its surroundings, doing work. As a result it loses energy and cools. The constant cooling of rising air creates a drop in temperature of about 4°F for every thousand feet of altitude gained (6.5°C per kilometer, Figure 119). On a miserably hot

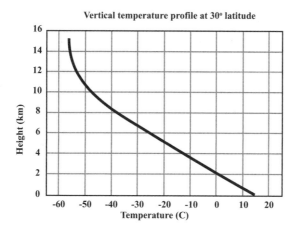

Figure 119 A typical temperature profile with altitude.

summer day, I'll look up in the sky at the tops of billowing thunderheads and picture the subzero temperatures up there, as if they were frosted mountain peaks towering above the scorched plains. The lower reaches of the atmosphere where this air cooling takes place is called the *troposphere.*

Water is a huge player in weather and climate. Its importance cannot be overstated when you stop to consider that the violence of tornados, thunderstorms, hurricanes, blizzards, and other phenomena is directly related to the properties of water. Its ability to retain heat is called *heat capacity.* Oceans, lakes, and ponds will stay much warmer in the autumn when the air around it is cooling off and will be much colder in the spring when the surrounding air is warming up. It takes a lot more heat to change the temperature of water than for air. In contrast to water, air has a relatively low heat capacity. The difference in the heat capacities of water and air create substantial differences between maritime and continental environments. For example, the city of Sitka, on the Gulf of Alaska, receives moist winds from the west that moderate its temperature swings. Sitka has January lows in the 30s (°F) and July highs in the 60s — an average temperature variation of 30°F (17°C). In contrast, Edmonton, Alberta, is located in a continental zone and has an average January low of –2°F and an average July high of 73°, a variation of 75°F (42°C).

The next important property is *thermal conductivity,* which is the ability to carry heat (or cold) from one place to another. Wet air is very effective at transmitting heat or cold from one place to another. Oftentimes on sultry summer days, people will lament, "It's not the heat; it's the humidity." A strong breeze in cold rain can rapidly drain energy away from the skin of hikers, causing hypothermia.

The third, and probably most important, property of water for weather is *latent heat.* This is the amount of heat required to change water from a liquid into a vapor. In the reverse process when water forms droplets they release heat. If water vapor condenses in a rising column of air, it releases heat, which fuels a more rapid rise. This process is responsible for the explosive growth of thunderstorms on hot humid days.

Sea Breezes

Daily wind patterns along coastlines are related to the heat capacity of water. Land heats faster than the ocean during the day and loses heat faster at night. Water temperature is relatively constant, while the temperature of the air over land shows a larger day–night variation. During the day (Figure 120) air over land heats up faster than air over water and forms a rising column of hot air. The departure of rising air creates a partial vacuum that pulls cool air in from over the ocean to take its place. The warm air at the top of the column blows out to sea. Once over colder water, the warm air cools, becomes denser, and sinks toward the ocean. Throughout the day, the resulting circulation of air coming off the water toward the land along the surface is called a *sea breeze*.

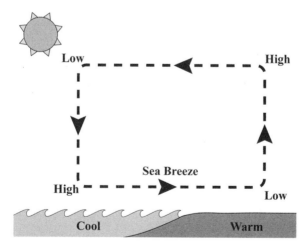

Figure 120 Sea breeze created by heated air over land rising during the day and falling over the ocean. Image credit: A. Scherlis.

At nighttime the trend is reversed (Figure 121). The temperature of the ocean remains largely the same, but the land cools down rapidly. Air over the ocean will now become more heated and rise. As it leaves the ocean's surface, its departure will suck in air from the land. The air that rises from the ocean will be drawn inland at a high altitude, cool, then sink. The cool air will come off the interior and blow along the surface and back out to sea. The cool air coming down off the mountains of the interior of a tropical island is often called a *land breeze* or a *land wind*. These terms are largely archaic and are found in the era of sailing when a land breeze aided a ship's departure from an island. On islands such as Jamaica and Cuba, the land breezes can be a substantial relief at nighttime as the cool air wafts down from the hills, allowing for a blissful sleep.

Three atmospheric processes transfer heat: conduction, convection, and radiation. *Conduction* occurs when two objects are in contact with each other. *Radiation,* a transfer of heat, occurs when electromagnetic radiation, typically infrared, travels from one place to another, carrying energy with it. When air moves from one place to another, it carries heat with it in a process called

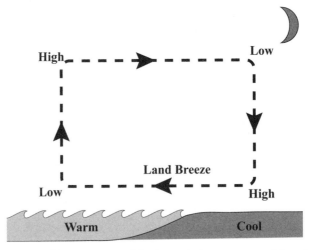

Figure 121 Land breeze created by heated air over water rising during the night and falling over land Image credit: A. Scherlis.

convection. Convection is the most efficient way heat gets transferred from one place to another in the atmosphere. The land–sea breeze pattern of air circulation at the boundary between land and water is an example of a *convection cell*. Convection cells arise when a circulation of air causes a conveyor-belt transport of heat from one region to another.

You create convection cells in your kitchen in a pot of boiling water. If you put peas into boiling water, you can use them as markers of the water flow. At a low heat, convection cells can form where peas boil up around the edges of the pot and sink in the middle, being transported across the surface. If you crank up the heat under the pot, you might see multiple convection cells form as the water boils harder. Figure 122 illustrates the formation of convection cells in a fluid. The circular motion is efficient at transporting heat from the source to cooler regions. You can see that the fluid in adjacent convection cells rises or falls in the same direction as the cell next to it, so there is an alternation of rotation between neighboring cells.

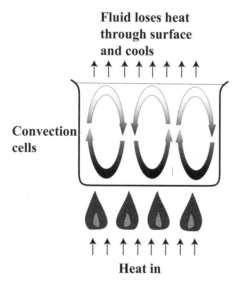

Figure 122 Convection cells arise when a gas or liquid is heated. The number of cells will depend in part on how much heat goes in.

GLOBAL WEATHER PATTERNS

Convection cells on a planetary scale create the main weather patterns we see. In each hemisphere there are three convection cells from the equator to the poles (Figure 123). The strongest of the three global convection cells is called the Hadley cell. It transports air from the equator to 30 degrees north and south latitude, where it cools enough to drop to the surface and return to the equator. The cell gets its name from George Hadley, an English meteorologist who first described wind pattern in the eighteenth century.

The region near the equator where hot air rises into the cell has very little wind and is called the *doldrums* by sailors. Here, air masses from north and south of the equator come together and rise vertically. Climatologists call this region the *intertropical convergence zone* (ITCZ). In the doldrums there is a nearly stationary band of thunderstorms created from the vertical projection of air that encircles the equator. In his book *Sailing Alone Around the World,* Joshua Slocum writes of his encounter with the doldrums:[1] "On the 16th the Spray entered this gloomy region, to battle with squalls and to be harassed by fitful calms; for this is the state of the elements between the northeast and the southeast trades, where each wind, struggling in turn for mastery, expends its force whirling about in all directions."

The region around 30 degrees N latitude is called the *horse latitudes*, where dry air descends from the Hadley cell and from the neighboring convection cell, the Ferrel cell. This area is also frequently becalmed and is associated with deserts, such as the Sahara, the Mojave, and the Arabian Peninsula in the Northern Hemisphere. The equivalent zone in the Southern Hemisphere is associated with the Kalahari Desert and the Australian outback. According to *The Oxford Companion to Ships and the Sea,* one possible origin of the term "horse latitudes" comes from the practices on southbound vessels sailing from England. Sailors were paid part of their wages before the start of a voyage. They then spent

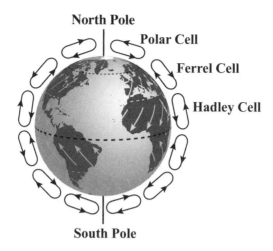

Figure 123 Three separate convection cells arise between the equator and the poles. The return air along the surface creates the prevailing winds. Image credit: A. Scherlis.

the advance in bars and brothels before setting out to sea, leaving them in debt. At sea they slowly paid off the debt, before reaching the break-even point, when they began to accrue money. To celebrate this moment sailors would bring out the effigy of a dead horse, which was ceremoniously beaten and thrown overboard. The break-even point was typically reached around 30 degrees N; hence, the name given to this latitude.[2]

The Ferrel cell, named after William Ferrel, who proposed its existence in the mid-nineteenth century, transports air along the Earth's surface northward in the Northern Hemisphere and southward in the Southern Hemisphere until it reaches approximately 60 degrees N and S latitude, where it meets the Polar cell. The Polar cell transports air from the poles along the Earth's surface toward the Ferrel cell. Where air rises at the convergence of the Ferrel and Polar cells, there is a zone called the Subpolar Low. This belt of low pressure is associated with strong storm systems in the North Pacific around the Aleutian Islands and in the North Atlantic around Iceland.

The Coriolis Effect and Global Weather Patterns

Prevailing winds are created from the path of air along the surface associated with the global convection cells. From the horse latitudes to the equator, the main surface winds flow toward the equator; beyond the horse latitudes the winds flow northward in the Northern Hemisphere. There is an additional factor that determines prevailing wind patterns: the Coriolis effect, which deflects winds in both hemispheres. The effect is named after the French scientist Gaspard-Gustave Coriolis, who wrote about it in 1835.

Figure 124 illustrates the basics of the Coriolis effect for two people riding on a rotating merry-go-round. One person throws the ball over to his friend directly across. The ball travels on a straight path, but since the merry-go-round is spinning, his friend has moved out of the way by the time it was supposed to reach him. From the point of view of an observer overhead, the ball travels in a straight line, but for the two observers on the merry-go-round, the ball appears to be deflected by what seems to be a new and mysterious force.

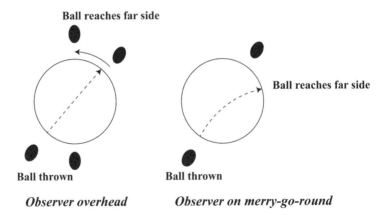

Figure 124 The Coriolis effect. When a ball is thrown across a rotating merry-go-round, it takes a straight path. From the point of view of observers playing catch on the merry-go-round, the ball seems to be deflected by a mysterious "force."

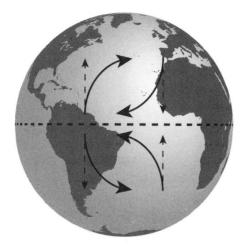

Figure 125 The Coriolis effect on the Earth. Air moving away from the equator gets diverted to the east. Air moving toward the equator gets diverted to the west. Image credit: A. Scherlis.

The Earth's rotation makes it similar to a merry-go-round. The main difference is that the Earth is a sphere. The rotational speed at the equator is faster than the rotational speeds nearer the poles. A northbound parcel of air from the equator will be moving faster than the surrounding air as it reaches higher latitudes and gets deflected to the east. Likewise, a southbound parcel of air toward the equator will be moving slower than the surrounding air at lower latitudes and gets deflected to the west. Generally, air parcels moving over long distances are deflected to the right in the Northern Hemisphere and to the left in the Southern Hemisphere (Figure 125).

The combination of the Coriolis effect and the global convection cells creates *prevailing winds* throughout the Earth (Figure 126). The prevailing winds are generated by the surface winds of the global convection cells. For the Hadley cells the air returning from the horse latitudes toward the equator gets deflected from the east to the west, creating the trade winds, which run from northeast to southwest in the Northern Hemisphere. The trade winds in the Southern Hemisphere run from the southeast to the northwest. The northern and southern variants of the trade winds converge at the doldrums.

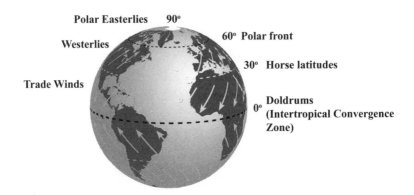

Figure 126 Patterns of prevailing winds on the Earth. The Hadley cell creates trade winds blowing from the east, and the Ferrel cell creates westerlies in the midlatitudes. Image credit: A. Scherlis.

Surface air in the Ferrel cell runs from 30 degrees to 60 degrees north in the Northern Hemisphere and will get swept to the east by the Coriolis effect, creating a prevailing wind called the *westerlies*. Wind is usually named for the direction of origin, so a "westerly" is a wind moving from west to east. Finally, there are the polar easterlies, which are formed as the surface winds from the Polar cell are deflected from east to west.

The trade winds and westerlies were important in the history of Europe and North America. On Columbus's first voyage to the Americas, he sailed south from Spain to the Canary Islands, where he caught the trade winds to take him west to the Caribbean. On his return voyage he sailed north to the latitude of the Azores, where he took the westerlies back to Europe. Similarly, ships from Europe sailed to West Africa to pick up a cargo of slaves, then sailed west on the trade winds to the Caribbean and North America, where the slaves were sold. After picking up goods from the New World, such as rum and sugar, the ships returned to Europe on the westerlies. This route is often called the triangle trade.

The Monsoon

The monsoon is a seasonal weather event that affects the Indian subcontinent and can be described as a large-scale sea breeze. Normally, the weather patterns in the Arabian Sea between India and the Arabian Peninsula are dominated by trade winds blowing from northeast to southwest. As summer approaches, however, the Sun heats the Indian subcontinent. This heating causes air to rise and draws in moisture from the Arabian Sea. As this air rises it condenses and creates a long rainy season that lasts a substantial part of the summer and can cause widespread flooding.

Navigators could use the monsoon winds to their advantage in timing their travel between Arabia and India to have following winds in both directions. The problem of sailing against the trade winds from Arabia to India may have prompted the development of a triangular kind of sail called the *lateen* that allowed sailors to sail partly against the wind.

Wind Speed

Wind speed is a clue to changes in weather and is a factor in travel decisions. Generally, we rely on forecasts to give us some idea of wind speed, but there are often clues from the way trees blow or the way waves behave in the open ocean. In 1805 Royal Navy officer Sir Francis Beaufort created a wind-speed scale to aid navigators. His scale, called the Beaufort scale, is still in widespread use. The wind speeds are classified in terms of "forces." By looking at the state of the sea, or the way trees blow on land, the navigator can make a decent estimate of wind speed and adjust plans according to the effect on the vessel. Table 2 is the Beaufort scale on land and sea.

Force	Strength	Knots	Effect
0	Calm	0–1	Smoke rises vertically
1	Light air	1–3	Smoke drifts slowly
2	Light breeze	4–6	Wind felt on face; leaves rustle
3	Gentle breeze	7–10	Twigs move; light flag unfurls
4	Moderate breeze	11–16	Dust and paper blown about; small branches move
5	Fresh breeze	17–21	Wavelets on inland water; small trees move
6	Strong breeze	22–27	Large branches sway; umbrellas turn inside out
7	Moderate gale	28–33	Whole trees sway; difficult to walk against wind
8	Fresh gale	34–40	Twigs break off trees; walking very hard
9	Strong gale	41–47	Branches break from trees
10	Whole gale	48–55	Trees uprooted
11	Storm	56–63	Widespread damage
12	Hurricane	Over 64	Devastation

Force	Strength	Knots	Effect
0	Calm	0–1	Sea like a mirror
1	Light air	1-3	Ripples without foam crests
2	Light breeze	4–6	Small wavelets, crests don't break
3	Gentle breeze	7–10	Large wavelets, some breaking
4	Moderate breeze	11–16	Small waves, frequently breaking
5	Fresh breeze	17–21	Moderate waves, many breaking
6	Strong breeze	22–27	Large waves, spray from tops
7	Moderate gale	28–33	Sea heaps up, streaking appears
8	Fresh gale	34–40	Big waves, large streaks in direction of wind
9	Strong gale	41–47	High waves begin to break
10	Whole gale	48–55	Very high waves, with long, overhanging crests, sea takes on white appearance
11	Storm	56–63	Extremely high waves, visibility diminished, large patches of white foam
12	Hurricane	Over 64	Sea filled with foam and driving spray, visibility substantially reduced

Table 2: Beaufort wind scale for land (top) and water (bottom). Wind speeds are in knots; 1 knot = 1.15 miles per hour.

AIR MASSES AND FRONTS

Although the surface of the Earth is three-quarters water, continental landmasses can play a significant role in weather patterns, particularly in the northern temperate zone (30 to 60 degrees latitude), where fronts periodically pass through on westerly winds. Continental air tends to be dry. It is more susceptible than maritime air to heating by the Sun in the summer and cools more rapidly than maritime air in the winter. Conversely, the heat capacity of water is large and moderates temperature swings with the seasons. Not surprisingly, air over oceans will be damper than air over continents. The capacity of air to hold moisture depends also on the air temperature: the warmer the air, the more water it holds. As a result summer air is usually humid, while winter air can be quite dry.

Air masses vary dramatically across continental North America. The southeastern regions of North America are strongly affected by moist tropical air from the Gulf of Mexico. The farther north one travels, the more significant a role dry continental air plays. In addition to the tropical maritime air, there is cool maritime air from the North Pacific and the North Atlantic. Maritime airflow moderates temperatures along the West and East Coasts of North America. Figure 127 shows the distribution of different air masses affecting North America.

A particularly dangerous area during spring in the central United States is known as "Tornado Alley": the boundary between the cold, dry continental air and the warm maritime air from the Gulf of Mexico. When the hot, moist air from the Gulf begins to move north in the spring, collisions with the continental air create intense storms. This boundary between the maritime and continental air masses in the United States is often called the *dry line*.

Continental Europe sees a similar pattern with air masses, and like North America, the shape of the coastline has a strong

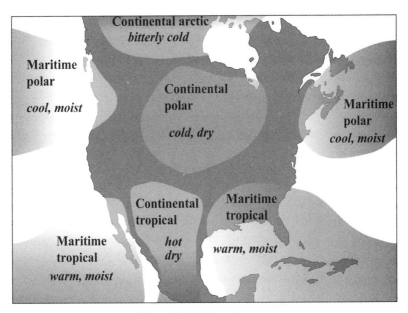

Figure 127   Air masses affecting North America.

influence on the patterns. Areas close to the Mediterranean are warm and moist, while farther inland air is dryer and more subject to seasonal variations, as in Ukraine. The westerly winds off the North Atlantic can bring in a substantial amount of cool maritime air, and the penetration of the Baltic Sea east to Russia plays a significant role.

Shifts in air masses are often reflected in the motion of *jet streams,* which flow west to east. The northernmost jet stream was discovered when pressurized cabins first allowed airplanes to fly above twenty-five thousand feet in the upper reaches of the troposphere. At those altitudes it was found that jets would sometimes fly a hundred miles per hour faster than usual when flying east and a hundred miles per hour slower when flying west. There are two jet streams in each hemisphere, corresponding roughly to the transitions between the circulation zones (Figure 128). The jet stream between the Polar and Ferrel cells is called the polar stream. The jet stream between the Ferrel and Hadley cells is called the subtropical stream. The polar jet stream meanders

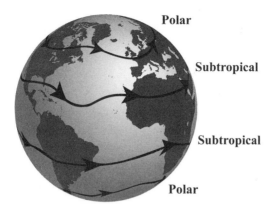

Figure 128 Subtropical and polar jet streams. The meandering of the streams is associated with weather fronts. Image credit: A. Scherlis.

frequently, creating tongues of cold air dipping to the south and warm air penetrating to the north.

On average, since cold air is denser than warm air, the northern stretches of continental landmasses create high-pressure systems, and the warmer maritime air creates low-pressure systems. In the temperate climes dominated by westerly winds, *fronts* develop at the eastward moving boundaries between the warm and cold air.

There are two distinct frontal systems in the midlatitudes: warm fronts and cold fronts. The fronts are named for the body of air that is moving in and replacing the existing air mass. For example, if you're in a region of cool, dry continental air and warm, moist tropical air is approaching, the boundary is called a *warm front*. On the other hand, cold air encroaching on warm moist air is a *cold front*. Warm and cold fronts have distinctive patterns of weather signs as they approach and can be forecast with some degree of accuracy by observation.

It is important to note that winds at the ground level are not necessarily the same as winds that blow at high altitudes, steering the fronts. The winds at high altitudes will usually be associated with the general flow of weather and can be found by observing high-level cloud formations. These high-level winds, called *steering winds*, are helpful in determining where the weather system is moving. Surface winds, on the other hand, can be the result of local forces. For example, thunderstorms moving from the west can pull air in from the east, so one could easily be misled by a wind blowing toward a thunderstorm.

WARM FRONTS

Warm fronts are the transition from cold, dry air to warm, moist air. These fronts are typically slow moving, and the signs of the approaching front may take twenty-four hours or more to develop. The front can cause long periods of rain with slight wind and long-lasting cloud cover. Once the front has passed, the weather can be unstable, with afternoon thunderstorms developing in the summer.

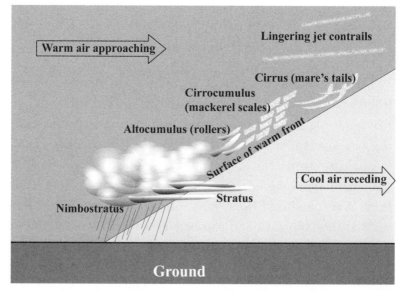

Figure 129 Characteristics of an approaching warm front. A lowering sequence of clouds appears as the wedge shape of the warm front overtakes the denser cold air in front of it.

Because warm air is less dense than the colder air it overtakes, it gets pushed upward, and the first cloud formations are seen at high altitudes. The resulting frontal boundary is wedge shaped (Figure 129). Because of its moisture content, the warm air cools and condenses as it gets pushed upward. The first signs of the front come in the form of clouds of ice crystals, then clouds of water droplets. The characteristic sign of an approaching warm front is a sequence of cloud formations that start from the upper atmosphere and gradually drop in altitude until they are at their lowest and bear rain.

Jet Contrails

Lingering jet contrails are the first harbingers of a warm front (Figure 130). These come from the moisture in the exhaust of engines in aircraft at high altitudes. If the atmosphere is relatively

Figure 130   A jet contrail lingering high in the sky can be one of the first signs of an approaching warm front.

dry, the contrails will disappear as the surrounding air absorbs the water vapor. If the upper atmosphere is saturated with water vapor, the contrails will linger and even broaden out, acting as seeds for further cloud formation.

CIRRUS CLOUDS

Cirrus clouds are often called *mares' tails* because of their wispy shape, which resembles the tail of a horse (Figure 131). The formations of ice crystals high in the sky slowly fall because of gravity. At higher altitudes wind speed is greater. As the ice crystals fall, they aren't swept through the sky as fast as those at a higher altitude, giving the characteristic shape for which they're named. The direction of the backs of the mares' tails is a good clue to the wind direction at high altitudes and the general trend of weather movement.

Figure 131 Cirrus clouds or mare's tails are formed by ice crystals in the upper atmosphere.

MACKEREL SCALES

An old weather rhyme goes:

> Mackerel scales and mares' tails
> Make lofty ships carry low sails.

"Mackerel scales" are one variation of a cloud formation called *cirrocumulus*. These clouds have a fine structure and do not show shading (Figure 132). These often have a mottled pattern resembling fish scales, hence the nickname. The combination of mackerel scales (cirrocumulus) and mares' tails (cirrus) are a strong sign of an approaching warm front.

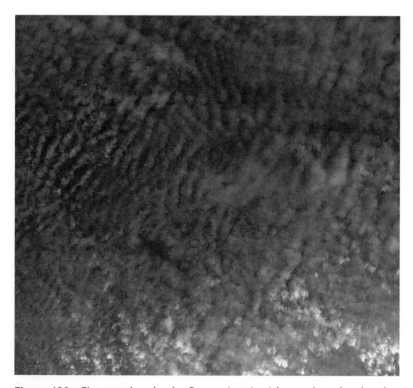

Figure 132 Cirrocumulus clouds. One variant is nicknamed mackerel scales. Source: Brian Klimowski. Copyright © Brian Klimowski. Reproduced with permission.

Altocumulus Clouds

Altocumulus clouds are puffy midlevel clouds. These typically are seen some hours after the first appearance of cirrus and cirrostratus clouds in the approach of a warm front. They are formed of water droplets and have a semitransparent appearance. Frequently, altocumulus clouds create a formation colloquially known as "rollers." Rollers are long horizontal bands of clouds (Figure 133) caused by faster air moving over slower air. When there's a layering of faster wind over slower wind, pockets of low-pressure eddies can form. Inside these eddies of circulating air, water vapor condenses, giving rise to rollers. The upper-level wind direction is generally perpendicular to the long axis of the rollers.

Figure 133 Altocumulus clouds. The clouds in the formation in the photograph are sometimes called "rollers."

STRATUS CLOUDS

Stratus means "flat" or "layered." Low stratus clouds are often the next-to-last set of clouds before rain in the sequence of an approaching warm front. If you've been watching the progression of lingering contrails, followed by mares' tails, then mackerel scales, then roller clouds, when you see stratus clouds, you might want to give some serious thought to setting up a good watertight camp, or find some kind of shelter, because rain is likely to be on its way. Low stratus clouds associated with rain are flat and grey and cover large sections of the sky. The clouds are low enough at this point that the sensation of humidity on the skin associated with the approaching warm front becomes quite obvious. The odor of damp soil may be intensified.

NIMBOSTRATUS CLOUDS

By the time you see nimbostratus clouds, it's probably too late, and you're getting rained on. They are relatively flat, rain-bearing clouds. Local patches of falling rain create some undulations in the cloud cover. The duration of rain can be quite long during a warm front, lasting up to several days, depending on how fast it is moving.

There's a general saying about weather systems:

> Short in coming, soon it will pass
> Long in coming, long it will last.

This aphorism refers to the speed of approach of weather signs. If the signs of an approaching warm front take a long time to build, it is likely that a long period of rain ensues. If the signs come through rapidly, it's likely that the period of rain passes quickly. During long rainy spells, I'll often look out of my tent, repeating this aphorism, and searching for signs of a letup and clear weather. When the nimbostratus clouds begin to blow apart and patches of blue sky appear, this is often a sign that the storm is abating. There's an old saying of Dutch extraction in Upstate New York:

> A storm's blown out when you spot a patch of blue sky
> Large enough to make a Dutchman a pair of britches.[3]

Once the front has passed, the weather accompanying the low-pressure zone of warm air is often unstable, with afternoon thunderstorms blowing up in the summer and cloud cover common in the winter. Conditions are typically warm and humid.

COLD FRONTS

Cold fronts have a different character from warm fronts. Because cold air behind an approaching front is denser than the warm maritime air, it creates a blunt leading edge (Figure 134). The cold front hugs the ground and pushes the warm air mass ahead of

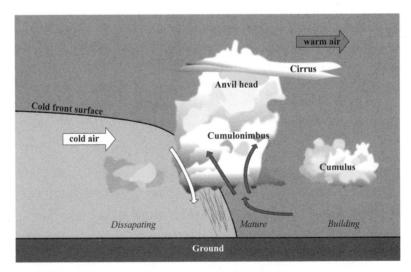

Figure 134 Shape of a cold front with dense cold air impinging on warm moist air.

it. The abrupt leading edge creates a much faster transition than warm fronts, often breeding thunderstorms.

As the cold front approaches, it pushes warm air in front of it to higher altitudes. This is a crucial factor in the development of thunderstorms. Warm moist air contains large amounts of latent energy that is released when water vapor condenses. As the warm air rises, it gains altitude and cools, forming droplets. The heat given off by the droplet formation causes the surrounding air to warm further and rise more rapidly, creating a vertical column. When this happens, a partial vacuum arises on the surface that sucks in more warm air. A sudden drop in atmospheric pressure often signals the approach of a cold front. Thunderstorms are a kind of engine that feeds off the energy of the surrounding humid air. The vacuum effect can pull in air from a hundred kilometers away, even if the front itself is localized.

Figure 134 also shows the characteristic sequence of clouds associated with the production of thunderstorms. Puffball fair-weather clouds are a kind of *cumulus* cloud. They begin to build into thunderstorms with updrafts and start to show significant vertical structure as heated columns of air push the cloud tops higher. As these heated columns of air gain altitude, the storm begins to mature. Rising air columns can reach beyond the upper level of the troposphere and bubble out, freezing into ice crystals. The ice crystals then spread, creating a large patch of high cirrus clouds that is blown ahead of the storm in strong upper-level winds. The high band of cirrus clouds creates a characteristic "anvil head" that often accompanies a mature thunderstorm.

As water droplets gain altitude from the updrafts, they eventually begin to cool and grow in size. At a certain point the droplets are heavy enough that updrafts can no longer keep them aloft, and they begin to fall. As more and more droplets fall, they entrain cold upper-level air, producing strong *downdrafts*. The downdrafts often precede the rain itself and can produce violent gusts in advance of falling rain. Strong downdrafts can be exceedingly dangerous to boats and airplanes. The ensuing rain is intense, accompanied by lightning that appears as both cloud-to-cloud and cloud-to-ground strikes.

The mature rain-bearing cloud is called a *cumulonimbus* cloud. Once the storm has passed or has blown itself out, the remnants of the dissipating cumulonimbus cloud is often gray in color, having a chaotic filamentary appearance.

The sequence of winds in an approaching cold front can be telling. First, one experiences winds from the east and southeast feeding the storm that draw moist warm air along the surface. In contrast to the surface winds, high-altitude clouds moving from the west can signal the direction of prevailing winds aloft. On the ground, as the storm approaches, the transition to vertical airflow creates a "calm before the storm." The easterly or southeasterly wind drops out as dark clouds move in overhead. Cool wind then starts to flow as part of the downdraft, often becoming violent, followed by rain and lightning. At this point the wind shifts into the southwest and west as the front passes.

The clockwise *veering* pattern of winds is typical for cold fronts in the Northern Hemisphere: the inflow of air from the east and southeast, followed by the westerly winds after it passes. In the temperate zones of the Southern Hemisphere, the wind will exhibit a counterclockwise or *backing* pattern, where the wind starts in the northeast, moves to the north as the front passes, then shifts to the west.

Large thunderheads are accompanied by strong vertical airflow. Hail forms when water droplets freeze in the upper atmosphere, fall, become coated in water, and freeze again when pushed upward by another updraft. The cycle repeats, and multiple layers of ice can create hailstones as large as golf balls.

Once the front itself has passed, the area under the zone of high pressure is typically relatively cool and dry, with blue skies and puffball cumulus clouds blowing up periodically. During the winter these high-pressure continental zones can produce bitterly cold weather.

Another dangerous formation associated with cold fronts is a *squall line*, effectively a long, stretched-out band of thunderstorms that form a single unit. The characteristics of the air inflow and strong downdrafts are similar to those of a cold-front thunderstorm but can be more violent. In my limited experience approaching squall lines appear to be heralded by a cloud sequence that looks like a sped-up warm front approaching, with the formations passing in a few hours rather than over the course of a day or more. As it approaches, the squall line appears as a smooth, elongated band of clouds across the sky in the distance. The initial downdraft stretches along the length of the squall line and is called a *gust front*, which can hit suddenly with great intensity, followed by pouring rain and lightning.

It is important to realize that thunderstorms don't need a cold front to form. If you are surrounded by warm, moist tropical air, the heating of the Sun during the day can create scattered thunderstorms, some of which can be violent. A small updraft can plant the seed of a thunderstorm on a hot summer day and be amplified into a large storm as it sucks in moist, warm air. These typically blow up in the afternoon and abate toward dusk. The updrafts creating storms can also be formed as part of the sea-breeze convection cells, typically on islands in the tropics.

In the absence of fronts sweeping by, wind over the course of a day often follows a predictable pattern, particularly during the summer months on continents. Still air accompanies daybreak. As the Sun rises, unequal heating occurs among parcels of land and bodies of water. Thunderheads may build in the distance, pulling in air from far away. As the day progresses wind picks up, often peaking in mid- to late afternoon, then dropping toward dusk. Explorers and fur traders traversing North America by canoe would arise well before daylight and paddle in the gray of morning twilight. As the day progressed along the shores of a large lake, wind and waves would build, forcing the paddlers to hole up and wait until the wind abated near sunset, when they could get back out and put in more miles before dark.

Navigators in the tropical Pacific used towering cloud formations as a way of finding land up to a hundred miles away. This is not an infallible method, but it works and is part of the navigational lore of Hawaiians in the form of a chant, "May the peaks of Havaiki be banked in clouds."[4] This sign of land comes, in part, from the processes that produce sea breezes: air gets sucked in from the ocean during the day, is heated over land, and rises. Water in the rising air column condenses, producing thunderheads that tower over land over thirty thousand feet high. Figure 135 shows the top of a thunderhead eighty miles from the camera, illuminated by the Sun that has just dipped below the horizon, casting the lower clouds in darkness.

The highest mountain in Jamaica is seventy-four hundred feet high, and for an observer at sea level, it would be seen at a distance of 86 miles. The top of a towering cumulus reaching a height of thirty thousand feet would be visible up to 173 miles away. One of the bigger issues in using this as a navigational technique is sorting out whether a towering cloud is truly representative of land.

Figure 135 Top of a thunderhead cloud illuminated by the setting Sun. The Moon is in the background. The photograph was taken eighty miles away from the cloud formation in the foreground.

Typically, the land cloud will remain stationary, whereas other clouds will move.

With lightning and thunder, distance to a thunderstorm can be estimated from the delay between the light flash and thunderclap. A flash from lightning arrives almost instantaneously, while sound travels at approximately 340 meters per second to reach the observer. If you start counting at the flash and stop counting when you hear thunder, the distance of the strike is one mile for every five seconds of delay. As a storm approaches, the delay decreases.

Most lightning strikes are cloud to cloud. Violent winds are more likely to be a danger to someone caught outside in a thunderstorm, but cloud-to-ground strikes can be deadly, albeit rare. There are a few simple rules that one can use to reduce the chances of getting struck if caught outside:

1. Never seek shelter under a tall, lone tree.
2. Crouch on the ground, in the lowest area you can reach.
3. Avoid shallow overhangs, such as shallow caves on a cliff: they can channel lightning.
4. Avoid beaches — current paths can run into water underneath the sand.

CYCLONIC STORMS

A *cyclone* (Figure 136) is a large, contained region of *low pressure* that pulls in surrounding air, accompanied by rotation about its center. Similarly, an *anticyclone* is a large, contained region of *high pressure* that sheds air to its surroundings, accompanied by rotation about its center. You may recall that the Coriolis effect deflects air moving over long distances. Air is deflected to the right in the Northern Hemisphere and to the left in the Southern Hemisphere.

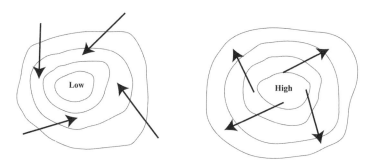

Figure 136 High- and low-pressure regions and associated airflow. These create cyclones and anticyclones. The above figure is representative of systems in the Northern Hemisphere. The sense of rotation gets reversed for cyclones and anticyclones in the Southern Hemisphere.

As air is drawn into a low-pressure pocket in the Northern Hemisphere, the right-hand deflection creates a counterclockwise rotation around the center. In the Southern Hemisphere, the left-ward deflection creates a clockwise rotation. In an anticyclone, air is shed from a high-pressure system and flows outward. The same Coriolis effect comes into play, diverting the wind, but since the air is flowing outward, it creates a clockwise rotation in the Northern Hemisphere and a counterclockwise rotation in the Southern Hemisphere.

If a storm becomes large, the Coriolis effect will influence its structure. A classic example of this is a hurricane. Hurricanes are born from low-pressure disturbances that move west off the coast of North Africa, carried by the trade winds over the warm waters of the tropical Atlantic. As they grow in size, they suck in warm moist air from far across the ocean, fed by the same latent heat effect found in thunderstorms. As rising air condenses into water droplets, it releases heat, creating more upward convection. The inflowing air from long distances creates a counterclockwise rotation. The circulation of wind around the low-pressure center intensifies, and it becomes a heat engine, bringing in more and more latent energy from the hot moist air and converting it into wind speed. In this case the airflow is not just vertical but has a strong rotational component.

Hurricanes and typhoons are two names given to storms more broadly known as *tropical cyclones*. The distinction between a hurricane and a typhoon is largely a matter of location, but they're effectively the same type of storm. As long as a tropical cyclone is drawing in warm moist air over the ocean, it grows in intensity, with an eye developing at the center of an upward spiraling column of air feeding the storm. The most intense winds are just outside the eye, diminishing in strength as one moves away from the center. Inside the eye, there is little or no wind and extremely low pressure.

The first sign of an approaching tropical cyclone is a large ocean swell. As we'll see in chapter 12, the longer the wavelength, the faster a wave travels across the ocean. Swells from hurricanes will have long periods of up to fifteen seconds between crests. Because of the high speed of the storm-induced swells, they outrun the tropical cyclone and arrive days in advance of the storm itself. As the storm approaches, the height of the swell increases and a pattern of clouds appears on the horizon that resembles a fast-moving warm front: cirrus and cirrocumulus clouds, followed by altocumulus clouds, and, finally, bands of rain accompanied by strong winds.

Tropical cyclones present an obvious navigational hazard, even to the largest ships. Prudent navigators can note the appearance of long swells, even aberrant ocean currents, and take evasive action. A sense of the likely position of the eye of the storm can help guide the navigator to seek the best course. Wind speeds are the highest where the motion of the tropical cyclone combines with its rotation. This zone is to be avoided. The other side where the motion of the cyclone goes against rotation has diminished wind speeds and is safer to traverse. The direction of oncoming swell and motion of clouds aloft can give some clues about the location of the faraway center of the storm. Fortunately, in the era of satellite imagery, a lot of the guesswork has been taken out of tropical cyclone forecasting, but this was not always the case.

Columbus on his fourth voyage to the New World in 1502 was sailing near what is now Martinique and saw signs of a hurricane brewing. It is not entirely clear how he became acquainted with the warning signs of an approaching tropical cyclone, but he evidently could detect its presence in the distance. Columbus sought shelter in the harbor of Santo Domingo on the island of Hispaniola (the island that is now home to the Dominican Republic and Haiti). He warned the governor of Hispaniola, Nicolás de Ovando, of the oncoming storm, but de Ovando refused Columbus the shelter of Santo Domingo and was in the process of sending a fleet of twenty-eight vessels laden with treasure back to Spain. Columbus tried to dissuade him from sending the fleet, but the governor persisted. By sailing along the coast, Columbus found shelter in an estuary, where he rode out the storm. Twenty-four of the twenty-eight ships of the treasure fleet sank.

Tropical cyclones move in the direction of *steering winds.* Their formation in the tropics creates an east–west motion because of the trades. As the storms mature they will eventually veer or back toward higher latitudes, then get caught in the westerlies, where they'll get swept back to the east. Once they are over colder water, the storms usually lose the driving power from latent heat, and the remnants diminish in intensity.

Cyclonic storms are also known in temperate zones and can be quite powerful. Along the East Coast of North America, there are nor'easters that begin as a low-pressure pocket on the southeastern coast of the United States, moving northward. As counterclockwise circulation about the low begins, warm, moist air flows off the ocean inland. The warm moist air meets cold continental air drawn in from the north, and the storm increases in intensity. Nor'easters typically occur in the winter and are often associated with blizzards, although they are known at other times of the year.

The 1991 Halloween Nor'easter, described in the best seller *The Perfect Storm,* was an unusually intense cyclonic storm in which a nor'easter absorbed a northbound tropical cyclone, Hurricane

Grace. A high-pressure system to the north of the low created a huge pressure gradient, resulting in a long channel of sustained high wind. This, in turn, produced giant waves that lashed the Eastern Seaboard.

A variation on temperate-zone cyclones is called *explosive* (or *bomb*) *cyclogenesis*. This is the very rapid creation of an intense cyclonic storm outside the tropics. These storms often form in the North Atlantic and North Pacific, developing over a period of twenty-four hours or less. Although most common in winter, they can arise throughout the year. A notorious example of explosive cyclogenesis occurred during the August 1979 Fastnet yacht race, which eluded forecasters. The resulting storm took the lives of fifteen sailors and stranded many more.

Nick Ward was a crewmember on the sailboat *Grimalkin* during the 1979 Fastnet race. He wrote of his experiences in the book *Left for Dead*. As the rapidly deepening low-pressure system of the cyclone approached, he noticed a strange coloration of the sky to the west: "It was then that I noticed the combination of colours in the sky — reds, oranges and ochres, weird but exquisite, unlike anything I had seen before. The reds reminded me of colours I'd seen in a Rothko oil painting. I joined Matt, Mike and Dave on the windward rail, looking intently to the west, all of us transfixed by the beauty of the skyline. With the sunset not due until 8.30pm, I was baffled by the colour scheme this deep orange sun created in the sky so early in the day."[5]

Strange sky colorations are often reported before severe storms. A PhD thesis was written on a green color that sometimes accompanies severe thunderstorms in the U.S. Midwest.[6]

Winds as Direction Indicators

As mentioned in chapter 4, winds can be used for orientation. Winds may persist for some time or change direction, depending on conditions and seasons. Over very short time periods, if you know the direction of the wind, it is usually a reliable assumption that it will continue for at least the next hour. Over longer periods of days, wind direction can shift, depending on weather. Over a period of months, winds and weather follow seasonal shifts. For example, along the East Coast of North America, prevailing winds are the westerlies, but seasonal shifts create discernable patterns. During the summer months, on average, winds tend to be more out of the southwest, while in the winter months, they tend to come more out of the northwest. Passing storms can alter these patterns for a period of days, however.

Wind coming from an area assumes the distinctive character of the region that bred it. In New England a southwesterly wind from the Gulf of Mexico is hot and humid, while a northwesterly wind, coming from the northern part of the continent, is cold and dry. Air off the Atlantic, from the northeast, is cool and damp.

Both the seasonality and the character of wind lend themselves as a natural compass. The Bugis, a people in Indonesia, use a compass based on the seasonal wind patterns.[7] David Lewis documented the wind compasses of a number of Pacific Islands navigators in his book *We, the Navigators*.[8]

Anthropologist Rick Feinberg analyzed variants of wind compasses on the Pacific island Anuta in his book *Polynesian Seafaring and Navigation*.[9] The names of the compass points often appear in multiple roles. For Anutans the name *tonga* can be associated with (1) the season of trade winds blowing out of the southeast or (2) the archipelago of Tonga, which lies to the east-southeast of Anuta. Names for winds and directions often merge together in wind compasses, and variations of *tonga* appear in many Pacific Islands wind compasses. Typically, the direction "tonga" indicates the southeast quadrant. An example of a wind compass from Anuta is shown in Figure 137. *Tokerau pare* can mean "landward," or "north." "Tonga" is generally agreed to be just south of due east, and the direction *tonga maaro* is to the "right" of tonga, or south-

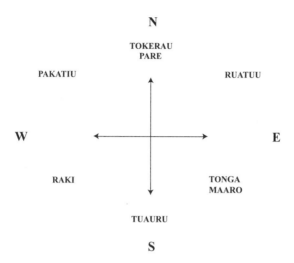

Figure 137 Anutan wind compass. Adapted from Feinberg, *Polynesian Seafaring and Navigation*; Kent: Ohio, 1988; 92. Copyright © 1988 by The Kent State University Press. Reprinted with permission.

east. *Raki* is "very big" and fills the southwestern quadrant. *Pakatiu* is northwest, and *tuauru* is south.[10] Another of Feinberg's informants, Pu Nukumanaia, drew a slightly different wind compass, but the names and directions are roughly the same as in Figure 137.

Navigators using natural compasses typically rely on multiple sources for orientation, whether they rely on stars, wind, or other sources. A single source can fail or be misinterpreted, but the redundancy of multiple sources builds confidence. A prevailing wind may persist for some time, followed by a shift signaled by storms or a change of season. Knowledge of the conditions creating wind shifts is based on experience with local weather patterns.

WEATHER FOLK WISDOM

Before there was a clear understanding of the causes of weather patterns, people relied on aphorisms to guide their forecasts. The aphorisms were based on experience but when examined closely often have a valid underlying basis for the poetry. Two of these have already been described above:

> Short in coming, soon it will pass
> Long in coming, long it will last.

and

> Mackerel scales and mares' tails
> Make loft ships carry low sails.

Another aphorism that appears to work is:

> If dew is on the grass
> Rain will never come to pass.

Dew usually forms when the temperature drops at night, causing water vapor to condense on surfaces such as grass. The falling temperature is the sign of clear skies, when radiative transfer transports heats away from the ground, unlike cloudy nights, when the hot air is trapped. This implies a clearing trend.

One well-known saying is:

> Red sky at night, sailor's delight
> Red sky at morning, sailors take warning.

Over the years I have tried to understand this one, and I have to confess that I do not put a lot of faith in it. The best explanation that I can come up with is the following: Storm systems in the midlatitudes move from west to east. If you see clouds at sunset illuminated by the Sun, there must be a clear region along the western horizon for the sunlight to reach the clouds. When it does, it illuminates clouds to east. This implies that the western sky is clear but there are still lingering clouds in your vicinity. The opposite would be the case in the morning, where the sky is clear near the Sun in the east, but there are clouds illuminated to the west. In both cases the "red" in the adage refers to the color of clouds in the sky, not the color of the Sun itself. My experience is that this adage is not very reliable. I've seen big storms following red skies at night and clear weather following a red sky at morning. This piece of weather augury is featured in the Gospel According to Matthew:

> The Pharisees and Sadducees came to Jesus and tested him by asking him to show them a sign from heaven.
>
> He replied, "When evening comes, you say, 'It will be fair weather, for the sky is red,'
>
> "And in the morning, 'Today it will be stormy, for the sky is red and overcast.'
>
> "You know how to interpret the appearance of the sky, but you cannot interpret the signs of the times.
>
> "A wicked and adulterous generation looks for a miraculous sign, but none will be given it except the sign of Jonah." Jesus then left them and went away.
>
> — Matthew 16:1–4

While I'll leave the interpretation of this passage to biblical scholars, it's clear the red-sky-at-night lore was common at least two thousand years ago.

For thunderheads there is:

> When clouds appear like rocks and towers,
> The Earth's refreshed with frequent showers.

The tower structure is an allusion to the vertical structure of growing thunderheads.

For fair-weather puffball cumulus clouds, there is this:

> If wooly fleeces deck the heavenly way,
> Be sure no rain will mar a summer's day.

For veering wind patterns before a cold front:

> When the wind is from the south
> the rain is in its mouth.

After a front passed through, this was said:

> The wind in the west
> Suits everyone best.

The above sampling of weather proverbs is not meant to imply that English is the only language where the folk art of forecasting is practiced. Far from it. Weather systems are a shared combination of semipredictable trends with random surprises thrown in that give strangers something to talk about in all languages.

12. Reading the Waves

. . . .

ON THE OCEAN, waves are usually a by-product of the weather: wind over water. It could be a gentle breeze creating a cat's paw pattern on a pond or giant swells kicked up by a typhoon. To most the ocean surface may seem random and inscrutable, yet there is almost a boundless amount of information hidden in plain view, if only the meaning can be deciphered. The existence of distant storms is betrayed by ocean swells racing over thousands of miles with undiminished power. The presence of an island thirty miles away will reveal itself through the pattern of reflected waves. Experienced navigators can deduce meaning in wave patterns.

Waves are disturbances that travel through a medium, such as air or water, and come in many forms. Sound and water waves are two familiar examples. The medium itself doesn't move, but the disturbance travels. Waves are described in science textbooks by a wavelength, and amplitude, as in Figure 138. These are often called *sine waves* after the name of the mathematical function. The *wavelength* is the distance between one crest and the next.

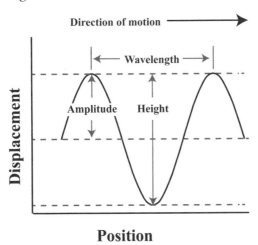

Figure 138 Typical sine wave, showing its amplitude, height, and wavelength.

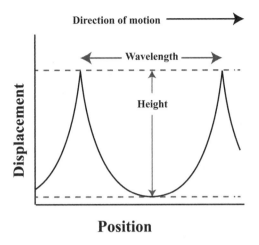

Position

Figure 139 Shape of a trochoidal wave. The wave height goes from the trough to the crest.

The *amplitude* describes the size or height of the disturbance as measured from the middle to the top of the crest.

A *water wave* is a disturbance at the boundary between water and air. Water waves can be much steeper than the textbook sine wave in Figure 138. If you ask a child to draw waves in the ocean, you would probably get something with sharp peaks, as in Figure 139. Mathematicians call this shape a *trochoid* — a curve traced out by a point on the rim of a wheel as it rolls along the ground. For trochoidal waves the textbook amplitude isn't easily defined. For these peaked waves the distance from the trough to the crest is easier to measure and is called the *wave height*. Most descriptions of ocean waves are in terms of heights rather than amplitudes.

Walks along the beach can be soothing and sometimes romantic, but can the average beach walker really see what's happening on the water? Water waves are formed from circular orbits of water on top of smaller orbits, as in Figure 140. When a wave reaches the beach, the orbit is broken and part of the circle continues as it curls over and crashes. Wind creates the largest orbits right at the surface, with their size diminishing with depth. Submarines

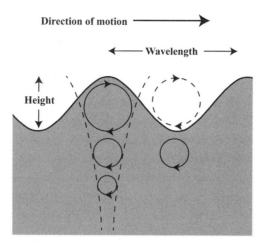

Figure 140   Water waves are circular orbits of water molecules. Each orbit gets smaller with increasing depth. Depicted here are deep-water waves.

can take advantage of this effect, diving down below storms to ride out conditions that would capsize vessels on the surface.

The waves shown in Figure 140 are called *deep-water* waves, in which the orbits are circular with radii that diminish with depth. Typically, "deep" means a depth of one-half of a wavelength or more. In shallow water waves orbits of water scrape against the seabed (Figure 141), where the friction creates an elongated, elliptical orbit. The friction against the bottom also slows down the wave. The shallower the water depth, the slower the wave moves. Again, the meaning of "shallow" is the depth compared to the wavelength. A water wave will be shallow if the depth is less than one-twentieth of its wavelength.

What is deep for one wave may be shallow for another. Tsunamis are created as the result of underwater earthquakes and landslides (rapid movements of solid material). As water moves to fill in the void, water waves with wavelengths more than a hundred miles long can be created. With an ocean depth of approximately four miles, tsunamis are typically "shallow water" waves but can travel at speeds of hundreds of miles an hour across the ocean, which is what makes them so unpredictable and dangerous.

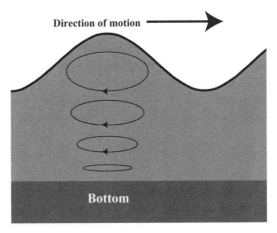

Figure 141 Shallow-water wave characteristics. The friction of the water molecules moving against the bottom creates flattened orbits of the water molecules.

For many kinds of waves, such as sound and light waves, their propagation speed doesn't depend on wavelength. This isn't the case for deep-water waves. Figure 142 shows the relationship between propagation speed and wavelength. The longer the wavelength, the faster the wave moves. Long-wavelength water waves can move at a considerable speed. From Figure 142 you can see that waves a hundred feet long can travel at roughly fifty miles per hour.

The first signs of a hurricane are the long-wavelength swells that race ahead of the storm. While a hurricane might move across the ocean at speeds of a few miles per hour, the waves generated

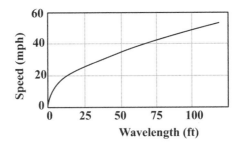

Figure 142 Relationship between wavelength and speed of deep-water waves.

by the storm are far faster and will signal the approach of a hurricane days in advance of other signs. Although they move rapidly, the long wavelength makes the time between crests fairly long compared to shorter waves. For normal waves this time between crests, called a *period,* may be three or four seconds. For the waves from hurricanes and typhoons, the period between crests may be as long as fifteen seconds. The arrival of long swells accompanied by high cirrus clouds that seem to radiate from a point over the horizon are strong signs of a hurricane approaching. When Columbus first became aware of a massive hurricane approaching Hispaniola in 1502, he may have been tipped off by these long, languid swells. Far from land the hurricane-induced swells can be so smooth and peaceful that it often seems like oil has been poured on the water, belying the violence to come.

CREATION OF WAVES

Nearly all ocean waves are generated by wind crossing water (tsunamis are a notable exception). Waves will build slowly as wind first skims the surface, creating what's called a *cat's paw*: very tiny ripples. Once little ripples are created, wind gains traction on the vertical faces, causing the waves to build to progressively larger heights. Waves that are building have more of a trochoidal shape, as shown in Figure 143, than the sine shape of Figure 138. As it builds in height, a wave crest becomes steeper and steeper, finally becoming unstable and tumbling over. This instability occurs when the wave height is greater than one-seventh of the wavelength and also when the interior angle of the peak of a wave is less than 120 degrees (Figure 143).

The longer the wavelength, the more energy a wave can hold. As a short wave builds, its height reaches a certain critical point when it becomes unstable for its wavelength and crashes down on itself. Under windy conditions progressively longer and longer

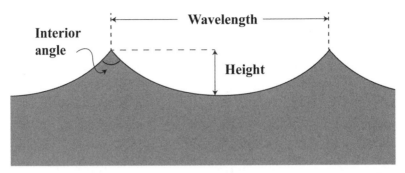

Figure 143 Instability condition for waves. When the height is greater than one-seventh of the wavelength or the interior angle is less than 120 degrees, the wave will break.

waves build up. Whitecaps seen on a windy day are part of this process. Typically, whitecaps first form when the wind is from eight to ten miles per hour. The condition of waves on the surface, characterized by average height and wavelength or period is called the *sea state* or just *sea*. As the wind continues to blow across the water, the wavelengths and wave heights increase until they reach a point at which the sea state is fully developed.

The term *fetch* describes the distance over which wind has time to build waves (Figure 144). The longer the fetch, the more developed the sea. For low wind speeds (ten knots), it can take only twenty miles for a fully developed sea to arise. For high wind speeds (above forty knots), a fully developed sea can only be realized after a fetch of over a thousand miles.

Fully developed seas have smoother waves that will be more like a smooth sine wave, as in Figure 138. Once a sea has developed and the wind goes away, the waves will continue in the form of *swells*. Swells have a smooth sinusoidal appearance and can travel thousands of miles without losing energy. Only when swells encounter obstacles, such as shallow water, current, or a contrary wind do they begin to lose energy, often breaking.

Fully developed seas under long fetches and high winds tend to be more "organized," meaning the wavelengths are more uniform

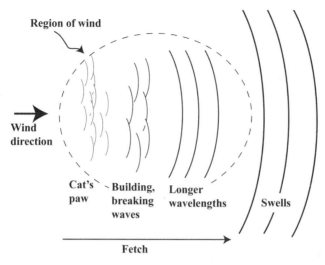

Figure 144 Fetch and the development of fully formed waves. Swells continue for long distances after departing the region of wind.

and the crests extend longer than the sea state under lighter winds. A twenty-knot wind is called a "fresh breeze" on the Beaufort scale, which states "moderate waves of some length, many white-caps, small amounts of spray." The resulting sea will be made up of a fairly broad range of wavelengths.

Thirty knots is a high wind. Beaufort states for this wind speed, "Sea heaps up. Some foam from breaking waves is blown into streaks along wind direction." The wavelength increases significantly, and the most probable wave period is around ten seconds, although there is still a significant amount of energy held in waves with periods as low as seven seconds.

Forty knots is a gale and has "moderately high waves with breaking crests forming spindrift." These waves will have long wavelengths for a fully developed sea, and most of the energy is carried in waves with periods of around twelve seconds.

Fetch and Sea State

Bigger seas develop with a longer fetch. When a major storm with high wind kicks up, the largest waves will be seen in regions where fetches of thousands of miles are possible. The Pacific Ocean is home to some of the longer fetches on the planet. Strong storms accompany the winter season, and accordingly, the largest waves are typically produced in the winters in each hemisphere. In the boreal winter (winter in the Northern Hemisphere), storms in the North Pacific will have fetches of up to four thousand miles, creating giant swells that crash against the western coast of North America.

Hawai'i, too, is home to some of the largest swells during the boreal winter and is a mecca for extreme surfers who will challenge waves in excess of fifty feet high. Surfers will often scan long-range weather charts for distant storms to take advantage of the large waves they kick up over long fetches.

One of the most notorious stretches of water is the *Southern Ocean*. This is not a name we're taught in school, but for mariners it is well known. Around 40 degrees south latitude, there is almost no land in the way of the westerlies. The longest fetch on Earth is ten thousand miles long, stretching from South America to Australia. During the austral winter (winter in the Southern Hemisphere — June through September), this region is home to some of the largest waves on the planet. In a stretch of the Southern Ocean to the southwest of Australia, there are average wave heights in excess of five meters (sixteen feet). An average wave height of a two-story house might already seem large, but waves can be much larger than this average, easily reaching thirty feet. When two or more of these waves combine, or a major storm runs over this fetch, waves in excess of sixty feet are not unheard of.

Joshua Slocum, author of *Sailing Alone Around the World*, contemplated crossing this region from Tasmania to South Africa in his first solo circumnavigation of the globe. He decided against risking fate and took a course around the north of

Australia, through a calmer stretch of the Indian Ocean. He also took pains to avoid the Southern Ocean in the austral winter. He wrote, "I had no wish to arrive off the Cape of Good Hope before midsummer, and it was now early winter. I had been off that cape once in July, which was, of course, midwinter there. The stout ship I then commanded encountered only fierce hurricanes, and she bore them ill. I wished for no winter gales now."[1] This stretch of latitudes is called the *roaring forties* by sailors, with good reason.

The Southern Ocean is a classic route for sailors aspiring to circumnavigation records. The westerlies blow in a favorable direction for a west-to-east voyage, and the higher latitude shortens the overall distance from South Africa to Australia. The danger lies in the storms and waves, which are particularly hair-raising during the austral winter with such a long fetch.

The story of Abby Sunderland is an object lesson in the perils of traversing the Southern Ocean during the austral winter. At age sixteen Abby attempted to become the youngest girl to do an unassisted circumnavigation. The International Sailing Federation (ISAF) collects records on circumnavigations and maintains standards but refuses to accept record attempts by anyone under age eighteen, to discourage reckless behavior.[2]

Abby's brother, Zac, had completed a solo circumnavigation in 2008–2009, unofficially becoming the first person under the age of eighteen to make the unassisted voyage. He was sixteen at the time of his voyage. Zac and Abby's father encouraged Abby's ambition to make her own attempt and helped train her. She had competition from Australian Jessica Watson, who was trying the same feat. If successful, either Jessica or Abby could lay claim to being the first girl under age seventeen to make the circumnavigation, despite the ISAF's refusal to accept such marks.

Jessica Watson already had a large head start, having left Sydney on October 18, 2009. Abby started her attempt on January 10, 2010, from Marina del Rey, California. Eight days into her first attempt, the power systems on Abby's boat, *Wild Eyes*, were not meeting the power needs of the vessel. After a rapid refitting Abby left Cabo San Lucas, Mexico, on February 6, 2010, sailing south along the coasts of Central and South America. After suffering a knockdown (in which the vessel is knocked 90 degrees to one side, with the mast parallel to the water's surface) while rounding Cape Horn, and a failure of her autopilot, Abby put in at Cape Town, South Africa, for repairs, ending the nonstop attempt.

Sailors consider Friday departures bad luck, but Abby put to sea again on Friday, May 21, 2010, with the intention of crossing the Southern Ocean directly to Australia during the austral winter. Jessica Watson had just completed her circumnavigation on May 15. During her passage across the Southern Ocean, Abby suffered repeated knockdowns. On June 10, the boat was dismasted by a large wave during a storm, and the *Wild Eyes* was left stranded at 41 degrees S, 75 degrees E, two thousand miles west-southwest of Perth, Australia. Fortunately, one of the emergency satellite beacons onboard her sailboat sent out a distress call. Australian authorities dispatched a Qantas Airbus from Perth on a four-thousand-mile round trip to assess the condition of the *Wild Eyes* and Abby. The French fishing vessel *Ile de la Reunion* made its way to the crippled *Wild Eyes*, reaching it on June 12. In the process of the rescue, the captain of the *Ile de la Reunion* had a close brush with death himself, falling into the churning frigid waters. He was eventually rescued. Abby was brought aboard and ultimately made her way back to California.[3]

Abby's experience in the Southern Ocean, although dramatic, is by no means unique. Sailors have been rightfully apprehensive about crossing this region. During the first global solo circumnavigation sailboat race in 1968–1969, British competitor Donald Crowhurst turned away before reaching the Southern Ocean. He kept a phony log and ultimately committed suicide rather than risk the passage. Such is the reputation of this ten-thousand-mile fetch.

WAVES APPROACHING SHORE

When waves approach shore they usually start out in the deep-water condition, in which the speed of the waves is determined by the wavelength. As waves get into progressively shallower water, the orbits of water molecules begin to feel friction from the bottom, and this creates a drag that slows down the waves. Figure 145 illustrates the transition from deep water to shallow water to breaking waves. As the sea floor gets shallow, the speed of the waves drops. As the speed slows the waves get closer together and tend to "heap up," meaning that their faces become steeper and the wavelengths decrease. Eventually, the waves become unstable and break. When this happens the contact with the bottom breaks the circular orbits of the water molecules. The top part of the wave curls over onto itself, expending its energy on the beach.

The character of the breaking wave depends on the wavelength of the swells approaching land and the slope of the seabed. If the slope is relatively steep, the waves break abruptly, creating a *dumping wave*. If the underwater slope is gentler, the waves will move toward shore with a steep face that may not break for some distance: these are good for surfing.

This process of waves breaking in shallow water is called *wave shoaling* by sailors. A *shoal* is a location where water is shallow. Even with no shore nearby, waves can break over a shallow

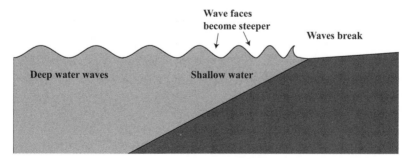

Figure 145   As waves approach shallow water, they feel the drag on the bottom. The wavelength decreases, and the waves get steeper and eventually break.

bottom, including submerged rocks. For mariners this is a vital issue. Navigation in relatively shallow waters can be dangerous and tricky. A submerged rock can tear a major breach in a hull, leading to disaster. When traversing shallow waters, a crew keeps a sharp lookout for shoals, surveying the shapes of waves, checking for telltale steepening of the faces, or listening for breakers at night.

If the shoreline is abrupt —that is to say, there is no gradual transition at all from deep water to land — waves will reflect and move in the opposite direction without breaking, like an Olympic freestyle swimmer making an underwater flip turn. A swell reflecting off a steep cliff that extends into deep water can be surprisingly calm, although it might seem as though it should be the perfect setup for a violent confrontation. On the contrary, a large swell meeting a cliff simply rises and falls as it reflects, with very little action. One sea kayak instructor I know calls this the "swellevator" and has students paddle right up to cliff faces to teach them how to control the kayak and conquer fear with knowledge. Canny lobstermen on the Maine coast will set traps in proximity to cliffs, knowing that others may be irrationally fearful of getting so close.

Wave Refraction

Waves on the ocean have crests that extend perpendicularly to the direction of motion, as in Figure 146. For a swell, the direction of motion in the open ocean will be a straight line, but if the swell reaches shallow water, the path will be bent.

In chapter 9 I discussed how light was bent (or refracted) by different densities of air, creating distortions and mirages. When the speed of a wave changes as it moves, it generally refracts. Water waves refract, too. As a group of waves approaches shallow water, the seabed will distort the shape of the waves and bend their overall path. In general water waves will bend so their crests are parallel to the shoreline when they break.

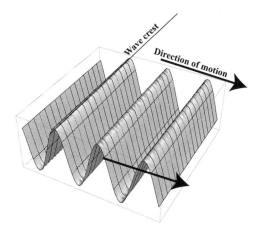

Figure 146 Wave motion is perpendicular to the crest of the wave.

A good navigator can "read" a shoreline and predict the wave conditions from its shape. Headlands that project out into the sea will concentrate wave energy as the crests bunch up and cause erosion. Embayments will spread out wave energy and collect much of the sediment eroded from the exposed headlands. One notorious headland in southern England is the Lizard, a peninsula projecting out from the coast of Cornwall into the English Channel. It's the site of a number of shipwrecks. If a ship doesn't give the Lizard enough sea room, a strong storm off the Atlantic can create dangerous breaking waves, pushing hapless vessels up against the jagged shoals that project out underwater.

Waves can sometimes refract so much around islands that they completely encircle them. Figure 147 shows how swells move around Baker Island, off the coast of Maine. In a southerly swell the waves will refract so much that they will be running both east and west on the north side of the island. I've traced out the wave crests in the figure and indicate the direction of motion with the arrows. Baker Island itself has a narrow neck of rock ledges extending to the north. In the condition of a strong southern swell from the Gulf of Maine, waves will break in opposite directions on the shoal, which can be quite surprising for the unwary.

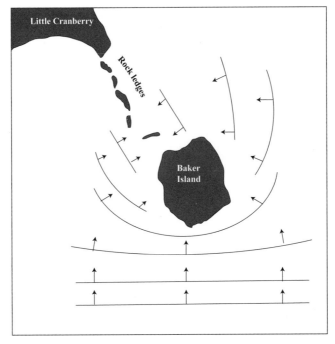

Figure 147 Wave refraction around Baker Island, Maine. Waves from the south can wrap around and break from both east and west on the north (lee) side of the island, breaking on the rock ledges.

Wave Interference and Reflection

Far out to sea it is rare to have only one swell or set of waves. Prevailing winds, storms, and local conditions can kick up swells and waves coming from different directions with different wavelengths. When two or more swells meet in the ocean, they will combine in a process called *interference*. Waves are able to pass through each other without becoming distorted, but when two waves meet, the displacements of the overlapping waves will add together. In Figure 148 I show two kinds of interference. In one case, called *constructive* interference (Figure 148, left), a crest from one wave will add to the crest of another wave, and the troughs will likewise add, creating a combined wave height larger than either of the two waves by themselves. In the case of *destructive*

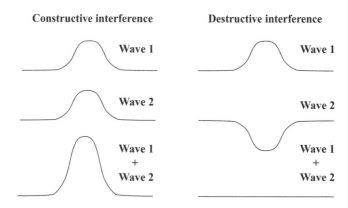

Figure 148 Constructive interference results when a crest adds onto a crest. Destructive interference results when a crest adds on a trough.

interference (Figure 148, right), the crest of one wave will add to the trough of the other wave, creating a smaller overall wave height than either of the two waves.

While Figure 148 shows a disturbance in one dimension, ocean waves are really two-dimensional and travel in different directions. Far out to sea, swells kicked up by two different storms can cross paths at some angle, producing what sailors sometimes call a *confused sea* that doesn't seem to have an underlying pattern.

When a series of waves encounters a steep cliff that extends deep underwater, instead of breaking, they're reflected, as noted above. The motion of the waves on reflection is much like light reflecting off a mirror. If the rays strike the cliff at a certain angle, the outgoing angle is the same, only flipped left to right. The interference of the incoming waves with the outgoing waves rebounding from a steep cliff at an angle will create a confused sea.

Wave Piloting

Land can make its presence known for quite some distance out to sea through its effect on waves. Swells can reflect off shorelines and refract around shallow regions. The use of wave patterns to navigate is sometimes called *wave piloting*. Pacific Islands navigators can often use the feel of the canoe's being pitched around on the waves as a way of identifying patterns that would help orient them with respect to land. Wave piloting has been documented among indigenous navigators from Taumako Island, Kiribati (formerly the Gilbert Islands), the Tonga Archipelago, Tikopia, and the Marshall Islands.[4,5,6] Perhaps the most developed system of wave piloting is found in the Marshall Islands, where representations known as *stick charts* are used to teach navigators how to find land using wave patterns.

Wave piloting begins with a thorough knowledge of swell patterns. The dominant swells are themselves the result of wind patterns. Some island clusters such as the Gilbert Islands experience trade wind swells from the east during certain seasons. Faraway storms can produce swells from different directions; for example, the product of a strong storm in the north Pacific.

While swell patterns are often seasonal, some are reliable year-round. The Southern Ocean with its global fetch creates a swell out of the southwest that penetrates most of the Pacific as far north as Hawai'i and is well known. The navigator Abera from Kiribati calls this *nao bangaki*.[7] David Lewis quotes Abera on the character of this swell: "It is big, long, and low and does not break; it is independent of the trade wind. If you are in a canoe bound from Onaotao to Tabiteuea [about north-northwest] you feel it as a slow heave that rolls the canoe a little from the port side. This swell can be detected over all the seas."[8]

Even in the absence of land, swells provide a natural compass that can supplement both star and wind compasses, providing redundancy. As stars, wind, and swells all show predictable seasonal variations, the combination of the three can form a consistent picture for the navigator. In many accounts of the use of swells for navigation, the Pacific Islands navigators feel the effect of the swell on the motion of the boat, rather than using a visual reference.

Land affects swells in three ways. First, when waves shoal and break on a shore, the swell is absorbed and a kind of "swell shadow" forms on the lee side of an island. Second, as waves reflect off a coastline, the outgoing waves will interfere with the incoming waves, which can create a distinctive pattern. The reflected waves, however, can often be quite weak, having lost much of their energy breaking on a shoreline. Third, as waves refract around islands, the lee side will show a crossing sea where the refracted swells from either side of the island will interfere with each other.

Figure 149 illustrates a swell shadow in the extreme case of an island with a steep drop-off underwater. The island, however, is shallow enough at the surface for crashing waves to extinguish the swell on the exposed side of the island. The steep drop-off will inhibit refraction of swells around either side of the island. A navigator will know he is in the shadow of an island or cluster of islands when the dominant swell drops out. If the island is known, the swell shadow can be used to create a line of position.

When Captain James Cook was voyaging in the Pacific in 1773, his log makes numerous references to the direction and strength of ocean swells. He often used the presence or absence of swells to infer whether he was near land. He rightfully feared the Pacific

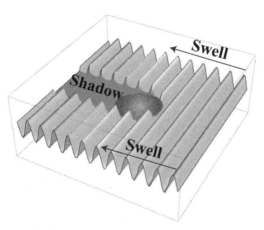

Figure 149 An island where waves break on shore can create the shadow of a swell on its lee side.

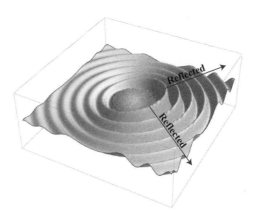

Figure 150   Pattern of reflected waves from an island.

Tuamotu Archipelago, which is often called the "Dangerous Archipelago" because of the numerous shipwrecks there. While negotiating a passage through the Tuamotus, he noted that the seas were calm, and he proceeded with the "utmost caution." Eventually, when a large swell from the south returned, he concluded that he was safely past the archipelago and out of danger.[9]

The reflections of waves from islands can create clues to their presence. When a swell reflects from the exposed coast of an island, it will create a radiating pattern of waves. The largest wave heights will be found rebounding into the main swell, as in Figure 150. The heights of the reflected waves will die out with distance but can be detected as far out as roughly thirty nautical miles.

If I overlay the incoming swell pattern and wave shadow in Figure 149 with the reflected swell pattern of Figure 150, the result is the pattern shown in Figure 151. This pattern arises from the interference of crest on crest creating a large disturbance and a crest on trough creating a small disturbance. Where the outgoing reflected swell has crests mostly parallel to the incoming swell, the resulting pattern is relatively smooth. In Figure 151 the incoming swell is portrayed as coming from the east, as it might in a region and a season dominated by trade winds. The

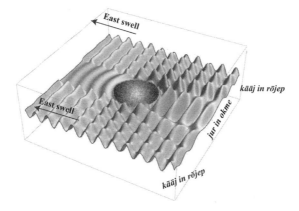

Figure 151   Pattern of wave interference around a circular island. The dominant swell is from the east.

smooth region is mainly to the east. To the northeast and southeast, the reflected swell is moving at roughly 45 degrees with respect to the incoming swell and creates a set of pyramid-shaped waves and a somewhat confused sea. Since the reflected wave spreads out over distance, eventually the interference pattern dies out.

On close inspection of Figure 151 in the region of pyramid-shaped waves to the southeast and northwest, the perceptive reader may discern a pattern where the peaks of pyramids form a kind of curved line that points toward land. In principle, if a navigator is able to discern the line of pyramid-shaped waves, he could use this as a way of finding land at some distance out to sea.

Joseph Genz made a recent study of wave piloting in the Marshall Islands as part of his doctoral studies at the University of Hawai'i. His primary consultant, Captain Korent Joel, was a native of Rongelap Atoll in the Marshalls. A major issue in studies of native navigational schemes in the Pacific Islands is the encroachment of Western European culture. Because the Marshall Islands are largely isolated in the central Pacific, the first major settlement of the Marshall Islands was a German outpost not established until the latter half of the nineteenth century. The Japanese took over in 1914 and established fortified garrisons for defensive purposes.

After their capture by the United States in World War II, the Marshall Islands were the site of extensive nuclear weapons testing by the United States. Although much of the indigenous navigational skills have been forgotten, some navigators who learned traditional techniques in their youth are trying to revive interest among younger generations. Korent Joel is keenly interested in resurrecting the navigational skills of his ancestors and has sought collaboration with anthropologists at the University of Hawai'i.

Joel's main work with Genz is the demonstration and understanding of wave piloting. The patterns of interference between an incoming swell and reflections are given names in the Marshallese scheme. The region of smooth interference seen to the east in Figure 151 is called *jur in okme*. This name is derived from the term for a pole used to pick breadfruit from trees that has a "V"-shaped curve said to resemble the constructive interference of the reflected and incoming swells. The region to the northeast and southeast with the pyramid-shaped waves in Figure 151 is called *kāāj in rōjep*, which is the name of a hook and lure for flying fish.[10]

To test Joel's navigational schemes, Genz worked with oceanographers to deploy a set of buoys sensitive to waves to the east of an atoll named Arno. Although Joel identified the reflected swell from the atoll back to the east, the wave buoys did not.[11] This was in the *jur in okme* region of Figure 151. Another test was made of Joel's navigation. While Joel was sleeping, Genz and crew drove the research boat twenty-five nautical miles to the southeast of Arno, well out of sight of land. When Joel woke up, Genz asked him to find the way to Arno based on what he saw in the pattern of waves. Joel detected the kāāj in rōjep and directed the crew northwest toward Arno.[12]

Another wave pattern that indicates land is produced when waves refract around islands, producing crossing swells on the lee side. Figure 152 illustrates this in the case of an eastern swell refracting around an island, where I show the crossing swells on the western (or lee) side. This region of crossed swell can also

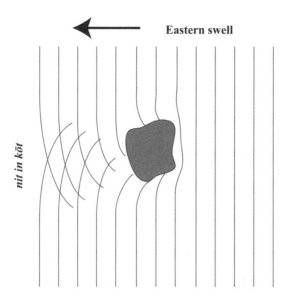

Figure 152 Refraction around island.

produce pyramid-shaped waves and creates a rather disturbed sea. Unlike the sometimes weak reflected swell, the refracted swell on the lee of an island is fairly distinct, even for westerners, and is often visible in satellite imagery. Joel calls this crossing pattern *nit in ḳōt*, which is the name of a small cage for capturing birds. The intersecting strips of the cage are said to resemble the crossing patterns of the swells in the lee of the island.[13]

Joel's explanation of the nit in kōt to Genz has a curious twist to it. Rather than referring to the two intersecting swells as transformations of the eastern swell through refraction, Joel calls them "northern" and "southern" swells.[14] In some sense the immediate direction of the swell gives it its name, regardless of its history.

One of the most curious concepts in Marshall Islands wave piloting is the *dilep*. The term means "backbone" or "spine" and is sometimes spelled as *drilip* or *rilib*. The initial consonant is a kind of rolled "r." The dilep is a wave path connecting pairs of islands. During Genz's work in the Marshall Islands, two informants, Korent Joel (Captain Korent) and Thomas Bokin,

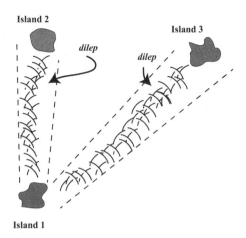

Figure 153 Dilep in Marshallese wave piloting.

explained dilep. According to them, it is a region where waves heap up through what appears to be constructive interference of two opposing swells. When a navigator sets out from an island toward another, he first establishes the heading, then searches for the dilep connecting the two islands. This is a path of disturbed waves. The word *booj*, or "knot," denotes the wave created by the constructive interference of two waves. The dilep is effectively a path of multiple booj.[15]

In Figure 153 I try to capture some sense of the concept as explained by Joel and Bokin to Genz. There are two dileps in the figure, one connecting Island 1 to Island 2 and one connecting Island 1 to Island 3. The goal of the navigator is to remain on this path. As long as the vessel is on the dilep, it has a symmetric motion under the influence of the waves. If he strays to one side or the other, the canoe will feel more of one swell than the other, which is manifested as an asymmetric rocking motion. The navigator's art is to keep the vessel on course by sensing the effect of the dilep through the behavior of the canoe.

Two other Marshallese interviewed by Genz gave a different explanation of the origin of dilep. Rather than describing it as the result of opposing swells, informants named Isao Eknilang and

Willie Mwekto claim that it comes from the reflected swell of the destination island.[16]

Anthropologist Marianne George describes a wave-piloting concept that is reminiscent of dilep in the Santa Cruz Islands in the southeast Solomons. Her source is a navigator named Koloso Kahia Kaveia, now deceased. A distinct path of waves indicates a course between the islands of Taumako and Nifiloli. In this case Kaveia explains the path as originating from the reflection of the dominant swell off the target island, much like Eknilang and Mwekto's explanation of dilep. As with the Marshallese dilep, the motion of the vessel is symmetrical while on course. But if the navigator goes off course, the motion of the boat changes. In Kaveia's explanation, if the vessel goes off course, it will exhibit an asymmetric elliptical motion rather than a direct up–down pitching motion while on course.[17]

What is dilep? The explanation of Bokin and Joel that it is created by intersecting and opposing swells doesn't work terribly well for our usual Western understanding. Their description implies that two opposing swells are somehow always oriented parallel to the path connecting pairs of islands. Westerners conventionally think of a swell as coming from one direction, and any reflection or refraction of that swell alters it, so it is no longer "the swell." In this Western sense, with perhaps one or two swells present in a region, it is impossible to have pairs of opposing swells parallel to the wave path between all pairings of islands.

When Genz and coauthors tried to detect the dilep on a north–south path between the atolls of Majuro and Aur with wave buoys, they could find the dominant eastern swell but found no trace of a swell from the west. Joel, however, claims to have felt a western swell.[18] It may very well be that Joel and Bokin are conceptualizing the patterns creating a wave path as the result of the intersection of swells. In Joel's description of the nit in kōt, his description of two swells was really of a single swell that was split into two and shifted by refraction around the island.

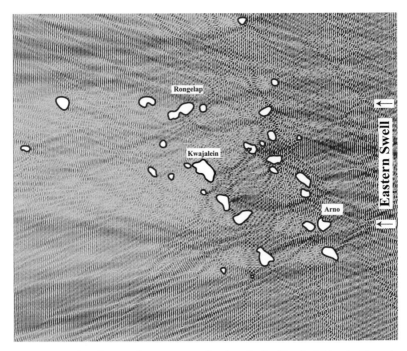

Figure 154 Simulation of pure wave reflections in the Marshall Islands. There is a single swell coming from the east (from the right in the figure). Darker shading represents an enhancement in wave energy. This was generated using a modified version of Paul Falstad's ripple tank applet, http://www.falstad.com/ripple/. Data input by author.

In Figure 154 I show the result of a vastly simplified computer model of an ocean swell from the east reaching the Marshall Islands. The islands of Arno, Kwajalein, and Rongelap are indicated in the figure. In the model I put in pure reflections off the atolls, with no refraction and no simulation of the loss of swell energy from swells breaking on the atoll shores. The darker regions have more wave energy and the lighter regions have less. Some clear features stand out. The most obvious feature is the swell shadows cast by the atolls on their lee side. The north–south orientation of the archipelago mostly extinguishes the easterly swell on the western side of the island chain. A more curious feature is the concentration of wave energy into what could be described as filamentary structures, which often link pairs of islands. Whether or not these patterns of wave energy represent what the navigators call dilep remains to be seen.

The reflections will be much weaker than those seen in Figure 154. The dominant swell and local wind-generated waves tend to mask reflections. Nonetheless, the interference between the incoming swell and the reflections do form distinct patterns. If indigenous navigators are able to mentally subtract out the contributions of the dominant swell and local wind-generated waves, what remains is a relatively unique signature of wave motion associated with each patch of ocean. With the proper training and a lot of experience, a navigator may discern these patterns, while they will elude the untrained. If true, this uncanny ability to detect weak signals in the presence of the "noise" would explain the absence of corroborating data from wave buoys in Genz's study.

STICK CHARTS

One component of wave navigation in the Marshall Islands is the use of *stick charts*. These are typically made from the midribs of coconut palms and pandanus roots lashed together into a latticework. There are several kinds of stick charts. The simplest are used as teaching devices to exhibit how waves reflect and refract around islands and form interference patterns. More sophisticated stick charts represent specific island clusters and the transformation of swells around those islands.

In 1901 a German naval officer named Captain Winkler published an article on stick charts titled "On Sea Charts Formerly Used in the Marshall Islands."[19] Winkler's article was the result of his effort to catalog the kinds and uses of stick charts from a number of sources. Joseph Genz also reports on stick charts in his dissertation and a recent paper.[20,21] There appear to be substantial divergences in the descriptions of the meanings of some charts. This is often the case for explanations of indigenous navigation

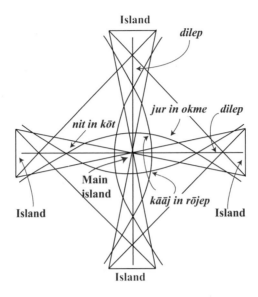

Figure 155 Marshallese stick chart, called a *mattang, wapepe,* or *meddo*. This illustrates the swell, reflection, and refraction patterns associated with several islands. Based on a photograph from Capt. Winkler, "On sea charts formerly used in the Marshall Islands, with notices on the navigation of these islanders in general," *Smithsonian Institute Report for 1899* 54: 487–508. Plate VIII (1901).

techniques. There are no textbooks to codify this kind of knowledge. Rather, information is passed by word of mouth from teacher to pupil, and the importance and emphasis may vary from one navigator to the next.

Figure 155 shows one class of stick chart called *mattang, wapepe,* and *meto*. The elements in common for explanations of this kind of chart are on the meaning of the center and four extremities. The center of the chart represents a main island of interest, and the termini at the top, bottom, and sides represent four neighboring islands. In addition, the fourfold symmetry reflects the existence of four swells emanating from the cardinal directions. The diagonal sticks running between the four neighbor islands are purely structural. In both Winkler's explanation and Joel's description, the straight sticks connecting the center island to the four islands are representative of the dilep. At this point, however, the commonalities end. The interpretations of the curved arcs and their intersections appear to carry multiple meanings, including the nit in kōt, the jur in okme region, and the kāāj in rōjep.

Wave piloting has many variations, depending on the navigators and local conditions; nonetheless, there appear to be systematic strategies underlying the art. These commonalities are likely from the keen observation of physical processes of wave motion on encountering land.

13. Soundings and Tides

• •

THE OCEANS ARE miles deep and have a varied bathymetry. In contrast, the ocean surface is highly uniform because of the fluid nature of water. Only close to shore do the continental shelves create relatively shallow seas with depths of two hundred feet or less. Tide and water depth both played important roles in early voyaging around Northern Europe. The character of Northern European seas allowed for a kind of navigation based on sounding. Fishermen often found their way by sampling the depth and character of the seafloor. With the discovery of productive fishing banks off the coast of North America, this practice spread. When sailors thought they were close to landfall, the crew proceeded to measure the seafloor depth to see if they were over the continental shelves.

A *sounding* is a measurement of the depth of the seafloor. The term derives from the Old English and Old Norse word *sund*, meaning "to swim." The practice of sounding was done with long canes on the Nile in ancient Egypt, presumably to avoid shallow water. While the earliest uses may have been for safety, knowledge of the depth and the composition of the seabed can be an aid to navigation.

Taking a depth sounding is simple: A sailor ties a weight on the end of a rope, throws it overboard, and pays out rope until it reaches the bottom, at which time the rope goes slack. The sailor then retrieves the rope and measures the submerged rope in outstretched arm lengths, or *fathoms*. In principle water depth and the composition of the bottom can act as a kind of underwater "landmark," guiding the sailor. Ole Crumlin-Pedersen, founder of the Viking Ship Museum in Denmark, argued that sounding was a primary component of medieval Norse navigation.[1] Others disagree, asserting that voyages mainly relied on other techniques, such as visual references and dead reckoning.[2] Sounding, however, was common practice in the shallow waters of the North Sea and the Baltic, particularly among fishermen looking for productive grounds.

By the 1600s a uniform system of sounding was developed by the British for the Royal Navy. A piece of lead, called a *lead*, appropriately, was tied onto a length of rope. The lead was standardized at seven pounds (half a stone) and had a hollowed-out bottom. A piece of tallow or beeswax was balled up in the hollow at the end of the lead. When the lead was dropped, a sample of the seabed adhered to the tallow, and the navigators would inspect the retrieved sample. To the uninitiated this seemed like magic. Sailors would inspect, sniff, and taste some of the vilest-looking seabed material to figure out their location. Fishermen also used this method to find their favorite fishing grounds.

This practice of sampling the seabed followed British seafaring culture to North America. There's an apocryphal story told in a number of versions about an old captain who had seemingly miraculous powers in using soundings. I relate below my own abridged version:

> The old captain was highly skilled at using the lead line to find his position, but many of the young sailors thought he was just faking this skill. On one voyage to the fishing grounds, the captain was laid low with a bout of rheumatism and had to shout commands up to his first mate from his cabin. After a successful fishing expedition, the captain gave orders to his mate for their return to Nantucket. He shouted up, "Now, sail north by northwest, and when you fetch up on shore, drop the lead and bring me the sample." The mate gave orders to the crew and dutifully followed the compass heading. A thick fog descended, and the mate proceeded cautiously. Upon hearing waves crashing on a nearby shore that he

took to be Nantucket, he took down his sails, dropped anchor, and began to make a sounding.

On the return trip some of the sailors decided it would be a joy to play a trick on the old captain. They convinced the mate to go along. Rather than heaving the lead and retrieving the depth sample on the tallow, he took some sand from the cooking box, and rubbed it over the tallow, then just wetted the lead to make it appear like a real sounding.

The first mate took the contrived sounding lead down to the captain. The old man looked it over at first with some consternation. He sniffed it. He wetted his finger and pinched a sample of the sand from the tallow and tasted it. He seemed quite puzzled by it and tasted it again. Finally, he turned to the mate and said, "Well, I'm afraid I have good news and I have bad news. The good news is that you drove the vessel north by northwest straight as an arrow, so you are to be congratulated on your seamanship. I'm afraid to say that the bad news, however, is that since we've been gone, Nantucket has sunk underwater, as we're directly above 'Sconset [Siasconset – a town on Nantucket]."

The British Royal Navy created a standard for marking lead lines. Intertwined with the rope were markers of various colors coded for the depth. Black leather was tied at two and three fathoms, white at five, red at seven, and black at ten, with knots for greater depth tied every five fathoms. After retrieving the lead a leadsman would report back the phrase, "By the mark," or "By the deep" followed by the depth in fathoms. So "By the mark, five," would report a sounding of five fathoms. This was also a common phrase in the United States. The American author Samuel Clemens took his pen name from this phrase based on his experiences as a pilot on Mississippi steamboats. "Mark twain" is a report of a depth of two fathoms.

TIDES IN HISTORY

Tides were relatively unimportant in the Mediterranean to the ancient Greeks and Romans. Travelers venturing into the North Sea or the Baltic, however, were often surprised by the strength of tides. According to Pliny the Elder, Pytheas of Massilia (Marseilles) observed a tidal range of eighty cubits (well over one hundred feet) in a voyage to Great Britain circa 325 BC.[3] This is likely an exaggeration by Pliny or Pytheas (or both), as ranges of perhaps fifteen feet in the North Sea are seen. Pytheas is also widely credited as first documenting the influence of the Moon on tides.[4]

Writers in the Middle Ages grasped that there was a connection between the Moon and the timing of the tides.[5] In some harbors a high tide occurs when the Moon is directly overhead, yet in other harbors a high tide coincides with moonrise. Nautical atlases from the fourteenth century onward included tide tables relating the height of tides in known ports and estuaries to the position of the Moon in the sky.

The *Prose Edda*, a thirteenth-century collection of Norse tales, contains a passage about the origin of the tides. The god Thor is challenged to a drinking contest by the giant Utgarda-Loki. Unknown to Thor, Utgarda-Loki connects his drinking horn to the ocean and tells Thor that he should be able to down the contents of the horn easily in one gulp. Thor drinks and drinks and drinks but is barely able to lower the height of the water in the horn. Utgarda-Loki mocks Thor, who goes at it again with renewed vigor but only seems to drain the gourd slightly. Thor gives up at this point. The next morning Utgarda-Loki reveals the trick he played on Thor and explains that this is the origin of the tides. While the story is fanciful, it has an interesting nugget of insight: the variation of sea level from tides is small compared to the depths of the oceans.

Explanations for tides before Isaac Newton developed his theory of gravity were scattered. Johannes Kepler understood that there was a link between tides and the Moon. Galileo tried to explain tides as the result of Earth's rotation combined with its orbit around the Sun. This would have predicted only one high tide a day, but most regions in Europe experienced two high tides a day.

TIDES, EXPLAINED

Isaac Newton first published his famous work *Principia Mathematica* in 1687. After the early work of Copernicus and Kepler on a Sun-centered solar system, there was much speculation on the nature of the forces acting on the Sun, Moon, and planets. Newton contemplated the idea that celestial forces were related to the gravity felt on Earth. He developed the *law of universal gravitation,* the concept that one force is responsible for all gravitational attraction.

In *Principia* Newton analyzed the origin of the tides in terms of gravity. According to Newton, lines of gravitational force spread out over space from its source in massive objects. When these lines spread across a sphere like the Earth, they create tidal forces in addition to the force of attraction. This theory is often called the *static theory of the tides.* Figure 156 shows the tidal forces on Earth caused by the Moon. This pattern emerges when one subtracts out

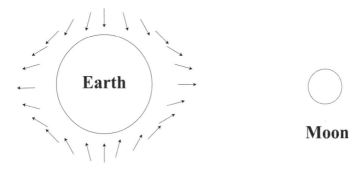

Figure 156 Tidal forces on the surface of the Earth created by the gravitational pull of the Moon spreading across the Earth.

the gravitational force averaged over the surface of the Earth from the variations caused by the field lines spreading out. There are tidal forces pushing away from the center of the Earth along a line connecting the Moon and Earth, and there are tidal forces pushing toward the center of the Earth perpendicular to the Earth-Moon axis.

The tidal forces illustrated in Figure 156 create a bulge pointing toward and away from the Moon (Figure 157). As the Earth rotates underneath this bulge, it will experience high and low tides in different regions. The average height of the bulge is about a foot spread over ten thousand miles. Land is not immune to tidal forces and will also show the effect of the passing bulge. The bulge is largely unnoticed as the entire surface of the Earth shifts up and down under the bulge and only small differences are discernible.

Every so often the shift on land has a demonstrable effect. In the 1980s in Geneva, Switzerland, a very precise circular accelerator, twenty-seven kilometers in circumference, showed a shift in its energy that was associated with the position of the Moon. After some sleuthing the accelerator physicists realized that the tidal bulge from the Moon distorted the shape of the accelerator ring every time it passed.[6] The real effect of the tides only becomes apparent when there is a difference in local displacements. The oceans, being fluid, are more easily distorted than

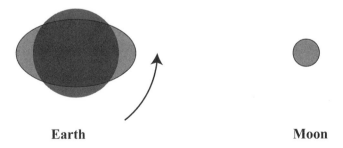

Earth   **Moon**

Figure 157 Tidal bulge caused by tidal forces.

land, and tidal differences are most apparent where oceans and seas meet land.

The tidal bulge is, effectively, a very long wave. When the bulge passes through the deep ocean, the effect is not noticeable, but when it impinges on the shallow waters surrounding the continents, the size of the tides can become substantial, as the tidal bulge heaps up like any wave encountering shoals. Here the difference in displacement between land and water under the bulge is the largest.

The Sun exerts a much stronger gravitational pull on the Earth than the Moon does, yet the Moon produces a stronger tide than the Sun. The reader might wonder why. Because the Moon is much closer to the Earth than the Sun, the lines of force from the Moon spread out much more dramatically over the surface of the Earth, creating stronger tidal forces. This explanation from Newton confirmed the experience of sailors, who linked the Moon to the tides. This is not to say that the Sun has no effect whatsoever. The solar and lunar tidal forces can combine to create larger tides or partially cancel to produce smaller tides.

For a new Moon or a full Moon, the tidal forces of the Sun and Moon are aligned (Figure 158, top), creating the largest tides, called *spring tides*. This alignment is called *syzygy*. For a half Moon the forces producing the tidal bulges for the Sun and the Moon partially cancel each other, reducing the size of the tide. The Moon still wins out, but the size of the tides is diminished. These tides are called *neap tides*, when the Sun and Moon are in the alignment called *quadrature* (Figure 158, bottom).

One of the most perplexing aspects of the tides is the seemingly random variation of tide heights and timing from one body of water to the next. A completely enclosed body of water, such as the Caspian Sea, would see no tide whatsoever. On the other hand, a body of water partly open to the ocean could show major tides, or smaller tides, depending on the size of the embayment.

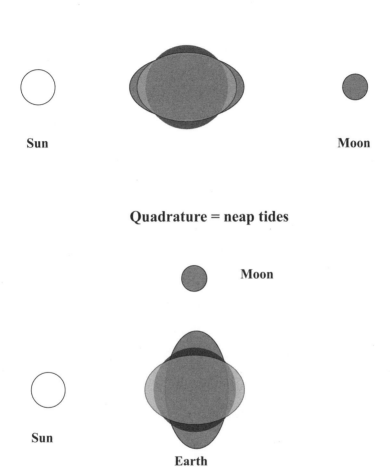

Figure 158 Spring tides occur when the tidal forces of the Sun and Moon add together. The Earth is represented as the black circle. The darker grey is the distortion due to the Moon, while the lighter grey is the distortion due to the Sun. These are higher than normal. Neap tides occur when they tend to work against each other, reducing the tidal range.

You can think of a partly trapped body of water as a bell that will ring with a distinct pitch or frequency if you hit it. If a large meteor hits a bay, the water will slosh back and forth with a unique frequency that depends on the size of the bay. On average, the larger the bay, the lower the pitch or frequency. A small harbor will show this kind of ringing in a phenomenon known as a *seiche*. Under the influence of the wind blowing across the surface of the harbor, water will slosh back and forth. If you take a time-lapse sequence of moored boats in a harbor, you can see that they rise and fall every five minutes or so. The larger the harbor, or lake for that matter, the longer it takes for the water to slosh back and forth and the lower the natural frequency of the trapped body of water.

In the case of tides, the lunar bulge is responsible for perturbing the water in the embayment. There is an analogy with musical instruments that describes the effect of the lunar bulge on partly trapped bodies of water. Some string instruments such as the Indian sitar have *sympathetic strings*. These are tuned to a specific frequency or pitch. When a musician plucks another string at the same frequency, the sympathetic string will vibrate on its own as the energy from vibrations gets transmitted through the instrument body.

In a way you can think of the embayments that line the continents as sympathetic strings that will vibrate under the influence of the passing lunar bulge. It is as if the Moon were a musician plucking the Earth as a string. If the size of an embayment is 'tuned' to the same frequency of the passing lunar bulge, it will show a huge tidal range. If it is smaller or larger than this ideal size, it will show some tidal range but not as large as that for perfect tuning. The lunar bulge passes once every twelve hours and twenty-four minutes.

Two areas with the highest tides on the planet are in Ungava Bay in northern Quebec and the Bay of Fundy in the Canadian Maritimes. These embayments have natural frequencies very close to the frequency of the passage of the lunar bulge (about twelve and a half hours).[7,8] This effect, called *resonance,* produces a tidal range of forty feet in these "tuned" bays. In contrast, Nantucket Sound, just south of Cape Cod in Massachusetts, is much smaller than either the Bay of Fundy or Ungava and has a higher natural frequency. As a result Nantucket Sound experiences a substantially lower tidal range of five to eight feet, depending on whether it's a spring or a neap tide.

Timing of the Tide

You might naively always expect a high tide when the Moon is directly overhead and the bulge is the largest. This is usually not the case. Often the highest tide occurs at other times during the lunar cycle and is often related to the timing of the water flows entering an embayment from the ocean. The time when the Moon is at the highest point in the sky is the *lunar transit* or *meridian passage.* As the tidal bulges point toward and away from the Moon, it is often important to look at the timing of the tide with respect to the *near transit,* when it is at its highest point in the sky, and the *far transit,* when it is 180 degrees opposite.

Figure 159 shows the timing of tides in Boston Harbor and in Newport Rhode Island relative to the transits of the Moon. The high tides in Boston nearly coincide with the near and far transits of the Moon when the peak of the bulge passes. In contrast, for Newport *low* tides are associated with the lunar transits. The plots themselves come from data obtained from the U.S. National Oceanic and Atmospheric Administration (NOAA).

Figures 160 and 161 show the flow patterns of tidal waters around southern New England. When a tide comes in, it is called a *flood tide;* when it goes out it is called an *ebb tide.* On the flood, water enters Boston Harbor directly from the Atlantic Ocean. The tidal flow is more complicated around Newport than in Boston. It tends to flow north toward Newport, then split both to the east into

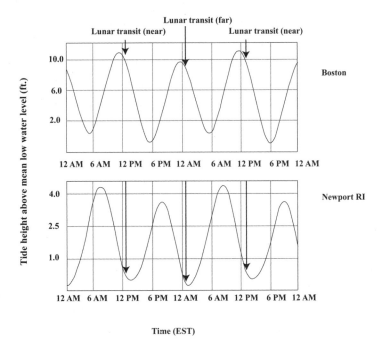

Figure 159 Heights and phase of the semidiurnal tides at Boston and Newport relative to the transits of the Moon. Data from NOAA, Nov. 3–5, 2009.

Nantucket Sound on one side and west into Long Island Sound on the other side. The speed of the flow will vary between one and four knots, so the lag time in filling these basins is substantial. The ebb tide simply reverses the patterns.

This flow pattern is responsible for the difference in the timing of high and low tides in Newport and Boston with respect to the lunar transit. One confusing consequence of these flow patterns is an ebb flow coming in from the Atlantic on the southeastern tip of Cape Cod and flooding back to the east, which naively seems backwards. At constrictions the tidal current can be quite high. One of these constrictions is off Race Point, on the north side of the narrow entrance to Long Island Sound. The difference in time between the lunar transit and the high tide is sometimes called the *tide lag.*

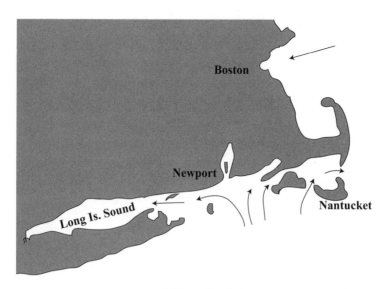

Flood tide

Figure 160   Flood tide in the Newport–Long Island–Nantucket Sound complex and Boston Harbor.

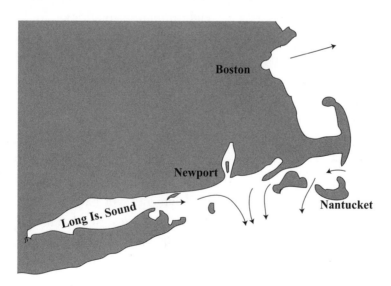

Ebb tide

Figure 161   Ebb tide in the Newport–Long Island–Nantucket Sound complex and Boston Harbor.

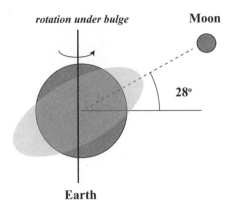

rotation under bulge   **Moon**

28°

Earth

Figure 162 The declination of the Moon creates a bulge that varies with latitude.

On the East Coast of North America, there are two high and two low tides in one tidal day, called *semidiurnal tides*. A *tidal day* is the length of time it takes for the Moon to make successive transits (about twenty-four hours, forty-eight minutes). In addition to the twice daily "push" of the Moon, there is a component of the push that happens once a day. In Figure 162 I've sketched a diagram of the Earth and Moon when the Moon is at a high declination in its orbit (28 degrees), creating a bulge that is not lined up with the equator. As the Earth rotates under the bulge, an embayment in the Northern Hemisphere will experience more of the bulge in a near transit of the Moon than in a far transit. This gives a tidal "push" that has a component that happens once a tidal day, as well as the semidiurnal push. Some embayments will have a natural frequency that will experience a *diurnal tide*, meaning that it has a high tide only once during a tidal day. There are also regions that experience a *mixed tide*, which has some components of diurnal and semidiurnal tides.

Puget Sound in Washington State experiences a mixed tide. In Figure 163 I show tides in Port Townsend, Washington. On top of this I've indicated the near and far lunar transits. The transits both occur at roughly the same time in the tide cycle over this period.

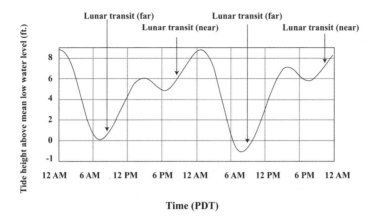

Figure 163 Mixed tide at Port Townsend, Washington (Puget Sound). Near and far transits of the Moon are shown. The Moon had a declination of approximately −12 degrees during this period. Data from NOAA.

Over the course of a lunar month, the shape of the tide cycle can vary substantially in Puget Sound.

Rotary Tides

Some regions on Earth experience a *rotary tide*, which is particularly common in the shallow seas of northern Europe. When tidal currents flow over a sufficiently large distance, the Coriolis effect caused by Earth's rotation comes into play. Scientists in the eighteenth century combined Newton's original static theory with the effects of fluid flow and the Earth's rotation to explain rotary tides. The Coriolis effect diverts current in a counterclockwise direction in the Northern Hemisphere, much the same way cyclonic storms circulate counterclockwise around a hurricane. At the center of a rotary tide is a point of zero tidal change, called an *amphidromic point*. The high tide circles around this point once in the lunar cycle of twelve hours, twenty-four minutes. The height of the tide increases the farther away you move from the amphidromic point.

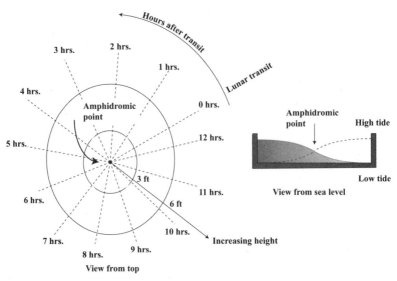

Figure 164   Illustration of rotary tides. These circulate about a fixed amphidromic point. The tidal range gets larger farther from the amphidromic point.

In Figure 164, I illustrate a rotary tide. I arbitrarily chose the starting time of zero hours to coincide with the local lunar transit and high tide. As time progresses, the high tides sweep around like the spoke of a bicycle wheel, with the position of high tide indicated at each hour. On the opposite side of the high-tide spoke, you find the low-tide spoke. Since the "wheel" of the tide completes its rotation in twelve hours, twenty-four minutes, there is net creep of the high tide later each cycle (forty-eight minutes each day).

Figure 165 shows the shape of a rotary tide system in the North Sea between the United Kingdom, France, Belgium, and the Netherlands. The line of high tide associated with the lunar transit is indicated as the "zero" hour, with progressively later high tides indicated by the radial lines. The dashed lines indicate the range of tides in feet for a region away from the central amphidromic point. The tidal range can be as large as sixteen feet at the narrowest point of the English Channel.

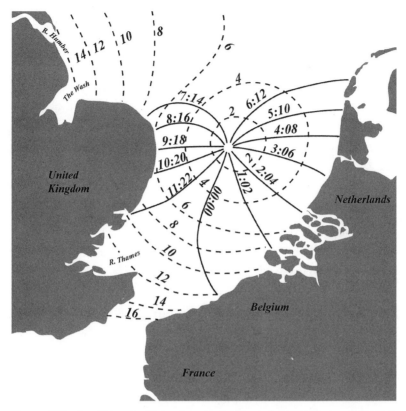

Figure 165 Amphidromic tidal system in the North Sea. The times of the high tide are indicated with the radial lines, and the heights of the tide are indicated by the dashed line encircling the amphidromic point. Image after L. R. Jones, *The Geography of London River* (London: Methuen and Co., 1931), 85.

Figuring the Tides

Although the theory of tides developed by Newton and his successors works quite well, the details of the seafloor and size of embayments drive the kinds of tides you'll find in any location on Earth. Modern computer simulations that incorporate the undersea landscape will predict tidal flow reasonably well, but oftentimes, it's easier to just fit data to a model that incorporates knowledge of the position of the Sun and Moon over the course of the day.

If you arrive at a coastal location and want to know the tides, you can often pick up a tide chart at a local bait-and-tackle shop. On the other hand, I find it's a worthwhile challenge to figure out the tides on my own. It takes a bit of time and patience, but if you work on it, there's a sense of satisfaction and connection to the ocean.

When you first arrive at a shore, you'll find lines of detritus washed up on the shoreline at various heights. Often, the highest spring tide will have its own line of driftwood and seaweed perched high up on the shore. You may also see lower lines of washed-up debris indicating the position of high tides closer to neap. From the position of the lines of debris, you can get a sense of the local tidal range. The structure of docks also gives a clue to the range. A dock made from fixed piers indicates a relatively low tidal range while a dock on a floating raft connected to a gangway, allowing it to rise and fall, indicates a higher tidal range.

Over the course of ten to twenty minutes, you can tell whether the tide is coming in or going out by looking at the features that are uncovered or covered up by the water. If you know the position and phase of the Moon, over the course of twelve hours, you can figure out the tide lag by comparing the time of the lunar transit to the time of local high tide. Once you know the tide lag, by advancing the time of high tide by twelve hours and twenty-four minutes each cycle, you can figure out the timing of high tide for every day. Finally, the timing of spring and neap tides can be figured over the course of a month. If you use the height of the detritus pushed up by high tide as a guide, you can find the phase of the Moon associated with spring tides, usually around a new or full Moon. There may be a shift of a few days for the highest high tide, called the *age of the tide*.

How did sailors figure tides before the advent of tables? Tables of a sort were printed in atlases along with portolan charts and sailing instructions as far back as the fourteenth century. The tables were based on a combination of local information and phases of the Moon. The high tide was reckoned from the position of the full Moon in the sky at any given port. For example, in Amsterdam, high tide on a full Moon happens three hours after its transit (meridian passage). This can also be seen in the rotary tide chart in Figure 165. Knowing the port city and the number of days after the full Moon, the tables give the mariner the relative timing of the high tide to lunar transit. Time was kept with hourglasses on most vessels, so once the timing of the full Moon transit was known, a navigator could keep track of the tides using these tables, knowing the rough time and date.

SOUNDINGS, AGAIN

Toward the end of the eighteenth century, sounding information began to appear on nautical charts, often in feet. Modern nautical charts display soundings to the seafloor in feet (United States) and meters (most other countries). It's crucial for a navigator to know something about the seabed in shallow waters to avoid running aground and sinking.

The depth of the ocean is great compared to range of tides, for the most part. However, when the seabed is shallow, chart makers have to make some decisions about how to represent depths on a chart. There are a number of choices. If we chose the mean sea level and think that the depth of a rock is twelve feet and the bottom of our hull is eight feet under water, that's all well and good, but what if the tide is low by an unknown amount? We'd risk hitting the rock.

There is a term used by modern nautical chart makers called *chart datum* that gives a standard reference for all depths. In the United States the National Oceanic and Atmospheric Administration uses something called the "mean lower low water" (MLLW) for the chart datum, which is the average of the lowest water height each tidal day over a period of approximately nineteen years. Are you completely safe using MLLW as the chart datum? Well, not entirely. If you are at the ebb of a spring tide, the water level will certainly be lower than this, and you could risk hitting a rock if you didn't take precautions.

In charts from the United Kingdom and elsewhere, there's a chart datum that is often used called "lowest astronomical tide" or LAT. This is the lowest tide that can be predicted under any possible set of astronomical conditions, although it doesn't take into account the effect of storm surges. You might ask, "How do you figure out the lowest astronomical tide? What conditions does this correspond to?" There is not an easy answer, as so many factors must be taken into account.

A spring tide occurs when the Moon and Sun are aligned. If they happen to be aligned when the Moon is over the equator, it can create a larger spring tide. If these events combine when the Earth is at its closest approach to the Sun, the tides are even larger. Finally, if all this happens during an equinox, it is the biggest tide you could ask for. Now, you might wonder, "How often does this alignment occur?" The answer is "exceedingly rarely." In fact, it's so unusual that it only occurs once every three thousand years or so. Do chart makers really use this condition, given that the seabed will look different in three thousand years? No. There's a more practical answer.

You might recall from chapter 8 that the Moon's orbit has a nineteen-year period called the Metonic cycle; that is, the time where its maximum declination shifts over one complete phase. Barring the broader issues of the points of closest approach, there will be a time in the Metonic cycle when the Moon's orbit contributes, in concert with the Sun, to a lowest astronomical tide during that cycle. The working definition of the lowest astronomical tide is the lowest that can be predicted for the period of the particular Metonic cycle that the chart is valid for.

Most sailors will tell you that the most dangerous part of a voyage is making landfall. When navigating close in to land, the art of guiding a vessel through relatively shallow continental waters is often called *piloting*. Skillful piloting requires a good sense of the depths of local waters and the timing and range of tides, along with the currents produced on different cycles. All of the discussion above comes into play for piloting. While tides generate currents close to land, large-scale ocean currents can be important for deep-sea navigation, as they can move a vessel up to a hundred miles a day. Part of the navigator's skill is the ability to reckon and adjust for deep-sea currents, which can be fickle in some parts of the world.

14. Currents and Gyres

· ·

WESTERNERS HAVE KNOWN about ocean currents since the days of Ponce de Leon, who fought against a powerful flow in his voyages near Florida. In these waters the Gulf Stream sweeps northward at speeds of four miles per hour, nearly as fast as, or faster than, a sailing vessel itself. On April 8, 1513, de Leon's vessels were pushed backward as they were trying to sail farther south off the coast of Florida. This baffled de Leon, who noted in his log: "A current such that, although they had great wind, they could not proceed forward, but backward and it seems that they were proceeding well; at the end it was known that the current was more powerful than the wind."[1]

Benjamin Franklin was fascinated by the ocean and in particular by the Gulf Stream. On his many transatlantic crossings, he amused himself by taking temperature readings and noting the daily progress of his ship. He eventually created the first map of the Gulf Stream.

Pacific Islands navigators were able to read ocean currents and make necessary adjustments in their long-distance voyaging. A journey between islands could cover hundreds of miles, lasting many days. The net effect of a modest current of one mile per hour could be huge as it adds up through one hundred hours. The currents in the equatorial Pacific are notoriously fickle, and it requires great skill to adjust the heading of a vessel properly.

ORIGINS OF OCEAN CURRENTS

Oceanographers broadly classify currents in two categories: *surface* and *thermohaline*. Surface currents are the ones most familiar to us, yet they are important only in the upper four hundred meters of the ocean. Thermohaline currents are associated with ocean depth and are driven by differences in both temperature and salinity. We'll only consider surface currents here, as they're far more important for navigators.

Four main factors create and influence surface currents:

1. Wind
2. The Coriolis effect
3. Solar heating
4. Gravity

Wind blowing across the surface of the water will create a current. Recall from chapter 12 on water waves that the effect of wind on the water created circular orbits of water molecules as waves form. In reality a small effect called *mass transport* occurs when water molecules slowly move in the direction of the wind. The circular orbits of water molecules forming the waves are more like a forward-moving spiral (Figure 166). As a rough guide, with a wind blowing for ten hours, the surface waters will flow at about 2 percent of the wind speed.

Solar heating warms water nearer to the equator. Expansion of water caused by warming raises the average sea level around the equator by about eight centimeters relative to that in the midlatitudes. This is enough of an effect to create a downhill slope, causing water to flow north and south away from the equator.

As with weather and tidal systems, the Coriolis effect is also a factor in ocean currents. The Coriolis effect will divert wind-induced current to flow to the right of the prevailing winds in the Northern Hemisphere and to the left of the prevailing winds in the Southern Hemisphere. This constant push of wind-induced currents by the Coriolis effect is the core of the modern theory of ocean currents, called *Ekman transport*.

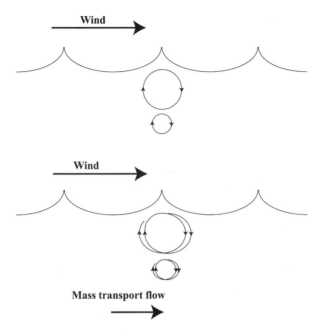

Figure 166 The effect of the wind creates a slow mass transport. The circular motion of waves under the influence of the wind creates a spiral motion of water molecules, which in turn generates a current.

EKMAN TRANSPORT

The modern theory of ocean currents developed through a twisted series of events. The quest for the Northwest Passage from the Atlantic to the Pacific via the waters of the Canadian arctic spawned a large number of expeditions and a tremendous amount of speculation. Ships and crews were routinely lost. The most famous of these was the expedition led by Sir John Franklin. After departing London in 1845, their last contact was with a pair of whaling ships near Baffin Island. They were never heard from again, but their fate was eventually discovered. The ships were crushed in the ice, and all hands perished in a vain attempt to hike to civilization. This spawned a massive search effort that charted large reaches of the Canadian arctic. A number of odd theories about the "polar sea" developed during this period. A common notion held that if one ventured north far enough past the ice pack, one would encounter a body of warm water stretching to the Pole. This notion was reinforced by periodic sightings of large gaps of water in the midst of the ice pack called *polynyas*. The term "polynya" (pronounced *puh*-**lin**-*yuh*) is borrowed from the Russian полынья, meaning a natural ice hole. Polynyas can form when the polar ice cap separates from the ice that is rigidly attached to land ringing the polar sea, creating large gaps of open water. In the summer months heating can cause a partial disintegration of the ice pack, producing large gaps of open water between floating slabs of ice.

One of the groups that searched for the Franklin expedition, led by Elisha Kent Kane, was financed by the United States and departed in 1853. Kane sailed north along a channel between Greenland and Ellesmere Island, where his ship was frozen in. One of his crew, William Morton, traveled overland to a cape in northwest Greenland. Morton saw a huge patch of open water with not a speck of ice in sight and hundreds of birds milling around and concluded that he had seen the fabled "Open Polar Sea." It was just a polynya, but Morton's sighting fueled the concept of a warm polar sea.

Silas Bent, an oceanographer working for the U.S. Navy under Commodore Matthew Perry was the first Westerner to chart the Pacific's version of the Gulf Stream, the Kuroshio (black current), which flows along the eastern coast of Japan. Like the Gulf Stream, the Kuroshio is a warm current that flows north, carrying tropical waters to higher latitudes. Bent believed that the combined effect of the Gulf Stream and the Kuroshio could be sufficient to warm the Arctic Ocean. He wrote:

> Now, since these streams [the Gulf Stream and the Japan current] possess such a wonderful power of retaining their heat, so long as they do not touch the land, as to raise the climatic temperature 30 or 40

degrees over half a continent lying eight thousand miles distant from the points in the Tropics from whence they spring, and from which they derive their heat, it does not seem unreasonable to believe that those portions of the streams which pursue their courses direct into the Arctic Ocean, carry with them warmth enough not only to dissolve the ice they encounter, and keep their pathways open all through the year, but also, to raise the temperature permanently above the freezing point of a large area of the sea around the Pole, and thus prevent this extremity of the Earth becoming locked in eternal ice, and overburdened, in the lapse of ages, with the accumulations of snow precipitated from the winds, loaded with moisture taken up by evaporation, and carried thence from more southern and warmer regions of the Earth's surface.[2]

The German geographer August Petermann championed the concept of a warm polar sea and suggested that a more viable route to the North Pole could be found by sailing through the Bering Strait that separates Alaska from Siberia and following the current northward. Morton's observation of what he perceived as a warm ocean, combined with Bent's observations, lent credence to this idea, although we now know that Morton sighted a polynya, not an open ocean.

In 1878 James Gordon Bennett Jr., the flamboyant publisher of the *New York Herald*, purchased a Royal Navy gunboat, the HMS *Pandora*, and renamed it the *Jeannette*. Bennett had previously financed Henry Morton Stanley's famous expedition to find Dr. David Livingstone in central Africa and was eager for new fodder to publish in the *Herald*. An amateur enthusiast of arctic exploration, Bennett was intrigued by the theory of a warm polar ocean that could be reached via the Bering Strait. He convinced the U.S. Navy to sponsor an expedition to follow his proposed route. In

March 1878 Congress authorized the expedition, with Lt. George DeLong in charge. The *Jeannette* was refitted for polar exploration and left San Francisco on July 8, 1879. DeLong sailed through the Bering Strait, making good time, but got frozen into the ice pack near Wrangel Island, off the coast of Siberia in September.

The *Jeannette* was frozen solid in the ice pack for the better part of two years, drifting in the polar sea toward the northwest. In June of 1881 the pressure of the ice pack closed in, slowly crushing the *Jeannette*. DeLong and his crew rescued a substantial amount of supplies, sleds, and three lifeboats from the fragmented *Jeannette* before abandoning it. He mounted a forced march over the ice ridges to reach the limits of civilization along the delta of the Lena River in Siberia, seven hundred miles away. The crew dragged the boats huge distances over the fractured ice, finally reaching open water. In September a storm blew the boats away from each other. One of the boats eventually made it to safety. DeLong's boat landed in the delta, but his crew was severely weakened, and eventually, all in his party perished.

Three years later wreckage from the *Jeannette* was found in Greenland, two thousand miles away from the site where it sank. The probable drift of the wreckage in the pack ice is shown in Figure 167. The large-scale movement of ice in the polar ocean may at first be surprising, as it might seem that ice would be frozen fast to the continents. There is a zone of ice that is rigidly attached to land, but some miles from shore the ice transitions to the open flow of the Arctic Ocean. In fact, the arctic ice cap and water underneath is constantly moving and changing its form, causing polynyas where trays of ice part and *pressure ridges* where they collide. The net flow of water and ice will form distinct currents, although that wasn't appreciated until the discovery of the wreckage of the *Jeannette*.

The discovery of the *Jeannette*'s remains in Greenland prompted Norwegian meteorologist Henrik Mohn to develop a theory of large-scale movements of the polar ice cap. Mohn believed that

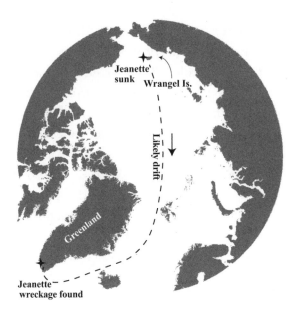

Figure 167 Drift of the wreckage of the *Jeannette*. Image credit: A. Scherlis.

the ice drift simply followed the direction of the prevailing winds. Fellow countryman and explorer Fridtjof Nansen set out to test this theory of polar drift with a large scientific expedition. Nansen reasoned that if he could get a suitably built ship frozen into the ice pack, he could simply drift over the North Pole and claim to be the first to reach it.

Knowing the experience of the *Jeannette*, Nansen set out to build a vessel that could withstand the forces of the ice pack. His vessel, the *Fram*, was built with a rounded hull shape that would cause it to rise up as the ice closed in on it, as opposed to being trapped in a vice, like many of the vessels designed for the open ocean. Nansen and crew left Norway in June 1893 and sailed along the Siberian coast until they reached the New Siberian Islands, close to where the *Jeannette* sank. Nansen sailed north until the *Fram* was frozen in. Nansen and his crew did depth soundings of the polar sea and marked the progress of the drift, all the while collecting data on wind speeds and directions.

The *Fram* drifted to 84 degrees N latitude, but it became apparent that it would miss the pole, some 360 nautical miles away. Nansen and companion Hjalmar Johansen, equipped with kayaks, dog sleds, dog teams, rifles, and provisions, left the *Fram* in 1895 to set out for the North Pole. The rest of the crew continued on with the *Fram*, drifting with the ice cap until breakup the following summer, when they reached open waters. Figure 168 is a map showing the drift of the *Fram* and Nansen and Johansen's sledge journey.

Nansen and Johansen made it only 2 degrees farther north, trudging over huge pressure ridges and fighting the drift of ice that carried them backward, as if walking on a conveyer belt traveling in the opposite direction. They finally turned south after reaching approximately 86 degrees N latitude, the farthest north for then. In the vast maze of arctic ice, they had no hope of finding the *Fram*, which was a mere speck on the ice. Instead, they headed for Franz Josef Land, crossing polynyas, feeding the dwindling pack of dogs on the meat of others they shot. When they reached open water, they abandoned their sleds, put down the last two dogs, and continued on with the kayaks lashed together under sail. By the time they reached Franz Josef Land, winter was closing in, but there was an abundant supply of walrus and polar bear.

The two shot enough animals to lay in a supply for winter and constructed a crude hut consisting of stone walls with moss chinking in the cracks and walrus hide for a roof. They passed the winter of 1895–1896 in the cramped hut. Nansen would read scraps of his almanac for entertainment. When spring arrived, they continued south through the archipelago of Franz Josef Land until they came upon the hut of British explorer Frederick Jackson, who was exploring a route to the Pole himself. With that meeting Nansen and Johansen were rescued. The *Fram*, its crew, and Nansen and Johansen safely returned to Norway in the summer of 1896.

-------- *Fram* **under sail**
———————— *Fram* **in polar ice**
·········· **Nansen and Johansen by sledge**

Figure 168 Voyage and drift of the *Fram*. It sailed along the coast of Siberia and was frozen in. Nansen traveled by sledge toward the North Pole, then to Franz Josef Land. Image credit: A. Scherlis.

The expedition had a wealth of scientific data, including information about the geology of the northern coast of Siberia. Most notable was a detailed collection of measurements of the drift of sea ice, depth soundings in the polar sea, and measurements of wind directions and speeds throughout the drift. It took years for Nansen to sort through these data, but some preliminary findings indicated that it was not wind alone that was responsible for the drift of the polar sea.

During the *Fram* expedition Nansen noted that when the ice pack was blown by the wind, it would systematically drift to the right of the wind direction. This was at variance with the initial idea that current simply followed the prevailing winds. He attributed this to the Coriolis effect. In an early account of the drift of the *Fram*, Nansen wrote: "But, according to my opinion, it is not merely chance winds which influence the drift of the ice;

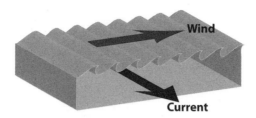

Figure 169 Ekman transport in the Northern Hemisphere. The current moves to the right of the wind direction because of the Coriolis effect. Image credit: A. Scherlis.

I thought, too, that at times there was evidence of a slight current in the water, under the ice, which also went in about the same direction. Nor do I think that the drift of the ice quite coincides with the direction of the prevailing winds. I had the impression that it often carried us a little further north than did the latter; but our abundant material is not yet calculated out, and before this is done it will not be possible to say anything for certain on the subject."[3]

Nansen was able to persuade a doctoral student, Vagn Ekman, to do a thorough treatment of this effect. Ekman worked out the mathematics of the combined effect of gravity, the wind, and the Coriolis effect to create the modern theory of ocean currents.

Nansen's original conjecture and Ekman's subsequent treatment of the phenomenon relied on a combination of surface wind and the Coriolis effect. Consider wind blowing across a large surface of water. If the wind is moving with any northerly or southerly skew to it, the Coriolis effect will create a systematic shift in the direction of the resultant current on the surface of the water (Figure 169). This shift will be to the right of the prevailing winds in the Northern Hemisphere and to the left of the prevailing winds in the Southern Hemisphere.

If we consider the Northern Hemisphere, you may recall from chapter 8 that trade winds generally blow from the northeast to the southwest and are found between the equator and 30 degrees N latitude. North of 30 degrees the westerlies blow from the southwest

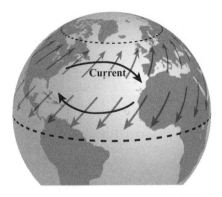

Figure 170   Influence of Ekman transport on current flow in the North Atlantic. Image credit: A. Scherlis.

to the northeast. This sets up a current flowing from east to west in the trades, then north, and from west to east in the zone of the westerlies.

The Coriolis effect will persist in depth. Water moving on the surface will pull along water underneath, but the next layer below the surface layer will be shifted systematically to the right of the surface layer in the Northern Hemisphere, just as the surface of the water was shifted to the right by the action of the wind. The layer of water flow below this second layer will be shifted farther still to the right and so on. The strength of the current diminishes rapidly with depth, dying out by four hundred meters. This is called the *Ekman spiral* (Figure 171).

Figure 171   Ekman spiral as a function of depth below the surface. Image credit: A. Scherlis.

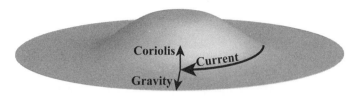

Figure 172 The combination of current, the Coriolis effect, and gravity give rise to large-scale gyres in the oceans. Image credit: A. Scherlis.

The main effect of the Ekman spiral is to create large-scale circulations of water called *gyres* in the oceans. Under the influence of wind, the upper surface layer of the ocean will deviate to the right of the driving wind, the next layer below that will deviate to the right of that still, due to the Coriolis effect/Ekman spiral. This will cause water to heap up in the center of a circular loop. The effect of gravity will balance the inward force created by the Ekman spiral, and there will be equilibrium between the two, as water circulates around the central bulge (Figure 172).

Because of the Coriolis effect, gyres will rotate clockwise in the Northern Hemisphere and counterclockwise in the Southern Hemisphere. The central bulge will vary in height, depending on the gyre, and even show variation over the course of years, often reflecting climate-induced changes; for example, shifting wind patterns.[4] The height of the central bulge can be as large as two feet relative to the periphery.

Figure 173 Circulation of gyres in the North and South Atlantic. Image credit: A. Scherlis.

Major Ocean Currents

The effect of the rotation of the Earth will cause the western portion of the gyre to pile up on the eastern coastline of continents, creating strong currents along those coasts. This effect is responsible for the strength of the Gulf Stream near Florida and is called a *western boundary current*. A similar current arises in the Pacific Ocean near Japan, called the Kuroshio (Japanese for "black stream/black tide").

The gyre of the North Atlantic can be divided into a number of distinct currents. Figure 174 shows a supercomputer simulation of the flow of current in the North Atlantic. The trade winds in the mid-Atlantic, just north of the equator, are responsible for the

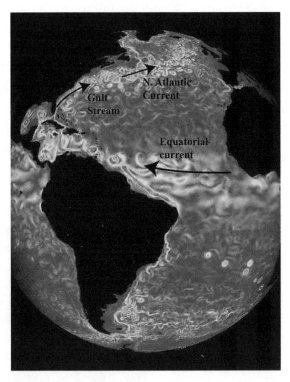

Figure 174 Computer simulation of currents in the North Atlantic. Adapted from NASA Sea Current Simulation, Los Alamos National Laboratory and Naval Postgraduate School. Credit: NOAA.

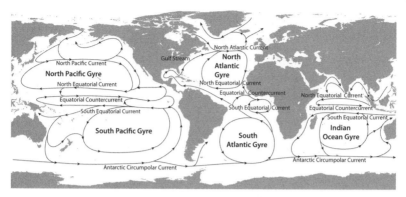

Figure 175 Major ocean currents of the world. Image credit: A. Scherlis.

North Equatorial Current, which flows westward from the Cape
Verde islands toward the Caribbean and the northern coast of
South America. When it enters the Caribbean, the Gulf of Yucatan,
and the Gulf of Mexico, it creates a number of eddies of circular
current in those regions but then forms a strong current in through
the Straits of Florida and forms the Gulf Stream–Florida current.
During the Deepwater Horizon oil spill, there was a strong fear
that the loop of this current in the Gulf of Mexico would begin to
transport oil around the Florida Keys and out to the Atlantic.

As the Gulf Stream passes Cape Hatteras in North Carolina,
it begins to move offshore. At this point it starts to meander and
form eddies around the edges. Nonetheless, there is enough flow of
warm water to keep Northern Europe temperate. The return flow
to the Cape Verde islands is considerably weaker and spread out
over a broader area. While the coarse details of the Ekman theory
explain the currents in the North Atlantic, much of the details will
be affected by the location of land and the onset of turbulence.

Figure 175 is a map of the major ocean currents. The doldrums
around the equator mark the dividing line between gyres in the
Northern and Southern Hemispheres. Depending on topography
and time of year, the actual currents can change substantially. The
dominant features are the gyres of the Indian Ocean, the North
and South Atlantic, and the North and South Pacific.

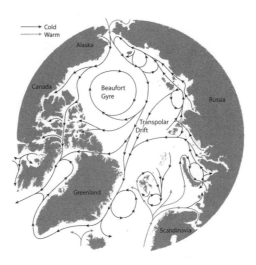

Figure 176 Currents in the arctic. The dominant features are the Beaufort Gyre and the Transpolar Drift discovered by Nansen. Image credit: A. Scherlis.

The Beaufort Gyre and Transpolar Drift dominate currents in the arctic (Figure 176). The latter current was responsible for the dispersal of the wreckage of the *Jeannette* and was the one Nansen took advantage of in his voyage on the *Fram*. If we look back on the theory of an open polar sea, we can note that there is a warm current flowing north through the Bering Strait, although it is hardly sufficient flow to melt the polar ice cap — yet. If the trends of global warming continue, we may be forced to reconsider the theory.

Complex currents near Iceland and Greenland include the Labrador Current that transports icebergs and cold water south, creating shipping hazards in the North Atlantic. When the cold Labrador Current meets the Gulf Stream in the North Atlantic, a very sharp transition forms. The transition will move north in the summer and south in the winter as the Poles and Tropics alternately assert their dominance.

In recent years some gyres have become big garbage dumps. Trash that enters a gyre has no exit. It stays there and orbits the center indefinitely. Recently, the North Pacific Gyre has been labeled the "world's largest garbage dump" as detritus from cities along the North Pacific rim collects there and never leaves. The North Atlantic Gyre is also experiencing extensive trash problems, for identical reasons. Other gyres with smaller adjoining populations aren't as plagued.

Near the equator there is a significant effect called the *equatorial countercurrents*. With a clockwise flow in the Northern Hemisphere and a counterclockwise flow in the Southern Hemisphere, when they meet in the vicinity of the equator, you might naively expect that all the current would flow from east to west. This isn't entirely the case. The water flowing west tends to pile up against the eastern shores of the continents. The westward flow in the Pacific pushes up against Asia and the westward flow in the Indian Ocean pushes against Africa. Although a large fraction of this flow will make the return path by traveling north or south, some fraction of it finds a return path close to the equator. Some of the return flow proceeds at low latitudes in a west-to-east direction, creating the equatorial countercurrents that vary with the season.

A seasonal shift in the Pacific equatorial countercurrent can be an important factor for navigation in some areas. Authors Father Ernest Sabatier and David Lewis wrote about the influence of currents on the voyaging season for inhabitants of the western archipelago of Kiribati in the central Pacific (Gilbert Islands).[5,6] During the period of October through March, the equatorial current is quite strong, running from east to west. The native navigators associated this time with the appearance of the Pleiades just after sunset. This is also a stormy period, and the combination of weather conditions and strong currents kept navigators ashore. In the summer months the weather moderates and the current dies down, sometimes showing a weak equatorial countercurrent running from west to east. The combination of weak currents and good weather is associated with the appearance of Antares just after sunset and is the prime season for voyaging.

Equatorial countercurrents also vary from year to year. This variability is related to the summer-winter heating differences in the two hemispheres. Strong equatorial countercurrents in the Pacific are also associated with the climate phenomenon known as El Niño (Spanish for "the child," corresponding to its formation around the Christmas season). Climatologists call this the El Niño Southern Oscillation (ENSO), which appears in cycles of roughly five to seven years and creates a global climate disturbance. Warm waters move toward the eastern coast of the Pacific, disrupting the colder Humboldt Current moving from the north. This in turn causes major shifts in weather patterns: northern Mexico and California see an increase in warm wet weather, while Canada and the Midwest experience warmer but drier conditions. The Winter Olympics in Vancouver in 2010 faced challenging warm conditions during an El Niño phase.

Determining and Accounting for Current

When charting a course, a navigator on the open ocean has to account for currents. Since currents will create eddies and vary with the seasons, this is as much an art as a science. As I've discussed before, a navigator can assess position from dead reckoning and from a fix based on celestial navigation. The interplay between these two can often be used to find local currents. If you know your speed and heading, by knowing a point of departure, you can use this information to figure out your position at some later time. If an unknown current is pushing your vessel in a certain direction, when you try to find your position using celestial navigation, the results of dead reckoning and the celestial fix won't match up. Discounting the possibility of an error, the difference is due to current.

In Figure 177 I show an example of this. From a point of departure, a navigator sails on a heading of 78 degrees true. After twenty-four hours, she figures her position from the heading and assumption that the sailboat has a speed of six knots. She then takes an observation of her position from sighting stars at twilight,

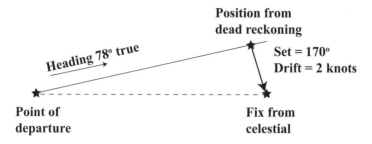

Figure 177 Using a comparison between dead reckoning and a fix to determine current set and drift.

and finds that her true position is substantially different than from dead reckoning. She accounts for this apparent discrepancy by assuming that a current has pushed her south. When a navigator talks about current, it has two characteristics: *set* and *drift*. *Set* is the direction of current, measured in degrees from true north. *Drift* is the speed of the current in knots. In the example of Figure 177, the navigator used the difference between dead reckoning and the celestial fix to figure out that the current had a set of 170 degrees and a drift of two knots.

Once a navigator knows the set and drift of current, she can make appropriate corrections to reach her destination. In Figure 178 I show the passage of a sailboat toward an island. With a current from the north, if the navigator holds to a heading directed at the target island, the current will carry her well south of the island. On the other hand, if she compensates for the set and drift of the current by choosing a more northerly heading, she will reach the target island directly. One of the daily routines of a navigator on the ocean is the estimation of current set and drift based on a comparison between dead reckoning and celestial fixes.

Seasonal shifts in currents and their unpredictability present a challenge to navigators, particularly in equatorial regions, where there can be a substantial seasonal variation. For the conventional Western navigator, charts and computer forecasts of currents are a primary reference, but given the tendency of currents to form

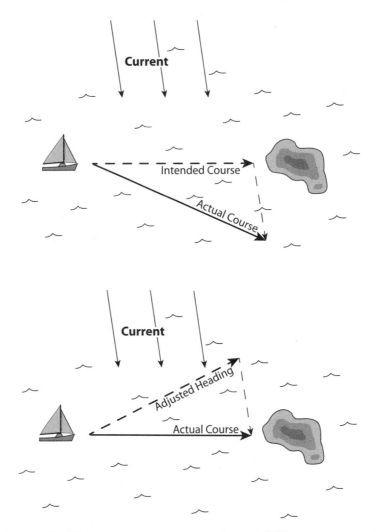

Figure 178 Effect of current on passage of vessel. Without a correction the vessel will miss landfall on the island. If the heading takes into account the current, the vessel can make landfall. Image credit: A. Scherlis.

eddies, even a navigator armed with this information cannot be assured of correct headings. In the modern era satellite imaging provides details of changing current patterns on a daily basis. The annual Newport-to-Bermuda race crosses a region of the Atlantic in which variations and eddies spun off the Gulf Stream can create conditions that favor some sailors and hinder others. The art of finding and following a favorable current can spell the difference between winning and losing. The following quotation comes from a description of the sailing in the race north of the Gulf Stream:

> BETWEEN NEWPORT AND THE GULF STREAM
> Sailing in cold water and often in fog, the navigator must select a route to the optimal position on the northern edge of the Gulf Stream, avoiding the bad side of warm eddies north of the stream, or taking advantage of the favorable side of the clockwise rotating warm eddies. Current in the eddies may reach 3 knots and the warm eddies can be 60 to 100 miles in diameter. Satellite photos and their interpretation are available so these days the navigator has a pretty good idea of the location of the stream and its major eddies. It is often difficult to stay near the rhumb line and navigators often worry about falling off to the east, however recent strategies have maximized VMG to the island in the early phases of the race wherever it leads.[7]

"VMG" means "velocity made good." The computer simulation in Figure 174 shows the many eddies associated with the Gulf Stream as it moves off the coast of North America.

Beyond racing, an understanding of currents is critical for successful ocean navigation. So how did one cope with currents in the absence of charts or satellite images? In some cases it was a matter of luck and sailing toward a long coastline, so the effect of drift in the current wouldn't make a difference. One simply sails, hits a coastline, then figures out how far along the coastline to travel.

In the case of voyages to or between islands, early sailors didn't have the luxury of satellite photos or accurate celestial position fixes. As mentioned above, the Pacific Islands navigators in particular had to cope with this problem, compounded by variations in the equatorial currents of the Pacific. The sailors had to be able somehow to measure the current and adjust their heading to take it into account. With their seasonal variation, the equatorial currents in the Pacific are some of the most challenging on the globe.

BACK BEARINGS

When setting out on a voyage, a navigator would sail some distance away from the island to clear any local currents caused by the island itself. The navigator would then look back at the island of departure and set a course using landmarks on the island to ensure that the vessel was heading in the correct direction. The direction the canoe was pointing would be different from the desired landfall, as the navigator had to compensate for current, but the landmarks on the island ensured that he was pointing in the right direction, taking into account the current. Caroline Island navigators called this practice *fatonomuir*, which translates into "facing astern."[8]

British anthropologist Raymond Firth documented the art of current finding on the island of Tikopia in the eastern Solomons. Much voyaging was done between Tikopia and another island, Anuta. Anuta is seventy miles to the northeast of Tikopia and is only half a mile across; its highest point is two hundred feet above sea level. Voyaging from Tikopia to Anuta is challenging, as Anuta is merely a speck in the ocean. Firth writes: "Anuta was termed figuratively *Te Fatu Sekeseke*—the Slippery Stone—since it is such a small spot in the ocean to be found, and so easily slid away from." Anuta can only be seen from about ten miles away. The Tikopian navigators had to set sail with careful corrections for the current. Firth writes about the practice of using landmarks on Tikopia to set sail to Anuta, noting that the canoe is "set carefully in the required direction by using marks of orientation on Tikopia. On the northern side of the island is a beach named 'Mataki Anuta,' 'Looking on Anuta,' i.e., facing in that direction. At the back of this beach a gully runs up the mountain side; this is known as 'Te

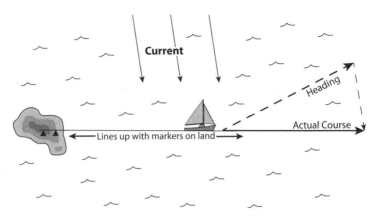

Figure 179 Use of back bearings to establish sailing direction in the presence of currents. Image credit: A. Scherlis.

Rua i Soso,' 'The Hollow from Soso' (this last being a taro plantation above). When setting out for Anuta the crew turn the stern of their canoe to this gulley and keep it in sight as long as they can."[9] This use of back bearings is shown in Figure 179.

WAVES AS CURRENT INDICATORS

If one removes all visual clues about current, the drift relative to land, or some other indicator, how can one figure out whether a current is flowing? After all, if I put you up in a balloon in the fog, the balloon would simply drift with the wind, and you couldn't tell whether you were moving relative to the ground. Much the same could be said about currents in the open ocean, except for waves and swells that travel long distances and enter a region where strong currents flow. When waves generated from far away reach a region of current, the waves' heights generally stay the same, the periods stay the same, but the wavelengths change.

If an ocean swell generated a far distance away reaches a current that's moving in the opposite direction, the wavelengths get shorter and the wave faces of the swells become correspondingly steeper.

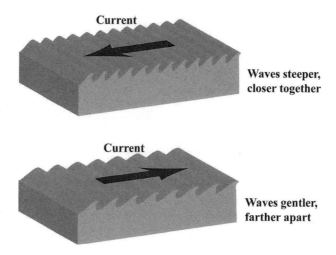

Figure 180   Effect of current on waves. Image credit: A. Scherlis

When waves move against currents, they can cause nasty conditions for the sailor. Conversely, if waves meet a current going in the same direction, the wavelengths will be stretched out, and the waves will become smoother as a result (Figure 180).

Experienced navigators can use the shape of waves as a means of determining the local currents even when out of sight of land. This was particularly useful for the Pacific Islanders who voyaged in equatorial regions with fickle currents.

Writer Steven Thomas apprenticed himself to the navigator Mau Piailug from the island of Satawal in the Caroline Islands. Piailug introduced Thomas to his techniques of navigation, and Thomas wrote of his experience in *The Last Navigator*. He writes of his introduction to the use of waves as a current indicator:

> "How can you determine the current when there are no swells?" I asked.
>
> "You look at the water and it is tight," he answered. "The small waves go like this [pushing in one direction] and then—how can I explain?" He extended both his

hands and pulled them back as if stroking the keys of a piano. He claimed the sign was now present and that it indicated a weak current from the west, flowing against the light northeasterly wind.[10]

Joshua Slocum also used the appearance of waves as a way of deducing currents. He wrote about this effect from his experience sailing south from the Cape Verde islands toward South America: "The sloop was now rapidly drawing toward the region of doldrums, and the force of the trade-winds was lessening. I could see by the ripples that a counter-current had set in. This I estimated to be about sixteen miles a day. In the heart of the counter-stream the rate was more than that setting eastward."[11]

CURRENTS INDUCED BY LOW-PRESSURE SYSTEMS

You may recall that low-pressure systems can produce cyclonic disturbances that circulate counterclockwise in the Northern Hemisphere and clockwise in the Southern Hemisphere. In the tropics these disturbances can turn into typhoons and hurricanes producing winds that get sucked into the ultralow pressure vortices that form the eyes. The wind will be pulled in over large distances, and there is often enough flow for the storm systems to create their own local currents.

The passage below is taken from the book *Storm Tactics Handbook* by Lin and Larry Pardey and relates their initial incredulity when an unexpected current created by a typhoon in the Bay of Bengal sent them far off their intended course. Only as more weather signs set in did they become aware of the approaching storm:

"Wonder if the shipping routes shown on our pilot chart are wrong," I asked Larry on our fourth night out of Galle, Sri Lanka. " I was thinking the same thing," he answered. "I saw three ships on my last watch. How many did you count?"

I'd seen the lights of four freighters on the horizon to the north of us as we reached along on an easterly course.

We'd headed southeast for the first three days of this voyage; our goal: three degrees north latitude, the normal lower limit of the typhoon zone. When our log reading and dead reckoning track showed we'd reached this point, we changed course to head directly east for 600 miles before working northeast into the Andaman Sea. We'd had cloudy skies for two days, so when our first sight this fourth morning out made no sense, we waited anxiously for our noon sight, while we both went over Larry's navigation figures, trying to find some mistake in the arithmetic. The noon sight seemed so out of whack that we wanted to ignore it. "We couldn't possibly be 110 miles north of our course, " was Larry's grumbled comment as he rechecked his figures and read the sextant again. Then I reminded him of the ships we'd seen. They should have been at least a hundred miles north of us. "If some sort of current was setting us north . . ."

"Can't be a two-and-a-half-knot current," Larry stated. "Pilot charts show a southeasterly current coming out of the Bay of Bengal at this time of year!"

But our first afternoon sight made sense only if we accepted our noon sight. That started us worrying. "If something has caused a current that strong, it must be a pretty big disturbance," Larry said. "Let's get headed south again." We tightened our sheets until we were sailing on a close reach 45 degrees away from our rhumb-line course. Within a few hours the last ship dropped below the horizon astern of us. Then we got the second sign.

A long, slow swell began to interrupt the steady pattern of the seas that came rolling toward us. Along with the normal wind waves, this rolling surge added an awkward motion to our 5-knot charge into the sea. I got out our stopwatch and counted the seconds between each crest of these obviously different swells. Four swells passed each minute, instead of the normal eight or ten. Their unusually slow spacing began to add to our certainty. There was a big weather disturbance somewhere ahead of us![12]

The above passage illustrates how an experienced navigator has to make sense of celestial fixes, allow for the possibility of unexpected currents, read signs in the waves, and try to create a coherent picture. Large storms will create waves with a very long period between crests. The combination of the unexpected current and the appearance of slow swells are strong signs of an approaching typhoon. In their narrative the Pardeys go on to describe how they took evasive action and safely rode out a harrowing storm on its margin. Unfortunately, the crew of a Canadian yacht wasn't so observant and sailed into the middle of the storm; the ship sank with all hands on board. In the case of the Pardeys, their experience and ability to synthesize the different pieces of information undoubtedly saved their lives, whereas others perished.

15. Speed and Stability of Hulls

• •

ALTHOUGH THE UNDERLYING physics describing how water-craft move through the water is the same throughout the world, marine environments are varied, and the needs of different cultures can dictate their design. The shape of a vessel's hull is a major consideration in navigation, particularly when combined with sails. A vessel's speed and ability to move with or into the wind has to be taken into account when calculating dead reckoning.

The use of vessels to carry humans over large stretches of water goes back as far as recorded history. There is circumstantial evidence of the use of seaworthy craft much further back. In January 2011 the Greek Cultural Ministry unveiled the discovery of stone tools found on the island of Crete that are believed to be 130,000 years old.[1] If this date holds up, it is evidence of the earliest waterborne journeys over substantial distances. Water gaps from Crete to major nearby locations include the closest island of Karpathos (thirty-two miles). Crete is visible from a high point on the western end of Karpathos. The visibility of Crete from Karpathos makes a deliberate, as opposed to accidental, voyage to Crete plausible. Radiocarbon dating puts the time of the earliest settlements in Fiji at least as far back as 1600 BC, marking the beginnings of long-distance voyaging out of sight of land in the Pacific.[2]

One of the earliest recorded stories of long-distance voyaging in the West comes from Herodotus, who reports a circumnavigation of Africa carried out by the Egyptian king Necho circa 600 BC by Phonenician sailors:

> Libya is washed on all sides by the sea except where it joins Asia, as was first demonstrated, so far as our knowledge goes, by the Egyptian king Necho, who,

after calling off the construction of the canal between the Nile and the Arabian gulf, sent out a fleet manned by a Phoenician crew with orders to sail west about and return to Egypt and the Mediterranean by way of the Straits of Gibraltar. The Phoenicians sailed from the Arabian Gulf into the southern ocean, and every autumn put in at some convenient spot on the Libyan coast, sowed a patch of ground, and waited for next year's harvest. Then, having got in their grain, they put to sea again, and after two full years rounded the Pillars of Heracles in the course of the third, and returned to Egypt. These men made a statement which I do not myself believe, though others may, to the effect that as they sailed on a westerly course round the southern end of Libya, they had the sun on their right — to northward of them. This is how Libya was first discovered by sea.[3]

Here, "Libya" can be taken as synonomous with "Africa." Whether or not this voyage really took place, the account is consistent with a knowledge of the length of time it would take an able-bodied crew to circumnavigate the continent.

Design Considerations

Watercraft are built to move people and goods from one location to another. Not only must they support their own weight on the water, but they must also support the weight of crew and cargo. Consider the forces on a block immersed in water (Figure 181). At any given point under water, pressure increases with depth because of the weight of the column of water pressing down from above. On the block in the figure, the pressure on the top of the block is lower than the pressure on the bottom. The pressure on

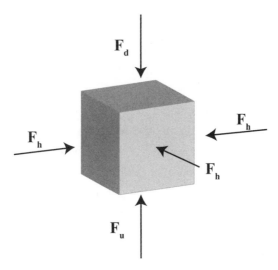

Figure 181 Buoyant force is created by a greater pressure at a lower depth of water on a solid object.

the four sides (F_h) is all equal and cancels itself out. The difference in pressures between the top and bottom faces gives rise to a net upward force called *buoyancy*. The force of gravity on the vessel with crew and cargo pulls it down.

As a vessel is placed in the water, it lowers and begins to displace water. As it gets loaded with cargo and crew, it sinks lower still, displacing more water. The upward buoyant force is equal to the weight of the water displaced by the vessel. The vessel will reach an equilibrium depth when the upward buoyant force and the downward force of gravity are equal and opposite. The more water displaced, the larger the buoyant force. Vessels are thus rated in terms of their *displacement* — the weight of the water their hulls displace, which is typically rated in tons for larger craft. The word *ton* or *tonne* derives from the Saxon word for a large barrel, which, when full, weighed about a ton.

Stability on water depends on the relation of two points in a vessel: the *center of gravity* and the *center of buoyancy*. If you add up all the forces of gravity pulling down on the vessel, the crew, and the cargo, you can approximate the sum of these forces as if they

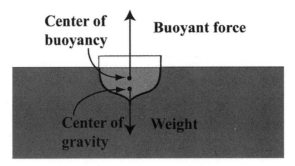

Figure 182 The buoyant force on a vessel is equal to the weight of the water it displaces (grey area).

were acting at one point: the *center of gravity*. Likewise, the buoyant force will act over the surface of the hull of the vessel. If you add up each of the buoyant forces, you can approximate the sum of these as if they are all acting at one point: the *center of buoyancy*.

The location of the center of gravity will depend on how cargo and crew are placed in the vessel when it's loaded. The location of the center of buoyancy depends on the shape of the hull, how much of it is immersed in water, and its orientation to the surface. Figure 182 shows a hull with both the center of gravity and the center of buoyancy indicated. Only the part of the hull under water gives a displacement contributing to the buoyant force. The amount of the hull protruding above the waterline is called its *freeboard*. If a vessel is too heavily loaded and has too little freeboard, it is dangerously susceptible to having large waves wash over the top of the hull.

Speed

The faster a vessel moves, the more efficient the transport, but there is a trade-off between speed and stability. As the wind fills the sails of a vessel, it begins to gather speed, but it also starts to experience forces that retard its forward motion. These forces are collectively called *drag,* which increases with speed. Vessels will reach an equilibrium speed when the driving forces, such as the wind in the sails, balance drag.

A moving vessel displaces water to both sides. The very act of pushing water aside requires force and creates a wave. As mentioned in chapter 12, the speed of a deepwater wave depends on its wavelength: the longer the wavelength, the faster the wave moves. If you drop a pebble into a pond, the splash creates a large number of wavelengths all at once. The longer waves expand more rapidly, followed by the shorter waves. In a sense the bow of the vessel is creating waves with different lengths all the time, but there is only one wavelength that moves at the same speed as the vessel and stays with it. This wave in front is called the *bow wave.* The back (stern) of the boat also produces a wave called the *stern wave.* Just like other waves, the bow and the stern waves add together, sometimes constructively and sometimes destructively.

At slow speeds the bow and stern waves will be relatively short, but as a vessel accelerates, the waves moving along with it lengthen. There is a critical speed called *hull speed,* when the wavelength produced by the bow and the stern are exactly the length of the vessel. At speeds below hull speed, it's relatively easy to increase the speed, but close to hull speed the drag becomes very large.

Figure 183 shows a sailboat at three different speeds. At top the boat is well below hull speed, with several wavelengths between the bow and the stern. In the middle the boat is just at hull speed, where there is exactly one wavelength between the bow and the stern. On the bottom the ship is moving just a bit above hull speed: the ship literally has to climb uphill over the bow wave to move faster than hull speed. It is rare for most sailing vessels to exceed hull speed. In some cases lightly loaded double-hulled vessels (e.g., a canoe with an outrigger) can exceed hull speed.

When a vessel exceeds its hull speed, this is called *planing*: the drag forces drop precipitously, and it skims across the water. One example is a vessel moving down a steep wave. The pull of gravity

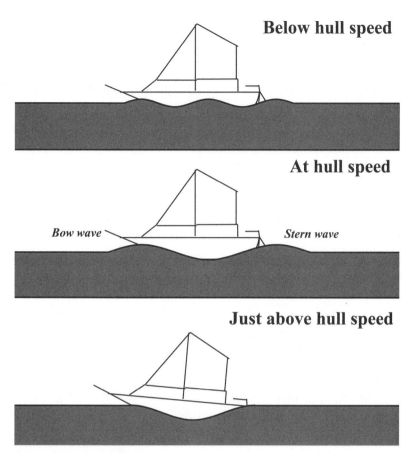

Below hull speed

At hull speed

Bow wave   *Stern wave*

Just above hull speed

Figure 183 A sailboat below hull speed (top), at hull speed (middle), and just above hull speed (bottom).

can accelerate it faster than hull speed, in which case it's called *surfing*. While this can be a sport for surfers or kayakers who paddle rapidly to catch a wave, then surf down the face, it can be a problem for vessels in a high sea. If a large wave catches a boat from the stern, the resulting surfing motion often makes it difficult to steer, and a sailor can lose control.

In addition to the effects of the bow and stern waves, other factors can determine the speed of a ship. The bow acts like a wedge, displacing water to both sides as it moves through the water. The narrower the wedge, the less force is required to move through the water. The roughness of the hull is also a factor. As water moves past the hull, the forces of friction between water and the hull itself can slow down motion. In addition, tiny swirls of water create turbulence which robs the vessel of speed. All of these considerations lead to some design criteria for a fast ship or boat:

- Make it as long as possible.
- Make it as narrow as possible.
- Make its hull surface as smooth as possible.
- Have a pointy bow.

STABILITY

Some of the fastest vessels are narrow. Racing shells are long and thin. This comes at a cost, however. Anyone who has sat in a racing shell for the first time knows that they feel unstable and prone to capsize. In general, a wider hull gives a ship more stability, but at the expense of speed. The stability of a vessel depends on how it responds to three possible kinds of rotation (Figure 184). These motions are given specific names, depending on how the ship is tossed around in the waves or under the influence of various forces:

1. **Pitch:** an up-and-down motion of the bow and the stern. A vessel generally pitches when it is encountering waves head-on or the waves are coming from behind (called a *following sea*).
2. **Yaw:** when the bow moves right while the stern moves left, or vice versa. This often occurs when waves hit a ship at an angle of 45 degrees (called a *quartering sea*).
3. **Roll:** when the ship rotates about its long axis. This condition often arises when waves hit one side of the ship, in what is called a *beam sea*.

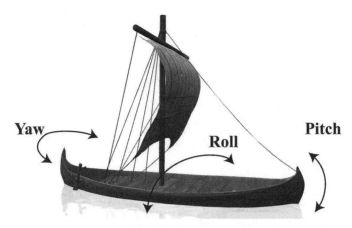

Figure 184 Illustration of the rotational motions of pitch, yaw, and roll for a Norse vessel called a knarr. Image credit: C. Mairs, copyright © 2011–12 President and Fellows of Harvard College.

A pitching motion of a ship is probably the most common and easiest to tolerate. The ship's heading is unaffected by pitching as it charges up and down the faces of waves. When a ship takes on a beam sea, it has a tendency to roll back and forth. Like pitching, rolls will not influence the heading of the vessel but can become unnerving if the roll becomes large. If the roll is very large, the ship may get pushed sideways, which is called a *knockdown*, or, worse yet, may turn upside down, which is a *capsize*. For the navigator, a yawing motion is the most problematic, as the ship will be changing its heading constantly, making it difficult to correct.

What determines the stability of a ship? Consider a spinning ice-skater. If the arms of the ice-skater are extended, he'll rotate slowly, but as he pulls them in, the speed of rotation increases dramatically. The shape of the ship and the way it is loaded will determine how it responds to seas. Think about the motion of a ship under the influence of pitching and yawing. If the ship is long and has a lot of cargo stowed fore and aft, it is like a skater with his arms extended. Any pitching motion will be slow. If the ship is short and loaded near the center, it will be much more susceptible to pitching and yawing under the influence of waves.

The way a ship is loaded also affects how easy it is to steer. If all the weight is in the bow and stern, it takes more effort to get it to turn and the ship is relatively sluggish, but if most of the weight is near the center, it becomes easier to turn. The same loading that makes a ship more stable also makes it more difficult to turn. When a vessel is loaded at a port, the captain has to pay attention to the way it is loaded fore and aft, which is called the *trim*.

Ship stability under the rolling motion is more complicated than pitch and yaw. The analogy with the skater and outstretched arms still works, in part. A ship with a wide hull will be more stable than a ship with a narrow hull. The trade-off between speed and stability is always a consideration for ship designers.

Gravity and buoyancy are also factors that influence the stability of a ship under the rolling motion. In Figure 185 I show what happens to the center of buoyancy and the center of gravity as a vessel rolls. A well-loaded vessel should have all its cargo tied down, so that it won't rock back and forth as the ship moves. The center of gravity remains at the same spot relative to the vessel as it rolls. On the other hand, as a vessel rolls, the center of buoyancy will shift. As it rolls to the right, more of the hull on the right-

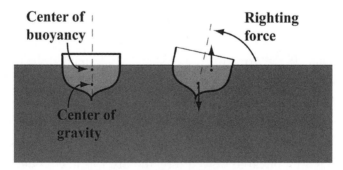

Figure 185 Righting force of a vessel under a rolling motion. The center of buoyancy is displaced laterally as the ship rolls. The combination of the force of gravity on the center of mass and the buoyant force on the center of buoyancy creates a force that tends to right the ship.

hand side is in the water and less is on the left, shifting the center of buoyancy to the right. The shift of the center of buoyancy relative to the center of gravity creates a righting force that naturally combats the roll of the vessel.

The stability of a vessel can be improved by *ballasting*: adding weight close to the keel to lower the center of gravity. Ballast was often a load of rocks, although in modern sailboats it is sometimes lead built into a deep keel.

The shape of the hull also plays a role in the stability. A wide, rectangular hull shape will shift the center of buoyancy dramatically and contribute to the stability of the ship. A narrow, circular hull will be far less stable. In Figure 186 I compare the stability of two Norse ships from the Middle Ages: the *longboat* and the *knarr*. The long, narrow longboats were typically used for swift attacks, while the knarr was used for long-distance transport. I have based the hull shapes in the figure on archeological reconstructions of vessels found in Roskilde Fjord, Denmark, at the Viking Ship Museum.[4,5] When rolled at the same angle, the combination of the width and hull shape of the knarr produces a larger shift in the center of buoyancy than the longboat, giving the knarr better stability under rolls. The knarr was better suited for voyages across the North Atlantic.

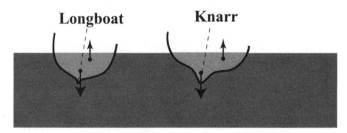

Figure 186 The displacement of the center of buoyancy for the same angle of roll will depend on the hull shape. The narrower Viking longboat has less righting force than the more seaworthy knarr.

Keels and Leeway

In the figures above, you can see that the bottoms of the hulls have a "V" shape to them. The *keel* helps the vessel track in one direction. The shape is particularly important for sailing vessels. The ability to sail in any direction the navigator wishes is limited by the wind and the capabilities of the vessel. One of the most important developments in sailing technology is the ability to sail *into* the wind, aided by the shape of the keel.

As a ship gathers speed, the V shape of the keel acts like a knife moving through the water, resisting forces that push the ship sideways. When combined with a sail that acts like a wing, the hull resistance allows the ship to sail into the wind. In the extreme case of modern racing yachts, the keels extend far under the hull to improve windward performance. The amount of sideways slip for a vessel under sail is called *leeway*.

Hulls of Selected Ships

Below, I discuss some vessels used by a number of cultures, their characteristics and construction. A detailed account of ancient vessels is a field unto itself: nautical archaeology. I can only touch on some of the design trade-offs that faced different cultures and how they may have approached them. Six representative hull shapes are shown in Figure 187. Next to the hulls, I've indicated representative figures for their typical lengths, displacements, and hull speeds. Needless to say, there can be a large variation of these parameters for any class of vessels. The lengths of the hulls are indicated in the figure; they are not to the same scale.

The Norse hull shapes and parameters are based on the archaeological remains in the Viking Ship Museum in Roskilde, the Skuldelev ships. The dhow hull is representative of a number of contemporary vessels and descriptions of vessels. The kayak hull

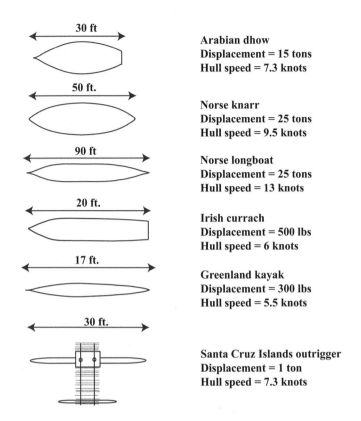

Figure 187 Hull shapes of vessels discussed in the text. Note that these are not to the same scale, but the hull lengths are indicated for each.

is based on an example brought to Europe by the explorer Fridtjof Nansen. The currach hull is based on a contemporary Aran Islands (off the western coast of Ireland) version. The Santa Cruz outrigger is a composite based on a number of descriptions dating from the time of Captain James Cook to the present.

Dhow

The dhow (pronounced "dau") is still widely used along the eastern coast of Africa, the Arabian Peninsula, and the Arabian/Persian Gulf. During the era of the spice trade, the seasonal character of the monsoon allowed traders to find favorable winds for voyages between India and the Arabian Peninsula.

The hull of the dhow shown in Figure 187 is designed for deep-sea trade. With a length of thirty feet, it has a hull speed of 7.3 knots. The broad hull gives it good stability and with a displacement of fifteen tons, it could carry a fair amount of cargo. Other dhows had tapered bows and sterns, while this version has a square stern, possibly influenced by Portuguese or other European designs. Dhows in use on the east coast of Africa have displacements up to sixty tons.[6] A common construction technique uses planks of wood stitched together with cords to make the hull, a technique that dates back to ancient Egyptian practices where the cords are countersunk into grooves. The edges of the planks lie flush against each other. Both the edges of the planks and the cords are treated with a waterproof combination of fiber and sap or tar.

Knarr

The knarr (the k is pronounced) was the workhorse of Norse trade, enabling the far-flung trade routes that flourished across Europe and the North Atlantic during the Viking Age (circa AD 800 to 1300). With a length of fifty feet, it had a theoretical hull speed of 9.5 knots and a displacement of twenty-five tons. It sailed with a square sail and had a V-shaped keel, presumably to minimize leeway. The broadness of its hull shape would have given it a decent stability against the storms of the North Atlantic. Even with good hull stability, an open-boat passage to Iceland or Greenland must have been ghastly, given the intensity of the weather in that region.

The knarr hulls were built in a style called *clinker*, in which wooden planks were placed from the keel upward with over-lap between adjoining planks. Iron nails joined the planks together. Archaeological work at the L'Anse aux Meadows site in Newfoundland uncovered the remains of a Norse repair operation for boats. The inhabitants mined local bog iron, smelted it, and fashioned nails, presumably to replace ones that rusted in the saltwater environment.[7]

Moss soaked in tar was jammed into the seams to make the hulls watertight. Cross bracing was added after the hull was constructed. This design allowed the knarr to flex a small amount under the influence of heavy seas, as opposed to reacting rigidly to waves.

The Norse used a *steering board*, or *starboard*, which is a long plank extended into the water from the right-hand side of the boat. By adjusting the angle of the steering board, the navigator could hold or shift the course of the vessel. In many ways the starboard acts like a wing, generating a force akin to lift for a wing to move the stern of the boat to left or right, depending on its orientation. The starboard was, by convention, on the right-hand side. When bringing the knarr to a dock or unloading onto land, the starboard side was kept to the water, and the left, or port, side was where the vessel was unloaded. This practice is the origin of the terms *port* for left and *starboard* for right.

Longboat

As with the knarr, our best knowledge of the Norse longboat is from archaelogical finds. As the name implies, the longboat can be quite long, up to a hundred feet, with a narrow hull. Because of its length and narrow beam, it had a high theoretical hull speed of up to thirteen knots. With both wind and human power, it was likely the fastest vessel plying the waters of Northern Europe during the Viking Age. The narrowness of its hull suggests that it wasn't terribly seaworthy in the storms of the North Atlantic and hence was likely used more as an attack vessel than a trading vessel.

Like the knarr, the longboat was built with clinker construction. Although the boats were quite long, they had a shallow draft, allowing them to sail in shallow waters, including estuaries, aiding their capabilities as a raiding vessel. Longboats are depicted in the Bayeux Tapestry of the Norman invasion of England.

CURRACH

The Irish currach (or curragh) is an older design than the knarr. Historical records point to an Irish tradition of long-distance voyaging. According to the Íslendingabók of Ari Thorgilsson, the first Norse settlers to Iceland encountered Irishmen already living there. A famous story of oceanic voyaging by the Irish is the tale of Saint Brendan (circa AD 500), who set out in a boat with pilgrims in search of the Island of Paradise. In the story of Saint Brendan, there are mentions of a number of features that would be unfamiliar to most Irish at the time but that suggest knowledge of the North Atlantic.

In the story he finds an island with a fiery mountain — possibly Iceland; he sees a "coagulated sea," possibly the frozen far northern ocean; he sees a "crystal pillar" — possibly an iceberg. These features are woven among mythical events, such as finding Judas on a cold rock, granted periodic relief from hell. From a Latin manuscript (circa AD 800) of Saint Brendan's travels, *Navigatio Sancti Brendani Abbatis*, there is a description of his construction of a vessel made of wood and covered with skin that coincides well with more recent descriptions of the currach:

> He [St. Brendan] pitched his tent at the edge of a mountain stretching far out into the ocean, in a place called Brendan's Seat, at a point where there was entry for one ship. Saint Brendan and those with him got iron tools and constructed a light ship ribbed with wood and with a wooden frame, as is usual in those parts. They covered it with ox-hides tanned with the bark of oak

and smeared all the joints of the hides on the outside with fat. They carried into the ship hides for the making of two other ships, supplies for forty days, fat for preparing hides to cover the ship and other things needed for human life. They also placed a mast in the middle of the ship, a sail and other requirements for steering a ship.[8]

The currach was a vessel of choice by the Irish, and it has a related cousin in Great Britain called the coracle. The earliest currachs had a wooden frame covered by skins and could be propelled by both human and wind power. Once assembled, the joints were covered with rendered animal fat. We can only look at modern versions of the currach to get some idea of a likely design. They are quite small and lack a keel, which indicates that a minimization of leeway was not a strong factor. With a somewhat narrow, rounded hull, the modern examples indicate that their primary use was in near-coastal navigation. A voyage to Iceland would have been extremely adventurous. Figure 187 shows the hull of a traditional currach from the Aran Isles. With a length of twenty feet, it has a hull speed of 6 knots and a limited displacement.

KAYAK

The kayak (or *qajaq*) is a small vessel used for hunting and transport by the Inuit, in a range from Alaska to Greenland. Norwegian explorer Fridtjof Nansen deserves much credit for popularizing the Inuit kayak in the West. Traditional kayaks were made of seal and walrus hide covering a frame fashioned out of driftwood and bone. The frame was first lashed together with sinew. The hides were stitched into one piece, and this casing was then stretched over the frame and sewn into place and finally waterproofed with seal oil. The construction technique is reminiscent of the currach, although the hull shape is quite different.

The narrow width of the kayak compared to the hull length reduces the yawing motion, making it easy to paddle long distances, and reduces drag. On the other hand, the narrow rounded hull makes it susceptible to rolling. Not surprisingly, the Inuit developed a tremendous skill in paddling. The "brace" is a maneuver that allows the paddler to right the kayak if it begins to tip over. In the extreme case of a capsize, the Inuit could right the craft with little effort. Some Greenland Inuit were so secure in their skills that they would have their covering garments sewn into the cockpit to minimize exposure to cold water.

The example hull shown in the figure comes from a Greenland kayak Nansen brought back to Norway. It's seventeen feet long and has a hull speed of 5.5 knots. Modern versions of the kayak come in all shapes and sizes and are used mainly for sport.

Pacific Island Outrigger

Many sailing vessels used in traditional Pacific Islands voyaging are double-hulled. This design has a number of advantages. Two narrow hulls reduce the susceptibility to a yawing motion, allowing the vessel to track in a straight path. In addition, the separation of the two hulls imparts good stability, with a strong righting force that combats the tendency to roll to one side under the influence of the wind. The outrigger shown in Figure 187 is a design from the Santa Cruz Islands in the eastern Solomons but is typical of many found throughout the Pacific.

When under sail, the outrigger float is kept to windward. Its weight counteracts the force of the wind on the sail, keeping the craft upright. Early accounts from Western Europeans witnessing the outriggers marveled at their capabilities. Estimates of speeds in excess of 18 knots were reported.[9] Although the hull speed of a thirty-foot-long vessel is 7.3 knots, a lightly laden double-hulled craft can readily plane with a strong wind, moving much faster.

These craft are made with several construction techniques. In some cases the hulls are fashioned out of a single log. In other cases the hull construction is reminiscent of dhows, particularly for larger craft. Breadfruit logs and other wood are used to make hull planking. The flexible outer husks of coconuts form caulking, and heated breadfruit sap is used to waterproof the hull. The planks are initially held in place with wedges and palm fronds. Once the sap dried, rope made from coconut husk, called *sennit*, is used to lash the hull planks together. Like the dhow hull construction, the planks do not overlap but fit tight up against each other. The main hull and outrigger are held together with a platform connecting them.

Although the double-hull design has distinct advantages, there are drawbacks, too. In heavy seas large stresses can be placed on the poles and lashings holding the hulls together, with the attendant risk of their separating. One compromise in long-distance voyaging craft is to use two larger hulls tied closer together.

With the exception of the kayak, the hulls described above were used in conjunction with sails to give the mariners substantial freedom to make long-distance voyages. The design of sails, discussed next, is as important in the story of voyaging as hull shapes. The human mastery of the fluid dynamics of wind blowing over surfaces is a remarkable story, particularly considering that a physical theory describing this interaction was elusive until the twentieth century.

16. Against the Wind

· ·

ALTHOUGH THE EARLIEST watercraft were surely powered by humans, this imposed a natural limit on voyaging. Humans can only paddle or row so long before needing rest. If a large crew is required to row a cargo vessel, the crew must be fed and some of its effort must be expended to transport the food that fuels its effort. On the other hand, if the wind can be harnessed to propel a vessel, the ship needs a far smaller crew: one that tends to the sails, the steering, and navigation. A natural limitation on long-distance sailing, however, is the nature of a round-trip journey, which might require a departure or return that goes against prevailing wind.

As I discussed in the first chapter, the Lapita people began an eastward migration into the Pacific Ocean at least thirty-six hundred years ago from the Bismarck Archipelago of Papua New Guinea. The migration goes against the prevailing trade winds, which means that the Lapita had vessels capable of sailing into the wind. Geoffrey Irwin notes that making reconnaissance voyages into the wind is sensible strategy. If a voyage of discovery starts out against the wind, the passage back to the home port goes downwind, and a safe return is more likely.[1] Norse voyages east from Iceland, Greenland, and North America on the other hand bucked the prevailing easterlies. In general, sailors can wait for favorable winds before embarking on a voyage to a known destination, but forays into the unknown without a viable return strategy is a risky proposition.

The ability to sail against the wind implies a sophisticated combination of hull and sail design. Although the implications of windward sailing for navigation might be self-evident, there are subtleties that might elude the casual reader. Even with the cleverest design of a sailing vessel, it will experience a sideways slip when making its way to windward. This is called *leeway*. When

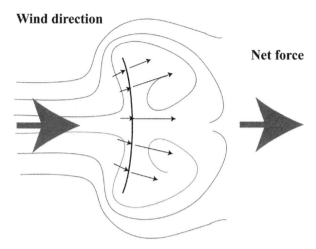

Figure 188 The force generated by a wind directly on a square sail. The reaction created by the wind diverted around the sail gives a net force that is transmitted down the mast into the boat itself.

a navigator on a sailing vessel calculates his position from dead reckoning, he must take the leeway into account to obtain an accurate estimate of position. Estimating leeway is something of an art unto itself. It depends on the craft, the angle into the wind, and the strength of the wind.

Sailing downwind is fairly easy to understand. Anyone who has held an umbrella aloft in a strong wind can appreciate the principle. In Figure 188 I illustrate a wind from behind impinging on a sail. The wind gets diverted around the sail. According to Newton's laws of motion, "For every action there is an equal and opposite reaction." Air has mass, and its diversion around the sail creates a reaction force in the sail that gets transmitted down the mast and into the hull. The reaction force propels the sailing vessel forward.

Leaving aside for now the question of how one designs a sail and vessel capable of sailing into the wind, it's important to note that one can sail only at a finite angle into the wind. The closest angle to the wind a boat can make will depend on a number of factors in the sail and hull design. Having said that, though, I note

Wind

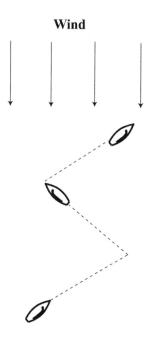

Figure 189   Tacking into the wind.

that the limitation on angle implies that a sailor often cannot take a direct path against the wind to a destination but rather a zigzag path in a process called *tacking* (Figure 189).

While the zigzag path is somewhat trying and time-consuming, it has advantages in reconnaissance. More territory is covered transverse to the overall direction of travel. If a small island lies to windward, a navigator is more likely to make landfall in one of his tacks rather than in a direct downwind path.

Lift

Wings, sails, and rudders have a strong kinship in the nature of the flow of air and water past their surfaces. In general they work by generating a kind of force called *lift* that is used for flight, sailing into the wind, and steering a vessel. To understand how they work, I'll focus on how wings generate lift, then turn to sails. The science of the flow of water and air past objects comes from a more general theory of *fluid dynamics*, where both air and water are considered fluids.

We usually talk about the flow of air as a stream where we can trace paths of air molecules in a predictable fashion. The paths taken are called *streamlines*. This kind of flow is generally called *laminar*. When wind moves rapidly across a surface, or there's a change in speed from one place to another, it sometimes becomes chaotic or *turbulent*: swirling around in eddies. In turbulent flow there are no well-defined streamlines, although we can try to capture some momentary sense of flow with continuations of streamlines. The eddies and churning of turbulent flow will constantly shift. In Figure 190 I show a photograph of a rising column of smoke. When

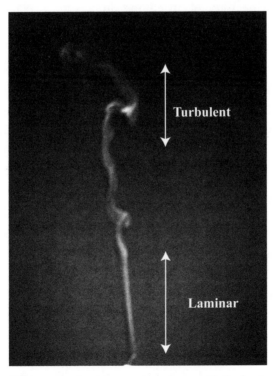

Figure 190   Photograph of a rising smoke column, with the laminar and turbulent flow indicated.

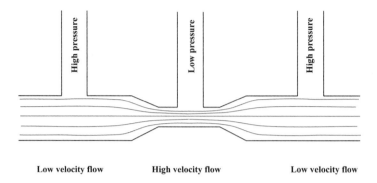

Figure 191 Bernoulli's principle in airflow: when air is flowing fast, it has low pressure; when air is flowing slowly, it has higher pressure.

it first rises, it's laminar; then the flow gets wavy and unstable in transition, and finally, it becomes turbulent.

The driving force of an engine propels an airplane forward. When a wing moves through the air, it generates lift. The airflow over the top and the bottom creates a low-pressure zone on the top surface of the wing and a high-pressure zone on the bottom. The pressure difference between the top and the bottom generates an upward force that lifts the wing. It is easier to first describe the principle of lift for a wing, then describe how it works for sails.

Lift in wings takes advantage of a property of airflow first described by the eighteenth-century Swiss scientist Daniel Bernoulli. In Figure 191 I show airflow through a tube that has a constriction in it. The flow is indicated in the form of *streamlines*. When the streamlines are far apart, the air is moving slowly, but as the streamlines are forced together in the constriction, the speed of flow increases. At the two wide portions of the tube and in the middle of the constriction, the pressure is sampled. Bernoulli was the first to point out that pressure is lower in regions where the airflow is faster and that pressure is higher in regions where the airflow is slower. According to Bernoulli's principle, the pressure in the constricted region of the tube will be lower than the pressure on the wider ends. If I were to attach a pressure gauge

Figure 192 Characteristic shape of a wing with a rounded leading edge and a sharp trailing edge. The attack angle is assumed relative to the direction of flight.

between the constricted and unconstricted regions, I would find a pressure difference between the two regions.

Wings have a curved leading edge and a sharp trailing edge, a design that is crucial in creating lift. When a wing starts moving through the air, a number of transient events occur that ultimately result in a flow pattern that gives lift. The airflow over the top surface of the wing eventually becomes faster than the airflow on the bottom surface. In this steady state condition, Bernoulli's principle tells us that there is a higher pressure on the bottom of the wing than the top and there is a net upward force or lift. The way this airflow develops in the first place is a result of the shape of the wing and other properties of air that weren't fully appreciated until the early twentieth century. To experience lift, a wing needs a slight upward tilt, called an *attack angle*, as shown in Figure 192.

The generation of airflow necessary for lift involves a property of air called *viscosity*. In simple terms viscosity is the "stickiness" of a fluid, and I include air as a fluid here. The higher the viscosity, the stickier a fluid gets; for example, maple syrup and oil have higher viscosities than water. Viscosity plays a crucial role in the way air blows around sails and creates lift in wings. The air right at the

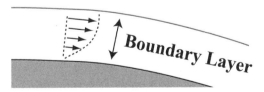

Figure 193 The velocity of air rushing over a surface creates a boundary layer, where the air right at the surface has no speed with respect to the surface, but the velocity grows as one gets farther away.

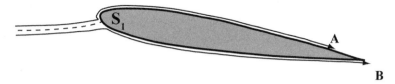

Figure 194 Airflow around a wing when it starts moving. The air moves with the same speed, but the air going over the top has to cover more distance to reach the trailing edge.

surface of the wing sticks to the surface, forming a very thin layer, called the *boundary layer* (Figure 193). Just beyond the boundary layer, the fluid can flow with some speed and form streamlines.

When a wing starts to move, the airflow around it is similar to what is shown in Figure 194. The point S_1 represents the dividing point between air flowing over the top and over the bottom of the wing. This is often called the *stagnation point*. Because of air's viscosity, the air in the boundary layer at the surface of the wing will travel at the same speed whether over the top or over the bottom. The path from S_1 over the top of the wing to the trailing edge is longer than the path from S_1 along the bottom of the wing to the trailing edge. This is just a consequence of the attack angle. We can compare the paths of two air molecules starting at S_1 at the same time, where one goes over the top of the wing and one goes along the bottom. Since they move at the same speed, the moment the air molecule on the bottom has reached the trailing edge of the wing (point B in the figure), the air on top is still somewhere on the upper surface (point A in the figure).

Two things happen to the air flowing over the wing at the trailing edge. If the air had no density at all, the flow on the underside could easily turn the sharp corner and meet the airflow over the top. Once they meet on the top surface, they trail off with another stagnation point on the upper surface, as indicated on the left-hand side of Figure 195. The finite density (or mass) of the air makes it impossible to completely turn the sharp corner of the trailing edge. It "tries to" turn the corner, but like a fast motorcycle rounding a

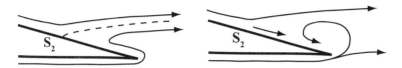

Figure 195 Airflow near the sharp trailing edge of the wing. If there were no inertia or viscosity, the air would turn the corner, but in reality it cannot and crates a vortex.

turn, it can only turn with a finite radius. The air flowing on the top surface likewise sticks to the surface, can't readily peel off, and moves toward the trailing edge, as in the right-hand side of Figure 195. The air from the bottom surface rotates into a *vortex*, like a miniature tornado, which falls behind the wing, allowing the air on the top surface to make it all the way to the trailing edge.

The sharp end of the wing, combined with the viscosity and finite density of air creates a condition where the air flowing over the top and bottom arrive at the trailing edge at the same time. For this to happen the air flowing over the top has to move at a higher speed than the air on the bottom. This is where Bernoulli's principle comes in. The pressure is higher on the bottom side of the wing where the speed of airflow is lower, while the pressure is lower on the top where the air speed is higher. The pressure difference creates the lift of the wing. For this difference in air speed between the top and bottom surfaces to exist, there has to be a net deflection of air downward from the wing. This is shown in Figure 196 on the right-hand side. The net downward displacement of air causes a reaction force of lift. On the left-hand side of Figure 196,

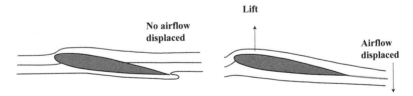

Figure 196 Left: airflow around a wing with zero air density and no viscosity. Right: airflow around the wing when viscosity and density are taken into account.

you can see what the airflow around the wing would look like if air had zero density and no viscosity: there is no displacement of the airflow and no resultant lift.

If lift is the angel, drag is the devil. Both are the products of the same microscopic forces at work that we collectively lump into such concepts as "viscosity" and "density." The stickiness of air flowing over the wing naturally creates friction, which retards forward motion. There's an even more pernicious force that creates drag: turbulence. As long as the boundary layer remains attached to the wing, the flow over the wing is laminar — that is to say, it is regular and can be described by streamlines. If, however, the attack angle increases beyond a certain critical point, the boundary layer will detach from the surface, and a turbulent froth of churning air will take its place. This churning froth is a region of very low pressure and will pull the wing backward. This condition is called a *stall* in airplanes.

SAILS AS WINGS

Sails can also act like wings, particularly if they have a curved leading edge and a sharp trailing edge. A leading beam holding the sail, called a *yard,* combined with the proper sail design can help create a winglike configuration as it takes shape under the influence of the wind and tension placed on it by ropes. The main difference between a wing and a sail is that the derived lift is horizontal. In addition, most airplanes are powered to create lift, while sailing vessels obtain their driving power from the wind itself. In Figure 197 I show the airflow around a sail that acts like a wing. The letter "S" designates the stagnation points where the streamlines split around the sail. The tension of ropes holding sails and resultant shape of sails is called *sail trim* by mariners. The ropes holding the sail ends are called *sheets*, from the Old English *scéata*.

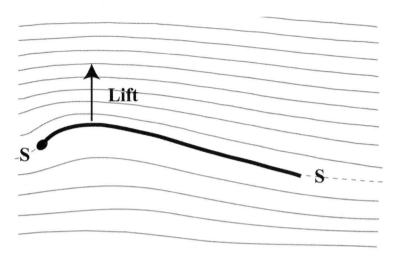

Figure 197 Sail acting as a wing with laminar flow around it. The stagnation points are at the leading and trailing edges of the sail, denoted by the letter S.

Leeway

Creating a winglike shape in the sail is a crucial factor for sailing into the wind, but it isn't everything. The vessel must also be able to resist the sideways force of the wind. In Figure 198 on the right, I show the forces at work on a cross-section of a knarr. The wind is coming from the left side in the figure. The force of the wind tends to roll the ship to the leeward side. You may recall from chapter 15 that the shift of the center of buoyancy with respect to the center of gravity provides a natural righting force. This righting force will counteract the roll caused by the force of the wind on the sail. The ship ends up sailing with some roll where the force from the wind and the righting force from the hull are in equilibrium. This equilibrium angle is called *heel* or *heeling angle*. In high winds the crewmembers are sometimes directed to sit on the windward side of the vessel to shift the center of gravity, reducing its heel. This is also the function of the outrigger on many Pacific Islands sailing vessels.

Figure 198 Forces acting on sailing vessels, seen as a cross-sectional view. Left image credit: C. Mairs, copyright © 2011–12 President and Fellows of Harvard College.

The keel of the ship plays an important role. With a "V" shape at the bottom of the hull, the keel will "grab" the water as it moves. This creates a kind of lock that combats the sideways motion from the wind. If the bottom of the hull were completely rounded, it would have far less ability to resist the sideways push of the wind, but the longer and more protruding the keel, the more it can resist the sideways push. This is indicated in Figure 198.

As I mentioned earlier, the resistance of the hull never fully combats the sideways motion of the ship, or leeway. On the left-hand side of Figure 199, I show the forces at work on a vessel as seen from above. These are:

- Lift from the sail acts roughly perpendicular to the wind.
- Hull resistance counteracts the tendency for the boat to be blown sideways.
- The force from the wind and hull resistance make the boat heel to leeward, but the forces of gravity

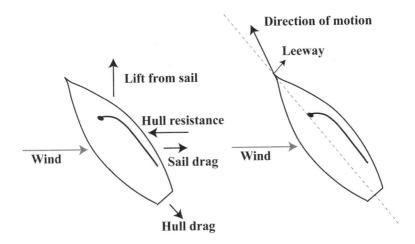

Figure 199   A view from above of the forces acting on a sailing vessel (left) and the resulting motion (right).

on the center of gravity and buoyancy on the center of buoyancy counteract this and keep the boat at a constant heeling angle.

• Aerodynamic drag from the sail and hydrodynamic drag on the hull tend to slow down the vessel.

All of these forces are in equilibrium in a sailing vessel at a constant heading in a constant wind. The resultant motion of the ship is shown on the right-hand side of Figure 199. Although the bow is pointed to one heading, the actual heading is shifted by the effect of leeway.

Like current, the determination of leeway is an important task for navigators. In one account author David Lewis describes how a navigator named Hipour from the Caroline Islands would gaze at the wake trailing a ship to estimate the leeway angle.[2] The use of the ship's wake to determine leeway is shown in Figure 200. The ship is sailing at an angle of about 80 degrees off the wind. Leeway creates a true heading that is 15 degrees off the direction the ship is

Wind direction

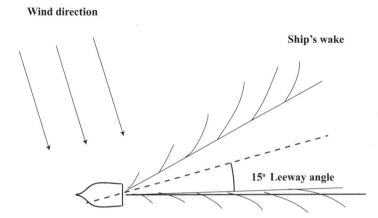

Ship's wake

15° Leeway angle

Figure 200 Determination of leeway angle from the difference between the boat heading and the wake.

pointing. To determine the leeway angle, the navigator gazes back at the wake and can "eyeball" the difference between the direction of the keel of the ship and the angle of the wake trailing behind.

In addition to the angle of the wake, it's been my experience that the sideways slip of leeway creates a region of "slick" water on the windward side produced by turbulence from the keel sliding through water. Both the wake angle and the slick water are noticeable with some careful observation.

Professor Ben Finney of the University of Hawai'i worked with replicas of Tahitian and Hawaiian voyaging canoes and concluded that they could sail at about 75 degrees into the wind. In part, one of the important characteristics that allowed sailing into the wind was the double-hull design, in addition to V-shaped keels that helped give the ships good hull resistance.[3]

Tacking at an angle of 75 degrees makes travel into the wind laborious at best. To gain one mile of true distance, the voyaging canoe would have to sail four real miles on its zigzag course. Often, it made more sense to take advantage of weather conditions and sail during seasons when a voyage could have more of a following wind. High and low pressure systems with the rotational wind pattern about the centers can be used to advantage in this case.

Sail Design

The windward capability of a sail depends on its design. As can be seen from the discussion above on wings and lift, the forces at work that create a sail with windward capabilities can be subtle to grasp, yet through trial and error humans have managed to find sails that do the job. During the eras of the ancient Greek and Roman cultures, most vessels supplemented square sails possessing limited windward capabilities with human power. Figure 201 is a photograph of a mosaic dating from the third century AD. The mosaic depicts the story of Odysseus and the Sirens, where Odysseus is seen lashed to the mast as the Sirens sing to him and his sailors proceed with wax plugging their ears from sound. The sails are rectangular in shape, with one attached to the mainmast and one attached to a smaller foremast. The oars, plainly visible in the mosaic, are used to augment driving power, particularly in times of little or contrary wind.

Figure 201 Representation of square-rigged galley in a mosaic of Odysseus and the Sirens from the third century AD in the Bardo Museum, Tunis. Image credit: Cain Maddern, www.wildfocusimages.com; used with permission.

Figure 202 Mosaic dating from the fifth or sixth century from Kelenderis, Turkey. Shown with permission of Professor Levent Zoroglu. Photo credit: Elif Musdu.

One variety of a sail with good windward performance is called a *lateen* sail. The name is derived from the French word for "Latin" (*latine*), implying a Mediterranean origin. A lateen sail either is triangular in shape or is a quadrilateral with a nearly triangular shape. The winglike characteristics are created by a forward-angled yard that allows the sail to fill with a curved leading surface, while sheets in the rear maintain a sharp trailing edge.

The path from square sails to lateen sails is not fully known, but it may very well be that sailors using square or quadrilateral rigged ships started to tilt the yards forward to gain a windward advantage.[4] Figure 202 is a photograph of a mosaic dating from the fifth or sixth century AD, excavated in Kelenderis, Turkey, by Professor Levent Zoroglu from Selçuk University. The mosaic shows a harbor scene with a large sailing vessel and two smaller vessels trailing behind, attached by ropes. The sail appears to be quadrilateral in shape. According to Zoroglu and nautical archaeologist Zaraza Friedman, this configuration was quite common in the era when the mosaic was made.[5]

An alternative explanation has been offered that this is actually an early example of a lateen sail.[6] Although there is some dispute among scholars about the precise configuration, the contrast between the mosaics in Figures 201 and 202 is striking. The yard in the Kelenderis ship appears to be angled forward, and there is no sign of oars for propulsion, only steering boards on the side. In addition, the bottom of the sail appears to be rolled up, which could be part of a practice called *reefing*, in which the sail area is reduced in high winds.

Another early development in the path to sails with windward capability was the use of a pole to stiffen the leading edge of a rectangular sail by the Norse, called a *beitass*. This is illustrated in Figure 203. The staff, when combined with the sheets holding the sails, created the winglike shape with better windward performance. The ability to sail into the wind appears in the *Saga of the Greenlanders*, when Leif Eriksson and his crew were returning from Vinland:

> When spring came they made the ship ready and set sail. Leif named the land for its natural features and called it Vinland (Wineland). They headed out to sea and had favourable winds, until they came in sight of Greenland and the mountains under its glaciers.
>
> Then one of the crew spoke up, asking, "Why do you steer a course so close to the wind?"
>
> Leif answered, "I'm watching my course, but more than that. Do you see anything of note?"
>
> The crew said they saw nothing worthy of note.
>
> "I'm not sure," Leif said, "whether it's a ship or a skerry that I see."
>
> They then saw it and said it was a skerry. Leif saw so much better than they did, that he could make out men on the skerry.
>
> "I want to steer us close into the wind," Leif said, "so that we can reach them; if these men should be in need

Figure 203 Knarr with leading edge of sail stiffened with a *beitass*. Image credit: C. Mairs, copyright © 2011–12 President and Fellows of Harvard College.

of our help, we have to try and give it to them. If they should prove to be hostile, we have all the advantages on our side and they have none."

They managed to sail close to the skerry and lowered their sail, cast anchor and put out one of the two extra boats they had taken with them.[7]

A *skerry* is a small rocky island protruding above water, coming from the Old Norse word *sker*. The phrase for "close to the wind" in this passage in Old Norse is *undir veðr*; literally, "under the wind." The phrase for "sailing into the wind" in Old Norse and the closest modern version, Icelandic, is "*beita undir veðr*," meaning "biting under the wind."[8] It is as if the sail is a tooth taking a bite into the wind. This phrase found its way into modern sailing English as "beating into the wind," where "beita" became "beat." The pole that stiffened the leading edge of the sail for the knarrs

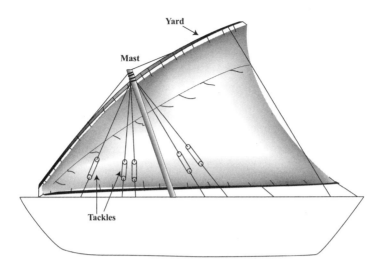

Figure 204   Lateen sail rig.

used by the Norse derives its name from the word "beita": the *beit-ass* is literally a "beat staff."

The above passage also sheds some light on naval tactics understood by the Norse. Leif talks about the advantage of approaching the island from the leeward (downwind) side by sailing into the wind. He didn't know if the people on the skerry would be hostile. If they proved to be hostile, he could simply turn downwind rapidly and sail away. In the British Royal Navy, this was called having the *weather gage*: a tactical positioning with respect to the wind to gain advantage over an enemy.

While the angling of a yard or the stiffening of the leading edge of a rectangular sail enhanced the windward performance up to 70 degrees into the wind, the fully developed lateen sail gave substantial windward performance of roughly 56 degrees into the wind.[9] Figure 204 shows a design typical of a lateen sail used in the Mediterranean and by Arab traders from about the eleventh century onward. The mast itself is sometimes angled forward, and the yard may be straight or curved like a bow. The sail itself is nearly triangular. A curved yard can enhance the leading edge necessary for a winglike shape. These are still used on dhows in the Indian Ocean and along the eastern coast of Africa.

During the Portuguese exploration along the western coast of Africa in the fifteenth century, lateen sails were used on vessels called *caravels*. The use of lateen sails by Western Europeans began to die out toward the end of the fifteenth century, however. One possible reason was their awkwardness. To tack (move from one windward heading to another), the sail had to be lowered, moved to the other side of the mast, and raised in a time-consuming operation. In Western European vessels caravels gave way to vessels with a combination of forward triangular sails and rectangular sails. Multiple sails often improve the windward performance of a vessel, as the flow of air through the sails is directed in ways that a single sail cannot manage.

Even with multiple sails, the square-rigged ships often didn't have as good windward performance as lateen rigs.[10] On long trade journeys ships could take advantage of shifts of prevailing winds with latitude to enjoy voyages that were largely downwind, as in the triangle trade in the North Atlantic.

This didn't spell the end of the use of lateen sails by any means. The Barbary pirates employed swift vessels called *xebecs*, which had multiple masts rigged with lateen sails. In addition, crews of slaves were employed using oars to enhance the speed in attacks. The xebecs had excellent windward performance and menaced merchant shipping off the coast of North Africa in the late eighteenth and early nineteenth centuries.

Pacific Islanders developed their own versions of lateen sails, often woven from the leaves of pandanus trees (a common tree in the central Pacific) and shaped by yards to give them the wing shape necessary for sailing into the wind. One version looks a lot like the Mediterranean/Arab lateen sail but was developed independently. One unique design is a sail called a "crab claw" (Figure 205), which resembles its namesake and has extremely good performance.[11]

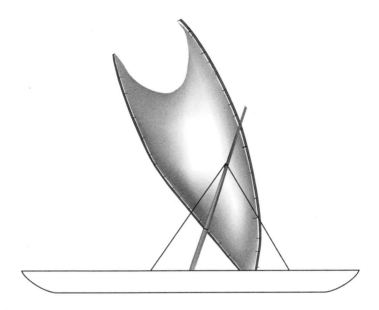

Figure 205 An example of a crab-claw sail.

Pacific Islanders solved the problem of tacking with a lateen sail in a unique way. Their outrigger canoes were built symmetrically so they could sail forward from either end. The outrigger itself always extends to windward to counteract the heeling force of the wind. When a navigator wishes to change direction into the wind, the sail is simply shifted from one end of the canoe to the other without having to go around the mast. This method of shunting the vessel back and forth against the wind is economical and allows for both the good windward capabilities of a wing-shaped sail and ease of use when changing heading.

The End of Sails

The nineteenth and twentieth centuries saw the slow extinction of the use of sails for nearly every purpose, whether it was for transport, fishing, or warfare. Western powers occupying Pacific islands forbade the use of indigenous sailing vessels, often out of a paternalistic concern for safety. There are recent attempts to revive native techniques of canoe building and navigation. In remote areas sailing in support of livelihood is rare but not unknown. In the present era, however, sailing is mainly a sport or hobby. The trend to more efficient sailing vessels carries on, but now with supercomputers that model the intricacies of fluid flow past sails and hulls; this was once the result of trial and error, when the stakes were much higher.

17. Fellow Wanderers

· ·

WE AREN'T THE only ones on a journey. We share paths with other travelers, even if temporarily. If we know their habits or even stop them to ask directions, it can help us find our way. This chapter is about those who wander: what they can tell us about where we are and what they can inform us about where we're going.

The modern English word for planets comes from the Greek πλανήτης (*planetai*), which is a derivative of the word for "wanderers." The brightest planets — Venus, Mars, and Jupiter — can act as temporary beacons for travelers, but they move against the fixed background of stars, as the name "wanderer" implies, and aren't as reliable as stars for navigation. Still, you can use planets for navigation, particularly if you have some understanding of their motions.

Planets often confuse navigators, and it's important to be able to distinguish planets from major navigational stars. Once, when Mars was in Taurus, I confused Mars and the star Aldebaran: they have similar coloration and, at the time, had roughly the same magnitude.

You can use a number of clues to distinguish planets from stars:

- Stars "twinkle," while planets will glow steadily. Refraction and small-scale turbulence in the atmosphere can cross the pointlike image of a star and temporarily obscure it, but planets present a finite-size disk, so they aren't as susceptible to "twinkling."
- Venus and Jupiter are usually brighter than most stars in the sky. The luminosity of Mars will depend on its position relative to Earth in its orbit.
- Planets will always lie close to the ecliptic. Some stars and constellations in the ecliptic include

Aldebaran (Taurus), Pollux (Gemini), Regulus (Leo), Spica (Virgo), and Antares (Scorpius).

The positions of the major planets used for navigation can be found in tables called *ephemerides*. The calculation of the positions of planets by astronomers and astrologers dates back at least four thousand years to the Babylonians. The use of planetary positions for celestial navigation was largely confined to the present era (circa 1600 onward) when accurate tables were available. The use of planets directly in navigation in earlier eras and in other cultures is largely unheard of.

Without an ephemeris a navigator can use planets roughly as direction-finding beacons over the course of a month or so, as they do not change their position in the sky so rapidly. At the start of a journey, a navigator can track the major planets. These can be added to the arsenal of stars used for orientation. Because of the wandering nature of the planets, their coloration, and their unusual paths, it is not surprising that human cultures personified them with the names of gods. Knowledge of the "habits" of the brightest planets can be helpful.

Venus

Venus can be the brightest object in the night sky. It has an orbit closer to the Sun than the Earth, with an orbital period of 225 days. Since it is closer to the Sun than the Earth is, it never strays more than 47 degrees from the Sun. The largest visible angular distance Venus makes from the Sun is called its *greatest elongation*. The proximity to the Sun gives it the names of the "morning star" or the "evening star," as it either precedes the sunrise in the eastern sky before dawn or follows the sunset in the western sky at dusk. When it's at its greatest elongation, Venus will appear to remain in the same relative position with respect to the Sun for weeks.

Rising Venus has been mistaken for airplanes, trains, boats, and even UFOs. These illusions are often the result of the thickness of the atmosphere at low angles. At altitudes below 6 degrees, the atmosphere substantially dims the apparent brightness of Venus, but as it rises it rapidly gets brighter. The visual image of the light on a boat, an airplane, or a train approaching from the horizon presents much the same characteristics: as it approaches, the light gets brighter as its height above the horizon increases.

Air Canada Flight 878 left Toronto on January 13, 2011, bound for Zurich, Switzerland. Some hours into the flight, the passengers experienced a sudden and violent buffeting of the airplane. A subsequent investigation revealed that the first officer had taken a scheduled nap, and when he woke up, he saw Venus in the distance and thought it was a C17 military transport and abruptly decreased altitude, believing he was on an "imminent collision course."[1]

Mars

The orbit of Mars is outside Earth's orbit, with a period of 687 days. Normally, Mars moves eastward along the ecliptic at a speed between 18 and 24 degrees per month, but there are exceptions. When the Earth overtakes Mars in its orbit, the effect is the same as the view of passing a telephone pole in a moving car: its apparent position with respect to the fixed background of stars seems to move backward for a period of time. This is called *retrograde* motion. When Mars enters and leaves retrograde motion, it appears to stand briefly still against the fixed background of stars.

The characteristic red-orange color of Mars gives it away, and it can be quite bright when it is at the point of closest approach to the Earth; I've seen it reflected off a lake at night during one of these close approaches. There is an urban myth that pops up from time to time that Mars, at one of its closest approaches, will look as large as the Moon. Although there's no truth whatever to this, every so often e-mail in-boxes will be cluttered with this "amazing celestial event."

JUPITER

Like Mars's, Jupiter's orbit is also outside of Earth's, but it is considerably farther out and has a period of 4,333 days, making it the most "stable" of the three wanderers discussed here. Jupiter will move eastward through the ecliptic at a rate of about 30 degrees per year. Jupiter has only a slight coloration and can be quite bright in the sky when the Earth is in its closest approach.

BIRDS

Many bird species have well-established habits and highly developed navigational instincts that sailors often took advantage of. Some birds use stars at night to find their way, others apparently sense the Earth's magnetic field. There are number of mythological references to the use of birds in navigation.

MIGRATORY BIRDS

Why make a *deliberate* voyage out of sight of land? It seems like a daring, even desperate thing to do. Author James Hornell speculated that the annual observation of migratory birds departing from land and later returning led navigators to venture off in search of their migratory grounds.[2] Observant mariners see flocks of birds depart from shore across the ocean in the spring, only to return again in the fall. This pattern is repeated year in and year out. Knowing that there must be an unknown land to which the birds flew, an adventurous mariner might be tempted to make the voyage by following their path.

The first settlers of Iceland were Irish monks who were found there by early Norse settlers. In an article, "The Role of Birds in Early Navigation," James Hornell advanced the idea that the monks found in Iceland followed the flyways of ducks that extend from Ireland to Scotland, then north to the Faroes, and thence to Iceland. At least one species, the Greenland white-fronted goose (*Anser albifrons flavirostris*), makes direct migrations between Ireland and Greenland via Iceland. Norse sagas also report the appearance of birds from Ireland showing up in Iceland, so these migrations were well understood by voyagers familiar with the two islands.

Migratory birds have a place in Polynesian folklore. Ethnologist J. Frank Stimson recorded a Tahitian poem called "Pathway of the Birds,"[3] which is about "sea kings" who follow the paths of migrating birds across vast oceans to unknown lands.

The discovery and colonization of Hawai'i by Tahitian sailors is thought to have occurred around AD 1000. The voyage from Tahiti to Hawai'i is roughly twenty-five hundred miles and takes about a month.[4] Anyone voyaging for a month has to be well stocked with provisions and relatively certain of his or her heading. It has long been believed that the Tahitians were guided to Hawai'i by observing the annual migrations of birds. A leading candidate for the migratory bird associated with the colonization is the golden plover (Figure 206), which breeds in Alaska in the summer and migrates south, over Hawai'i, then south to the islands around Tahiti.[5,6]

In Columbus's first voyage to the Americas, his crew was close to mutiny. He was nearing his first landfall in October 1492 at the same time that large migrations of birds were taking place from North America to the islands of the Caribbean. Columbus saw a large flock of migrating birds flying to the southwest. His navigator, Martin Pinzon, said to Columbus, "Those birds know their

Figure 206 Golden plover. Image credit: C. Mairs, copyright © 2011–12 President and Fellows of Harvard College.

business." Recalling that the flight of birds had helped Portuguese sailors locate some of the Azores, he altered course in the direction of the birds' flight, hastening his landfall.[7]

Shore-Sighting Birds

There are tales of the use of birds carried on sailing vessels for navigation. Mariners would let the birds fly free and follow them to land. Both ravens and doves are mentioned in a number of these.

A story of a great flood is common to many religions. One of the earliest recorded flood stories is the Gilgamesh Epic, as depicted on a set of tablets found in Babylon. On the eleventh tablet is a description of the great flood, where the hero releases three birds from shipboard: a dove, a swallow, and a raven. The first two birds circle back to the ship, then the raven departs, signaling the presence of land:

When a seventh day arrived
I sent forth a dove and released it.
The dove went off, but came back to me;
no perch was visible so it circled back to me.
I sent forth a swallow and released it.
The swallow went off, but came back to me;
no perch was visible so it circled back to me.
I sent forth a raven and released it.
The raven went off, and saw the waters slither back.
It eats, it scratches, it bobs, but does not circle back to
  me.
Then I sent out everything in all directions and
  sacrificed.[8]

In the Judeo-Christian version of the flood legend, Noah first sends forth a raven, then a dove. On the first departure the dove returns to the ark, having found no land. Seven days later Noah sends out the dove again. It returns with an olive branch. Seven days after this, Noah again releases the dove. When it doesn't return, it is taken as a sign that land is near.

Doves and pigeons are from the family *Columbidae*, and many are known for their homing abilities. Carrier pigeons in particular are impressive avian navigators, used to carry messages and for homing by humans for a very long time. Use of trained carrier pigeons persisted through World War I but tailed off with the development of cheap communication via telegraph and telephone lines. The origin of their homing abilities is still the subject of investigation; one leading hypothesis is that they are able to sense Earth's magnetic field.[9]

Ravens appear in Norse literature as shore-sighting birds. The settlement of Iceland is described in the *Landnámabók* (Book of Settlement). It is said that the first Viking to reach Iceland was named Naddoddr. Then a sailor, Gardar Svavarsson, was swept by a storm to Iceland, where he overwintered. The *Landnámabók* describes the subsequent, more purposeful voyage of Floki Vilgeroarson, who took three ravens with him. Floki earned the name Hrafna Floki (Raven Floki) in Icelandic legend: "Floki took three ravens with him to sea. When he set free the first, it flew aft over the stem; the second flew up into the air and back to the ship again; but the third flew forth straightway over the stem, in the direction in which they found the land. They hove in from the east at the Horn, and then they coasted the land by the south. But as they sailed west round Reykjanes, and the firth opened out to them, so that they saw Snæfellness, Faxi [a man who accompanied Floki] observed, "This must be a great land which we have discovered, and here are mighty rivers."[10]

Pliny the Elder describes a navigational practice of mariners from the island of Taprobane: they rely on shore-sighting birds to find land. In *Natural History* in the first century AD, Pliny writes: "[In making sea-voyages, the Taprobane mariners] make no observation of the stars, and indeed septentriones is not visible to them, but they take birds out to sea with them which they let loose from time to time and follow the direction of their flight as they make for land. The season for navigation is limited to four months, and they particularly shun the sea during the hundred days which succeed the summer solstice, for it is then winter in those seas.[11]

Taprobane was the name given in classical Greece and Rome to the present-day island of Sri Lanka. You may recall from chapter 7 that *Septentriones* was the Latin name for the Big Dipper but was also synonymous with "north." At roughly 5 degrees N latitude, the Big Dipper would be visible for some part of the time from Taprobane, although "north" is barely visible.

Unfortunately, Pliny is not an unimpeachable source on this point, as has been pointed out to me by Frank Reed.[12] Immediately following the above passage, Pliny goes on to describe the visit of three ambassadors from Taprobane to Rome. During their visit they marvel at shadows that actually point north and the Pleiades, which they say they cannot see from Taprobane. This is clearly nonsense. The Pleiades, at a declination of 24 degrees N, is visible as far south as at least 60 degrees S latitude. Pliny was evidently snookered by the "ambassadors" and reported this without any intellectual filter.

One of the earliest accounts of shore-sighting birds can be found in the Dialogues of the Buddha in the "Kevaddha Sutta" of *Digha*:[13] "Long ago ocean-going merchants were wont to plunge forth upon the sea, on board a ship, taking with them a shore-sighting bird. When the ship was out of sight of land, they would set the shore-sighting bird free. And it would go to the East and to the South and to the West and to the North, and to the intermediate points, and rise aloft. If on the horizon it caught sight of land, thither would it go, but if not, it would come back to the ship again."

According to author Harold Gatty, Polynesian navigators used frigate birds for sighting shore in much the same way Floki employed ravens to find Iceland.[14] Because of their lightness and tremendous wingspan, frigate birds can cover large distances over the ocean and can fly high enough to sight distant islands.

HOMING BIRDS

Many species of land-based birds will fly out to sea, fish, and return to land. Boobies, frigate birds (Figure 207), and pelicans are three species in the Pacific Islands that show this behavior. Sometimes, boobies are spotted as far as seventy-five miles away from an island. Very low atolls might be visible from only five miles away; the sighting of birds six or seven times that distance away dramatically increases the ability of navigators to find land.

These birds, sometimes called "homing birds," should be distinguished from pelagic birds, such as the albatross or the petrel. Although the pelagic birds ultimately find land to nest on,

Figure 207 Left: frigate bird. Right: brown boobie. Image credit: C. Mairs, copyright © 2011–12 President and Fellows of Harvard College.

they wander over large stretches of the ocean, and their presence is unlikely to be any clue to nearby land, unless it is during their nesting season.

As I have written in chapter 4, there can be an accumulation of uncertainties in position over the course of a long voyage. Small low-lying islands might be visible only five miles distant, making them tiny targets and relatively easy to miss. If, on the other hand, a navigator uses the behavior of known land birds as an aid, this can readily expand the effective "size" of the island target. The navigator proceeds toward his target destination based on all factors he can account for: heading, distance, leeway, and current. At some point, he comes across frigate birds feeding, but there is no land in sight. He drops sail and waits patiently while the birds feed. At some point he observes the birds flying deliberately in one direction, which he figures must be their home island, and sets a course in that direction until land is visible.

This concept of an expanded target of landfall is illustrated in Figure 208, where the visible range is shown as being substantially smaller than the range of feeding land birds. Author David Lewis estimates that both boobies and frigate birds will feed as far out as thirty to fifty miles from land.[15]

You may recall from chapter 2 that Caroline navigators divided a voyage into a series of *etaks*, which represent the position of a "reference island" against portions of the horizon at which certain

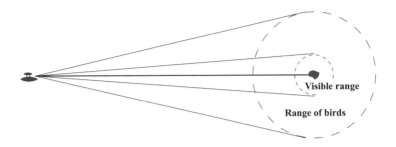

Figure 208 The principle of expanded target landfall. An inherent uncertainty in position arises on long voyages. By being able to increase the effective "size" of an island through the sighting of land-based birds, the navigators only have to get as close as thirty-five miles to an island.

stars would rise or set. As a voyage progressed, the navigator would mentally keep track of which etak he was in. The last etak in a voyage was called the "etak of birds," as this is where the navigator would expect to see birds associated with the target island.

In another Tahitian chant translated by J. Frank Stimson called "Land Sighted," references are made to both land (homing) birds and sea birds:[16]

> Land Sighted
> Lookout
> Watching, ever watching!
> Keeping a sharp lookout,
> a sharp lookout!
> Here am I upon the crosstrees
> watching for the sight of land birds;
> Keeping a sharp lookout!
> Captain
> Yours are the eyes of the goddess of the dark oceans,
> Yours are the eyes of the god of the sunlit skies;
> Theirs are the eyes keeping a sharp lookout!
> Lookout
> Darkened is the sky about me
> with the flocking of sea birds about the tall prow,

The sky is darkened about me!
Captain
The gods are watching through your eyes;
Theirs are the eyes keeping a sharp lookout!
Lookout
Now the far cries of land birds are heard as they
     swoop
into the troughs of the waves upon the horizon.
Ha! Now they settle upon a low-lying reef;
The cries of land birds are heard as they dive
into the troughs of the waves!
Now they come to rest upon the land
rising above the ocean's rim!
Captain
What is the land you have made out,
The land you have sighted?
Lookout
It is the Land of clear waters,
It is indeed our homeland, Land of peaceful waters.

Ships at Sea

Ships at sea can provide clues about position to a navigator. Imagine that a sailor is approaching an island and spots a small fishing vessel, ill-suited for the high seas. A navigator could wait until the fishermen had pulled up their nets, then follow the boat back to land, much like following a homing bird. More to the point, if you came across a ship, you might hail them and ask for directions.

In 1877 entrepreneur Thomas Crapo set out from New Bedford, Massachusetts, with his wife, Joanna, in a nineteen-foot-long boat, to become the record holder for a transatlantic crossing in the smallest boat. Crapo made no provision for navigation except to find ships on the crossing that he could hail and ask about his position.[17] The Crapos eventually made landfall in Penzance, England, after two months at sea. The custom of ships stopping each other on busy shipping routes was fairly common in the age of sail, and ships' logs are full of references to these stops, during which, among other things, information about positions at sea was compared.

Even the mere sighting of other ships at sea can give a navigator clues to position and orientation. Major transoceanic traffic is routed to give the shortest possible distance between ports. In many cases an oceangoing vessel follows a path called a *great-circle route*. A great-circle route can be distinguished from a rhumb line in the following way: A rhumb line is a course with a constant heading with respect to true north; a rhumb line, however, is not the shortest distance between two points. The shortest-distance path between two points on Earth would be what you would get by connecting those two points on a globe with a string and stretching it tight. In Figure 209 I show a great-circle route that connects New York City with London for air traffic.

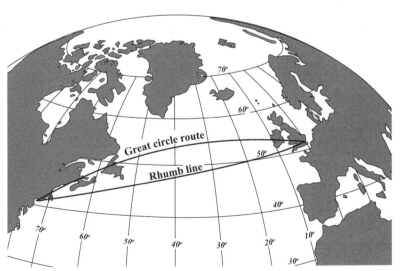

Figure 209   A great-circle route from New York City to London, compared with a rhumb line.

An airplane isn't constrained by coastlines, but shipping is. Figure 210 indicates some major oceanic shipping routes. A number of features come into play in these routes: connections between major ports, great-circle paths, the constraints of coastlines, large canals linking oceans, and deep-water straits. In the figure the great-circle routes between ports on the Atlantic and Pacific are readily seen. Canals create a number of choke points; notably, the Panama Canal and the Suez Canal. The Strait of Malacca between the Malay Peninsula and the island of Sumatra is another major choke point in oceanic traffic, and the Strait of Gibraltar likewise carries a high density of shipping in and out of the Mediterranean. Finally, the southern coast of Africa has a concentration of traffic between the Atlantic and Indian Oceans.

Not surprisingly, some of the regions where ocean traffic concentrates are the targets of pirates. The region around the Straits of Gibraltar was an area haunted by the Barbary pirates. The opening from the Red Sea to the Arabian Gulf along the Horn of Africa is home to present-day pirates from Somalia. Shipping in the Strait of Malacca is also vulnerable to piracy.

If a navigator has some sense of ocean traffic patterns, it can supplement his or her knowledge of position fixing, giving clues to either latitude or longitude. Here are three examples:

Lin and Larry Pardey   While crossing the Bay of Bengal, Lin and Larry deliberately sailed a course well to the south to stay out of the crowded shipping lanes between Sri Lanka and the Strait of Malacca. Unknown to them, a large low-pressure disturbance was creating a current that drew them back north. When they continued to see large ships, they realized that something was wrong and eventually used a combination of celestial fixes and the appearance of the vessels to deduce that an anomalous current had set in, pulling them to higher

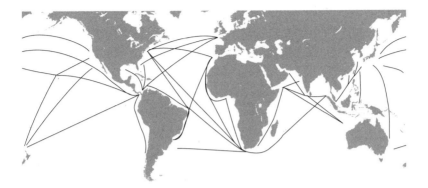

Figure 210 Major ocean traffic routes.

latitudes, and they took evasive action.[18] (See also page 361 in chapter 14.)

Steven Callahan   Steve Callahan set out on a solo sailboat crossing from the Canary Islands toward Antigua in January 1981. Several days out, during a storm, his boat was badly damaged in a collision with an unknown object. He was forced to take refuge in a life raft and drifted west in the North Equatorial Current for seventy-six days, making landfall near Guadalupe. During his drift he sighted a number of ships along the way, which he tried unsuccessfully to hail. Despite the lack of help, he was able to estimate his longitude and hence his progress to the west from the pattern of ocean traffic. He knew that he was approaching the Caribbean when he began to see oceangoing traffic on a north–south path between North and South America. He describes his ordeal in *Adrift: Seventy-Six Days Lost at Sea.*[19]

William and Simone Butler   William and Simone Butler set out on a global circumnavigation from Florida in 1989. After crossing the Panama Canal, they sailed west into the Pacific. Twelve hundred miles out, their sailboat was battered by a pod of whales and sank.

William and Simone escaped into a life raft and drifted to the east in the equatorial countercurrent, which brought them closer to land. Along the way they saw a number of shipping vessels heading to and from the Panama Canal along the great-circle routes across the Pacific. Like Callahan, they tried unsuccessfully to hail the vessels but were still able to use these as a measure of position. This is documented in their book *66 Days Adrift*.[20]

During the day oceangoing vessels are fairly easy to see at distances of up to several miles. On the other hand, a boat or a life raft is unlikely to be spotted from a large ship. Small vessels at distances beyond half a mile on the ocean are difficult to see. People on the bridge of container and fishing vessels usually aren't on the lookout for smaller craft on the high seas, and it is quite likely that their heading is directed by autopilot. To a large extent this explains why Callahan and the Butlers weren't seen by the ships they spotted.

At night the passage of oceangoing vessels can be identified by their *navigation lights*, which are based on an international standard. This is shown in Figure 211. The port, or left side, of the vessel is illuminated in the front and to the side by a red light.

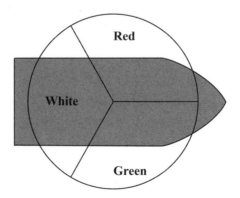

Figure 211 Convention for navigation lights on marine vessels.

The starboard, or right side, of the vessel is illuminated with a green light to the front and side. The rear sector is illuminated by a white light, although the white light is sometimes visible from 360 degrees if it is on top of a mast.

Oftentimes, the red, green, and white lights on a vessel can be simultaneously visible and give the viewer some idea of the heading of the vessel. If a viewer sees a red light on the right and a green on the left, the vessel is approaching. On the other hand, if a viewer sees red on the left and green on the right, the vessel is moving away. If only green is visible, the vessel is moving to the right, and if only red is visible, the vessel is moving to the left. Mnemonics abound for marine navigation. To remember the left/right orientation of navigation lights, one can use: "Port wine is red." This helps recall that the red light is on the port (left) side of a vessel.

Airplanes

Airplanes, like birds and ships, can be used in some circumstances as an aid to navigation. Airplanes can often be identified during the day by their contrails high in the sky, particularly if a warm front is approaching and the contrails linger. Jet airplanes at cruising altitudes of thirty thousand to forty thousand feet are difficult to spot by themselves, but even a short contrail will help find them. At night the same conventions for navigation lights on marine vessels are used for airplanes (Figure 212). An airplane's direction of motion can be identified at night by navigation lights on the wings. There are red and green navigation lights aimed forward on the ends of the wings (green on right, red on left) and flashing white strobe lights that point backward.

Planes can only be useful as directional indicators if you know something about their schedules and routes. In some areas, such as the central United States, the flights are so random that there's

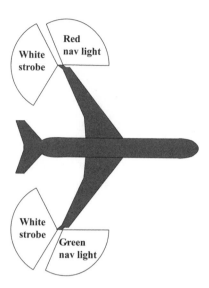

Figure 212 Navigation lights on a modern jet. White strobe lights flash off the rear of both wings. Normally, a solid green light is on the right side of the plane and a red light is on the left side of the plane.

little useful information when one is spotted. On the other hand, in less populated areas, the flights are more predictable. One region is New England and the Maritime Provinces of Canada. Planes leaving major population centers on the East Coast of the United States for Europe typically depart in late afternoon and early evening, following the kind of great-circle route shown in Figure 209. At this time there are large numbers of transatlantic jets that follow a southwest to northeast path on their outbound journey.

Conversely, incoming jets returning from Europe pass over the Maritimes and New England in the mid- to late afternoon. Figure 213 shows a satellite image from the U.S. National Aeronautics and Space Administration (NASA) of jet contrails appearing over Nova Scotia with the characteristic SW–NE orientation. I call this time the "international rush hour" because of the large volume of airline traffic overseas.

Other regions of the world where the paths of planes can be used for orientation are in the Pacific Northwest and the Gulf of

Figure 213   Jet contrails over Nova Scotia and the Bay of Fundy from airplanes flying from Europe to the East Coast of North America, captured in a satellite photograph. Image source and credit: NASA/GSFC.

Alaska, where flights to and from Asia on great-circle routes can be seen. In the northeastern Pacific flights between the West Coast cities of North America and the Hawaiian Islands are common and can also serve as directional indicators.

Knowledge of the missions of aircrews can be used to obtain a relatively good fix of location. This is akin to knowing the habits of birds. An example of the use of aircraft to find position was an episode during World War II. Louis Zamperini, Francis McNamara, and Russell Allen Phillips crash-landed in their B-24 bomber 850 miles west of Oahu, Hawai'i, escaping into a life raft. The three drifted west in the North Equatorial Current but at an unknown rate. They knew that they were drifting in the direction of the Marshall Islands but did not know how far these islands lay to the west.

One day a Japanese bomber appeared in the sky, made multiple strafing runs against the raft, and departed. Miraculously, none of the three occupants was wounded, although the raft required immediate repairs. Phillips and Zamperini set about finding their longitude from the timing of the attack. They surmised that the bomber came from a Japanese base in the Marshall Islands and probably departed soon after sunrise in the morning. The time of the appearance of the aircraft gave them a likely distance from the Marshalls, and a longitude. The difference between the longitude where they ditched and their location gave them a drift rate that they used to estimate the date of their landfall in the Marshalls. Zamperini and Phillips predicted their landfall with a precision of one day after 47 days adrift.[21]

THE MYSTERY OF UNDERWATER LIGHTNING

There is a strange phenomenon that goes by the name of *te lapa* in the Santa Cruz Reef Islands in the eastern Solomons. It has been described as "underwater lightning" and, by the reports of indigenous navigators, points the way to distant islands at night. David Lewis, author of *We the Navigators*, describes it: "It comprises streaks, flashes, and momentarily glowing plaques of light, all well beneath the surface. Exactly like lightning, it flickers and darts and is in constant motion. It occurs a good deal deeper down than common luminescence, at anything from a foot or two to more than a fathom."[22]

Lewis interviewed three navigators from widely separated Pacific regions who all described a similar phenomenon that points in the direction of land. It's usually found as far out as a hundred miles from land.

"Underwater lightning" was reported to Lewis by a navigator in the Gilbert Islands (part of the Republic of Kiribati). The Gilbert Islands are twelve hundred miles away from the Santa Cruz Islands. Abera, the navigator Lewis interviewed from the

Gilberts, called it *te mata* and gave a similar account to that given by the Santa Cruz navigators: "This is not the phosphorescence caused by a canoe's wake, but it is best seen when a canoe is traveling very slowly. Te mata moves; the longer movement away from land, the shorter toward it. It shoots out quickly in one or other of the directions rather than going back and forth. It is like lightning. We see it about eighteen inches below the surface and lower down. When land is nearby, there is a lot of phosphorescence about; this is nothing at all to do with te mata and is of no use whatsoever for indicating direction of land."[23]

Lewis also interviewed a navigator from Tonga, approximately fourteen hundred miles from either the Santa Cruz Islands or the Gilbert Islands, who spoke of a similar phenomenon, but called it *ulo aetahi*, the "glory of the seas." *Ulo aetahi*, like *te mata* and *te lapa*, is described as a flashing, darting light in the seas that points in the direction of land.

The light churned up in a canoe's wake is the result of bioluminescence from a single-celled organism called a *dinoflagellate*. If a large pressure disturbance, such as a canoe wake, or a wave crashing on a beach occurs among a dense concentration of dinoflagellates, a strong emission of light can be seen. Presumably, when the navigators interviewed by Lewis talked about "ordinary phosphorescence [*sic*]," they meant the bioluminescence emitted by dinoflagellates.

There is one well-known example of navigation with dinoflagellate bioluminescence. Astronaut James Lovell, who commanded the ill-fated Apollo 13 spacecraft, used the light from these creatures to find his way to safety. In 1954 Lovell took off from the aircraft carrier USS *Shangri-La* in the Sea of Japan. His electronic navigation systems failed, and he had no way of determining where the Shangri-La was located. After switching off lights in his cabin, he noticed a glowing trail kicked up in the wake of the aircraft carrier. He then followed this trail of light churned by the propellers and landed.[24]

Western navigators wrote reports of a phenomenon similar to underwater lightning when approaching land. In 1727 Captain George Shelvocke published *A Voyage Around the World by Way of the Great South Sea*, where he describes a strange lightning in the direction of the coast of Brazil that resembles the descriptions of *te lapa* and hints that this knowledge was widespread: "I must observe to you, that as you advance towards the coast of Brazil you'll see in the night a sort of faint lightning, flashing and playing (if I may so call it) in part of the Horizon extended over the coast; when you begin to observe this, you may assure yourself that you are not above 25 leagues from the land; thus I found it, and this is the general remark (as I was informed) of all the Portuguese Pilots."[25]

Marianne George, a collaborator with David Lewis, also reports seeing *te lapa* on a number of occasions, under the tutelage of a master navigator, Chief Kaveia.[26] George writes: "My eyes could see that there was a beginning and end of the line of light bolts coming toward me. It happens so fast — in just a fraction of a second — that it is not easy to see or describe."[27]

A recent investigation of te lapa was carried out by Professor Richard Feinberg from Kent State University in the Santa Cruz Reef Islands of Vaeakau and Taumako, where Lewis and George had encountered the phenomenon. Feinberg interviewed a large number of indigenous sailors in the region who largely corroborated Lewis's and George's descriptions of te lapa.[28]

In many of Feinberg's interviews, however, the character of te lapa and its use seems to vary from one subject to another. Feinberg himself set out in a boat with a master navigator, Clement Teniau, in the Reef Islands at night to see if he could personally document te lapa under what seemed to be good viewing conditions. Although Feinberg observed plenty of the normal bioluminescence, he did not observe te lapa. Feinberg says that he felt as though he had been searching for the abominable snowman[29] but does not completely dismiss it.

If te lapa *is* a figment of the collective imagination, it wouldn't be the first time that scientists chased after strange lights. There are other reports of inexplicable lights appearing at night, notably the will-o'-the-wisp or jack-o-lantern. This is a ghostly light that is reported to appear floating above bogs or marshes at nighttime and has a folkloric history describing its origin.

If, on the other hand, te lapa is real, what underlying processes create it? In a recent paper George offers some conjectures. Among the ideas she advances is the possibility that bioluminescence could be the original source of light but that the patterns of waves and swells between islands could focus the light like lenses do to produce transient patterns that point toward islands. Another possibility she advances is that te lapa is light associated with tectonic activity.[30]

Like Feinberg and George, I became curious about the origins of te lapa. My initial guess was that it might be associated with some form of bioluminescence that might readily be distinguished from the normal sources one finds in a canoe wake or crashing waves. I consulted with a colleague at Harvard University, J. Woodland (Woody) Hastings, who is an expert in the bioluminescence of dinoflagellates. I showed him the writings of David Lewis on the subject and asked him if it were possible that dinoflagellate light emission could produce such a result. Professor Hastings said that he had seen something that matches the description of underwater lightning when fish dart. A lot of times the "normal phosphorescence [*sic*]" reported by the navigators are the result of a slow churning of water, but when dinoflagellates are subjected to an abrupt pressure wave, they will emit light on a fast timescale.

Professor Hastings took me to his laboratory, where he grew large flasks of *Pyrocystis lunula,* a species of dinoflagellate. It took about ten minutes for my eyes to become adapted to the dark, and then I could see the light from the small creatures. Dinoflagellates will spontaneously emit flashes of light that look like tiny electrical sparks. In addition, if they are hit with a pressure wave, a whole population of them will flash light. Dinoflagellates have their own circadian rhythm: a sleep/wake cycle associated with day and night. They emit light during their night cycle.

It's not completely understood how the light emission helps these organisms survive, but one hypothesis is the so-called "burglar alarm" theory. In the burglar alarm model, if a dinoflagellate is near a shrimp or other small creature trying to eat it, the pressure wave causes light to be emitted. When this happens, it signals to nearby predators that might be interested in eating the shrimp that food is about. The shrimp is rapidly eaten by the larger predator, sparing the dinoflagellate. Whatever the mechanism, it is quite possible to get rapid bright flashes from dinoflagellates when fish dart underwater, and this matches a description of "underwater lightning." In my discussions with a number of divers and marine biologists from Woods Hole Oceanographic Institute, they confirmed to me the description of Professor Hastings of a kind of bright directional flash associated with fish darting in water rich in dinoflagellates.

I wanted to put this to the test in a controlled setting. Professor Hastings gave me a small sample of his dinoflagellates. I took them home and read some of the literature on their cultivation. Over the course of several months, I cultivated four liters of seawater populated by a large concentration of *Pyrocystis lunula* in my basement. To best imitate the motion of a darting fish in a controlled setting, I figured that a wobbling, segmented fishing lure might provide a reasonable approximation. Dr. Wolfgang Rueckner, at the Harvard University Science Center, graciously helped me devise a setup in a darkroom, where we used a long (approximately three meters) clear plastic trough, which I filled with the dinoflagellate-rich seawater. We waited for the dark cycle of the circadian rhythm of the dinoflagellates, then drew the lure through the trough. The resulting light emission indeed looks a lot like what I imagine "underwater lightning" might resemble based on the descriptions I read. The curious reader can examine the flash we recorded on YouTube, under the search phrase *underwater lightning project.*[31]

Now, the existence of a lightninglike flash of light from a darting fish does not, by any means, prove that this is the origin of te lapa, but it provides at least one plausible mechanism. The astute reader may now be wondering how darting fish could provide any directional information, and I must confess that I do not have a clear answer. Large predatory fish are known to swim toward islands at night, so it's possible that the flashes are associated with these migrations. The origins of te lapa are far from clear, but at least this is conjecture that can be tested. If te lapa is real, then it is something that presumably can be captured with sensitive imaging equipment. The spectral composition of the light can be compared with the known distribution of light frequencies emitted by local populations of dinoflagellates. This would be a major step in unraveling the mystery.

18. Baintabu's Story

• •

NAVIGATORS DON'T EXIST in isolation from society but are part of its fabric. The techniques of navigation don't exist as separate entities but have to be woven together during a journey. Below is a fictionalized account based on the legend of a female navigator named Baintabu from the Gilbert Islands. I try to show how the elements of culture, navigation, and voyaging might come together. Baintabu's story culminates on a voyage from Tarawa Atoll to Abemama in the present-day Republic of Kiribati.

French priest Ernest Sabatier recounted the legend of the woman navigator Baintabu from the Gilbert Islands in his book *Astride the Equator.* According to Sabatier, she accompanied a raiding party from her home atoll of Abemama to neighboring Tarawa. On the return journey from Tarawa, she was unceremoniously thrown off the lead canoe and rescued by the last canoe in the flotilla. Only the last canoe with Baintabu on board made it back to Abemama. Sabatier puts the date of this incident at roughly 1780. We don't know if the story as related by Sabatier is mythology or is based on real events.

A similar story was told by Capelle, a long-term resident of Jaluit Atoll in the Marshall Islands, to a Captain Winkler.[1] According to Winkler, around 1830 a flotilla of one hundred canoes set out on a voyage. All the canoes were destroyed, except for one, which had the chief's daughter onboard. Jaluit is five hundred miles northwest of Abemama.

The fictionalized version of Baintabu's story below is based on reports of the cultural and navigational traditions of the Gilbert Islands from a number of sources. By the time of Sabatier's reports (1938), much of the native culture had been encroached upon by Western influence, but Sabatier recorded some of the remaining vestiges and oral histories of the voyaging culture. Sir Arthur

Grimble (1888–1956) wrote extensively about the culture and navigational strategies of the Gilbert Islands in a series of articles. R. G. Roberts investigated the dynastic history and caste structure of the atoll of Abemama.[2]

Many of the navigational strategies attributed to Pacific Island traditions were put to the test in a series of experimental voyages starting in the 1960s, first by David Lewis, then by members of the Polynesian Voyaging Society, founded in 1973. In appendix 4 I relate some of the assumptions used to create a plausible voyage from Tarawa to Abemama.

BAINTABU'S STORY

The Gilbert Islands

The Gilbert Islands is the name of a cluster of atolls in the Republic of Kiribati that straddles the equator, formed on the tops of long-dead underwater volcanoes. The ocean is an overwhelming presence to the low-lying islands. At times large swells at high tides seem like an irresistible force that will take over an island by assault. The trade winds that blow through from the southeast can be as ferocious as the ocean, bending coconut trees over. In combination the sound of the waves crashing on the beaches and the wind whipping through the trees is like a monster tearing at the feeble atolls.

The Warriors of Beru

Around AD 1400 a group of warriors from the southern Gilbert Islands raided a group of atolls in the northwest. In the southern Gilberts one island, Beru, was overpopulated, putting a huge strain on their resources. Two men from Beru, named Kaitu and Uakeia, put together a fleet of thirty-seven canoes to conquer islands to the northwest.

The warriors from Beru eventually captured the islands of Abemama, Abaiang, Marakei, Tarawa, Maiana, Kuria, and Aranuka (Figure 214), and many inhabitants now trace their ancestry back to these founders. After the conquest of the islands, there was no central authority, however, and villages coexisted semipeacefully, although battles broke out between clans. For a long time the only authority on the atolls came in the form of local chiefs of the major villages.

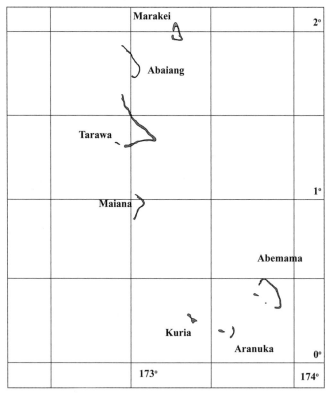

Figure 214 Map of the atolls in the Gilbert Islands conquered by the warriors from Beru.

Abatiku

Baintabu's clan lived on the island of Abatiku, the westernmost island on the Abemama Atoll (Figure 215). Abatiku has its own founding myth. According to legend, three sisters lived in a line of lands to the west at a time when the Earth had not fully separated from Heaven. The youngest of the sisters quarreled with her father and mother, left them in anger, and found the seed of a magic tree. She planted the seed, and after a short time it grew high and reached Heaven. The youngest daughter climbed the tree and came before a woman named Nei Ni-Karawa (Woman-of-Heaven), who took her in. The girl grew up, eventually married, and had a child. Her child was wandering about one day and saw a pandanus tree (a fruit tree found throughout the Pacific Islands).

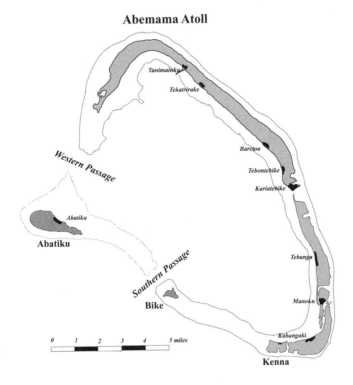

Figure 215 Map of Abemama Atoll. Dark regions are villages.

The child announced to her father that she was going to climb the tree to pick fruit, but her father admonished her, "A woman climbs no trees." But the child replied, "I shall indeed climb it." As she climbed the tree, she climbed out across a branch for fruit, and the branch broke. She fell to Earth, landing on Abatiku, where she met the only inhabitant, a man named Na Utonga, whom she married and with whom she had children. The children turned into great seafarers, and the tradition of seafaring was born on Abatiku.

In reality the first inhabitants of Abatiku were descended from the Beru warriors. The eastern atoll-facing side of Abemama was peppered with villages, large and small, up and down the coast. As the population grew, fights among rival clans in the villages broke out. Baintabu's ancestors, who were less inclined to fight, left the main islands for the more remote Abatiku. While the eastern side of the Abemama atoll was a relatively bustling metropolis, Abatiku developed a strong maritime culture in which the best canoes were made and the tradition of navigation far exceeded that of the more densely populated regions of the atoll. Abatiku became the Nantucket of Abemama.

With a smaller population density than the rest of Abemama, Abatiku had more natural resources and was blessed with abundant stands of trees, furnishing timber for canoe making. Abatiku was sheltered from the strong eastern swells in the lee of the atoll, giving it a peaceful air. In contrast to the narrow islets of the eastern side of the atoll, the island was wide giving an ample supply of fresh groundwater.

The culture and mood of Abatiku was more austere than that of eastern Abemama; the men were fishermen, sailors, canoe builders, and navigators. They did not think of themselves as warriors. Baintabu's grandfather honed the arts of navigation and made many journeys to the reefs of Maiana, Tarawa and Aranuka, returning with large numbers of fish and sea turtles. He was the best canoe builder on all of

Figure 216 The hut with no roof thatching constructed by Baintabu's father and grandfather.

Abemama and was intimately familiar with the waters surrounding all the atolls originally conquered by the Beru warriors.

Baintabu's father learned the arts of navigation from his grandfather, and together they constructed an enclave some distance from the main village that had two canoe-building houses, a strange-looking hut with no thatching on the roof (Figure 216), and arrays of stones laid out on the beach. Men from the village would come during the stormy season (November to March) and help construct canoes but could not divine the meanings of the strange hut, nor the stones on the beach. These were the secrets of the navigators.

THE RISE OF TETABO

When Baintabu's father was very young, a minor event occurred on the eastern islands of Abemama that had a far-reaching effect on the lands captured by the Beru warriors. A young couple from the village of Tabontebike (see Figure 215) traveled down to a sacred spot located at the curving channel between two islands. Strong currents accompanied the tidal flow in and out of the lagoon. The woman was several months pregnant, and the couple sought the magic in the channel. As the tide rushed through the channel into the atoll, they made magic beseeching the spirits to give their child a warrior's spirit and a huge size.

The baby was so large that he had to be delivered by cesarean section. They named the baby Tetabo, and he grew rapidly. Not only was he a giant, but he also was a ferocious warrior. When Tetabo was in his twenties, he united some of the villages of eastern Abemama with a gang that controlled much of that region. He soon imposed a reign that was half terror and half reward based on fealty to his dominance. He created a caste system throughout eastern Abemama. If a family or clan had something of value to offer and willingly aided him, they became *inaomata*. An inaomata was effectively landed gentry but could be called upon to help Tetabo in times of war. Freemen possessing land were *aomata*. People without land, but who were free to work for anyone, were *rang*, and slaves were *toro*. Toro and rang were frequently from the masses that dared oppose Tetabo. Toro [slaves] in particular were those captured on raids to neighboring islands.

As Tetabo was rising in power and coming to dominate the atoll, an outside threat began to materialize. Warriors from the northern atoll of Marakei began systematically to attack isolated villages on Abemama. This is where Baintabu's father and grandfather came into the picture. Tetabo realized that he could not rule Abemama merely as a bully and a tyrant, but he also had to be its protector. He could raise a war fleet to ward off the invaders from Marakei, but the war fleet needed a chief navigator. Tetabo, accompanied by his lieutenants, paid his first visit to Abatiku, which had, until that moment, been left out of his power games. Baintabu's father and grandfather had gained a different kind of notoriety across Abemama: they built the fastest canoes, took the longest voyages, and returned from trips to distant islands where others were lost.

When Tetabo approached a new village in his conquests, he demanded a tribute, usually in the form of land or concubines to seal the deal of loyalty. When Tetabo landed on Abatiku, however, he offered to grant the inhabitants the status of inaomata, or permanent possession of the island without the need for tribute of any kind, provided they rendered their services as navigators to his war fleet when asked. The council of elders of Abatiku convened, and Baintabu's father and grandfather agreed to the deal, realizing that refusal could create difficulties, while cooperation would ensure their defense. Abatiku's relative isolation made it vulnerable to attack, and it could benefit from an organized defense as much as the rest of Abemama. Tetabo's first assignment was a raid on Marakei.

Baintabu's father and grandfather helped organize a war fleet of thirty canoes against Marakei. By this time her grandfather had become too old for the battering he would take on the voyages, so Baintabu's father was left as the main navigator. He led a flotilla of thirty canoes through the western passage of Abemama Atoll and north to Marakei. Tetabo's warriors systematically plundered the outnumbered inhabitants of Marakei, putting an end to their raids. After this success Tetabo was called *matawniwi-iaonteabe*, Master of the Land, cementing his hegemony over all of Abemama as its ruler and protector. Baintabu's father in return became the royal navigator to Tetabo.

BAINTABO'S CHILDHOOD

Around this time Baintabu's father married. Men on Abemama were deemed of age when they were twenty-five years old, but women were married when they were fourteen or fifteen. Precautions against intermarriage forbid a couple to marry if they could trace their lineage back three generations to a common ancestor. Abatiku's population of fifty people made this problematic, so

Baintabu's grandmother and grandfather betrothed their son to a girl from the nearby island of Kenna. In preparation for marriage Baintabu's young mother had to spend many months confined to a dark hut called a *ko*, or bleaching house, as part of a ritual in which her skin was lightened. Daily her skin was massaged with coconut oil until she was deemed to be fair enough to be married. Baintabu's grandmother, father, and grandfather fetched the young, light-skinned maiden from Kenna without ceremony. This was the last time she saw her home island.

Baintabu's mother soon was expecting. During her pregnancy the local soothsayer and relatives made the ritual chants for a successful birth. Baintabu's mother went into labor and delivered Baintabu with the help of a midwife, but she started to hemorrhage. Despite all the incantations and herbs from the midwife, Baintabu's mother bled to death. Baintabu's father now had a young baby girl to care for but also had to support the village of Abatiku as chief navigator for the monarch Tetabo.

Normally, the mother takes a leading role in rearing young children, but this now fell to Baintabu's grandmother. Her father, grandfather, and grandmother combined as a family unit, along with the more distant cousins on Abatiku.

As a young girl Baintabu would accompany her grandmother on daily rounds. At low tide she would help dig shellfish out of the tidal flats. One of the early morning rituals was taking mulch to the *babai* pits. Babai is a giant taro that grows as a carbohydrate staple. As a water-loving plant, it is usually grown in swampy areas. Since the soil is quite poor on the atolls, the babai pits have to be enriched in mulch. Families would keep all their organic trash in a basket, and every morning the women would go out and mulch the pits.

Another daily practice expected of women was the production of *sennit* from the fibers of a coconut husk. Baintabu learned this from her cousins and grandmother. The fibers were separated and rolled against the thighs to create cordage. This cordage was used in everything from lashing together houses to making sheets to hold sails. On their daily trips to the babai pits, Baintabu and her grandmother passed the canoe houses and hut with no roof. Baintabu became fascinated by the work going on there. Sometimes, before her father set off on a long voyage, he would spend the entire night standing in the strange hut, looking up at the sky.

Baintabu begged her grandmother and grandfather to let her hang around the canoe house. Surely, she could make herself useful, bringing sennit or rendering breadfruit sap for caulking the planks. Her grandfather, who was in charge of building canoes, reluctantly agreed. The canoe houses functioned, in part, as a kind of men's club where they could discuss business out of earshot of the women, but because of her circumstances, "Old Graybeard," as the men called her grandfather, had a soft spot and let her wander around while they worked on canoes.

On the big islands of Abemama, the differentiation of genders was strong after the age of five, but on Abatiku it was not so important. On the larger islands there was a strict adherence to ritual; men were raised to be warriors, and women were raised to be mothers. The culture of Atabiku, however, was based more on seafaring skills than on warrior skills, because of its small population. Unlike the more metropolitan regions of Abemama, on Abatiku gender differentiation took a back seat to pragmatism.

At the canoe house Baintabu became friends with her cousin, Kimaere, who was learning to fish and build canoes alongside his father. Kimaere was five years older than Baintabu but was the only other child in the canoe house and had more in common with Baintabu than with the older men.

The canoe building season began when the November storms hit. This was signaled by the presence of *Nei Auti* (the Pleiades) just after sunset. Here, Old Graybeard could be assured of a regular supply of labor. The first task was wood gathering. Logs of different varieties were cut and allowed to dry out. Keels and

Figure 217 "Rocker" of the keel created with chocks and frame. After drawing by Arthur Grimble, "Canoes in the Gilbert Islands," *The Journal of the Royal Anthropological Institute of Great Britain and Ireland* 54 (1924): 106, with permission from Wiley Publishing.

stems were made from the hard wood of a tropical evergreen, *te itai* (*Calophyllum inophyllum*). Ribs were made from the malleable wood of *te buka* (lantern tree). Planking was made from breadfruit logs. The masts were made from coconut trunks. It took several weeks to fell these trees and drag them to the canoe house for drying. On the outings Kimaere and Baintabu would mostly tag along, but when it came time to remove the extra limbs, they both helped out and pretended to lend effort to the transport of the logs.

While the crew waited for the wood to dry, they set about working on the carving tools. The strongest material available for shaping the wood was fashioned from the shell of the giant clam, tridacna. The shell would be cut into an appropriate shape and lashed with sennit to a carved wooden handle to make an adze. Once lashed down, the adze was sharpened with pumice stone. All the parts of the canoes were hand-carved using this kind of adze.

The first order of business was to cut the te itai log into the keel using the adze to create a "V" shape. Once this was done, the curve of the keel (or rocker) was created by using a stick lashed to a frame to hold the middle down, while chocks raised the two ends (Figure 217).

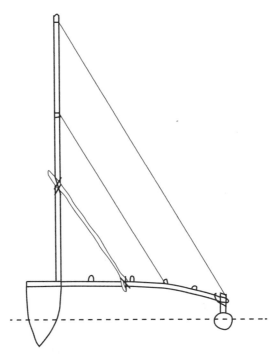

Figure 218 Cross-section of Gilbert Islands canoe hull, outrigger, mast, and bracing. After drawing by A. Grimble, "Canoes in the Gilbert Islands," 118, with permission from Wiley Publishing.

The shape of the hull was all-important, for speed and to resist leeway. Baintabu's grandfather kept the shape all in his head and instructed the men to shape each breadfruit plank to exacting specifications. The grandfather also had the name *te tiabaire*, the Measurer. Every detail of the shape of the boat (Figure 218) was based on his body: the fingernail width, the finger's length, the hand's length, the length of the arm from the elbow to the hand, and the fathom — the length of outstretched arms. Here is where Baintabu began to display the makings of a navigator: she had a head for figures. She would follow her grandfather around and compare the length of a certain plank to her body, creating in her mind an entire conversion table from her grandfather's measures to her own.

Once, much to her grandfather's surprise, Baintabu came running over to him in a panic, saying, "The men are lashing the hull so that the beam is too narrow; you must come quickly." Although he thought it was a boast, he went over to check. Baintabu showed that the distance between gunwales should be two lengths of her forearms but was a hand's width less. When her grandfather applied his own arms as measures, he concurred, scratching his head in disbelief, as did the men working on the canoe. Over the years, during the stormy season, Baintabu gradually took over more and more of the supervision of canoe building, as she had assimilated her grandfather's knowledge.

The season for fishing and voyaging was when *Rimwimata* (Antares) appears just after sunset and ends when Nei Auti (the Pleiades) appears just after sunset. The season of Rimwimata is roughly May through November, when the trade winds blow steadily from the east and southeast. At the same time a one- to two-knot current flows from the southeast to the northwest. The current and winds shift around substantially, even during this period, but they are notoriously unreliable and dastardly during the winter months.

By the time Baintabu was thirteen years old, she began regularly to accompany her cousin, Kimaere, on fishing trips to the nearby reefs during the season of Rimwimata. As long as they confined themselves to the islands of Abemama, they were relatively safe. Kimaere used a fishing trick he learned from his father. If he found an isolated pool, he would dive into its waters while Baintabu tended to the canoe. He searched for exotic creatures in his dives and sometimes would find a *te ntabanin*, a kind of sea slug. Back on the surface he then grated the slug on a piece of pumice they carried in their canoe as ballast and dumped the ground flesh back into the pool. The sea slug has a stupefying poison that makes the fish lethargic, and they float, unconscious. Kimaere and Baintabu would scoop up any that were floating and spear the remaining fish swimming lazily or blindly in the pool.

Often they would return to Abatiku with more fish than the hauls from the larger canoes with more crew.

Initiation into the Navigational Arts

The age of fourteen was a crucial time in Baintabu's life. Abatiku was an isolated island in the atoll, and her grandparents' house and the canoe houses were isolated even from the main village on Abatiku. Matters of gender were largely swept under the rug, but the onset of menses presented a major decision point. Her father was only present during the stormy season and spent most of his time as a navigator for larger clusters of sailing canoes during the season of Rimwimata (Antares). Baintabu had not been betrothed, but this was the age at which a woman entered a ko (bleaching hut) for a time of isolation prior to marriage. The elders met to discuss her case. Normally, women were excluded from village decision making, but this meeting had the extraordinary inclusion of her grandmother and other matrons of the village. Some argued that because Baintabu was a girl, the only right thing to do was to seek a suitable husband outside the island and have her enter the ko in preparation for marriage.

Baintabu's father was silent, so her grandmother spoke up. "Here on Abatiku, we have the privilege of being inaomata [free land owners], but we have this title only because we provide King Tetabo with his navigator for his warring fleet. Who will inherit the job of principal navigator for Tetabo and Abemama?" Much discussion ensued. Only Kimaere and Baintabu emerged as likely candidates, as they showed the most aptitude and spent much time in the construction of canoes. Some of the men spoke in favor of Kimaere, saying that only a man would command respect as navigator in a war fleet.

Finally, Baintabu's grandfather spoke up. His age and wisdom in these matters commanded the respect of the others. Understanding this, he cleared his throat and let a proper silence elapse before speaking, as if to underscore his thoughtfulness in the matter. "A navigator is only good if the canoes can be guided safely back home. The training of a navigator takes many years of patient and rigorous study in these arts. I have known both Kimaere and Baintabu from the time they were very young and can speak well of their capabilities. I can tell you that, while Kimaere is bold, nothing escapes Baintabu's attention. This ability to see and take in everything around you at a glance is a critical skill for a navigator. I can tell that this is in her blood. If we are to continue in our privilege as inaomata, if our canoe fleets are to make it out and back safely, we have only one clear choice: give Baintabu the training for navigator." After another pause the elders, including the matrons, nodded their assent.

Baintabu was told of the decision of the elders and given her first challenge as the heir to the role of navigator. Clothing may seem trivial in such matters to the reader in our era, but in Abemama this was a delicate matter. Upon reaching adolescence men and women dressed differently. Men wore loincloths, and women wore a skirt of coconut leaves split down the middle. Baintabu was spared the agony of the bleaching house but had to decide how to represent herself now that she had come of age. She decided to weave a kind of skirt out of barkcloth. This defined Baintabu in a manner unique but acceptable to the Abatikuans as both woman and navigator.

Up to this time Baintabu's grandmother and grandfather were the major forces in her life, but now her father took a leading role in her education as navigator. One evening her grandmother prepared the ritual dinner that was given to all navigators before departing on a long voyage. Both Baintabu and her father shared in this. As the Sun began to set, her father led her to the strange hut with no thatching on it.

In the deepening twilight Baintabu's father began. "If you are to become a navigator, you must know the stars. These are your primary guides on the seas. When you look up at the night sky, you will immediately be able to orient yourself and use the stars as a reference on all voyages out of sight of land. This hut is how we divide the sky. Most people look up at the sky and call it *karawa* [heavens], but navigators call it *uma ni borau* [roof of voyaging]." Baintabu immediately understood the significance of the hut and her father's mysterious presence there on nights before a long voyage. It was both a teaching tool and a kind of star observatory. She realized that the hut's long axis was oriented along a north–south line. It started to make sense. Her father observed the look of growing recognition in her face.

Baintabu's father continued, pointing at the construction of the hut (Figure 219). "I can tell that you now understand that the hut aligns with north and south. The plank on the east is called *te tatanga ni mainiku* [the roof plate of the east horizon]. All stars rise along this plate. All stars will set on *te tatanga ni maeao* [the roof plate of the west horizon]. In between the rising and the setting points, all stars will follow a path in the sky that looks like the arcs of the *oka* [rafters]. Your grandfather and I made this hut with curved rafters to show exactly how the stars move across the roof of voyaging. To make the rafters we took the strongest wood for making keels and bent them with steam."

"But do the rafters represent certain stars?" Baintabu asked. Her father positioned her with her head underneath the crest of the middle rafter. "This middle rafter is the path of *Taubuki* [Rigel, in Orion]. Do you see the ridge pole?" Baintabu nodded. "The ridgepole is also called taubuki. The highest point in the sky is where the path of *Taubuki* crosses the middle ridgepole, called *taubuki ni karawa*. As a star rises it follows its own rafter higher into the sky, and it is at its highest point when it crosses this middle ridgepole. As the star sets it follows its rafter back down into the western horizon.

Her father went on. "The southern rafter is the rafter of Rimwimata [Antares]. Can you see it now?" Baintabu stepped back and saw Antares, as most islanders were familiar with the

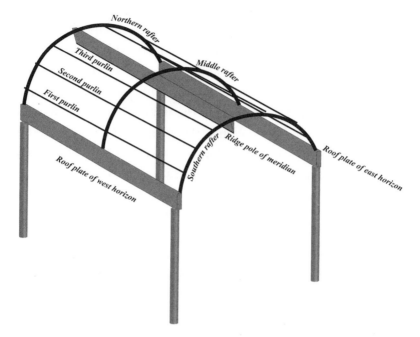

Figure 219   Names of the major features of the star hut.

star that marked the voyaging season. She saw that it was almost crossing the meridian ridgepole. "The northern rafter is the rafter of Nei Auti [the Pleiades]. All stars in the sky are referenced to these three rafters and their rising, setting, and meridian positions. You must be able to visualize them all across the sky at any time of night at any time of year."

"And what of the purlins in between the ridgepole and the horizon plates?" Baintabu wondered aloud. Her father answered, "These subdivide the rising and setting heights of the stars. You will have to memorize 168 stars in all. You will have to know where they are in relation to the purlins, the horizon plates, the meridian, and the rafters at any time of night and in any season."

"But this will take years," Baintabu protested.

"Yes, it will, but this is the way of the navigator. It took me seven years to learn the skills from your grandfather, and so it will take you as much time."

Navigational Skills

And so Baintabu made nightly trips to the star hut and learned one by one the names of the stars. First, she learned the major ones. *Te botonaiai* [Aldebaran] is the base of the canoe rib, named because it is part of a "V" in the sky. *Tanikaroa* [Orion's belt] are "the Three Fishermen" and reach the ridgepole of the sky directly overhead in Abemama. *Kameang* [Capella] was "make north" because it was beyond the northern rafter, and *Kamaiaki* [Canopus] was "make south" because it was beyond the southern rafter.

Gradually, her knowledge increased to encompass all the major stars. Her father showed her the meanings of the stones on the beach. They were lined up in the sailing directions of the major islands: Tarawa, Abaiang, Marakei, Maiana, Kuria, and Aranuka. She learned how to orient a canoe toward each of these islands using the star hut. She learned to associate a region along the horizon with the domain of each of the major stars rising and setting (Figure 220).

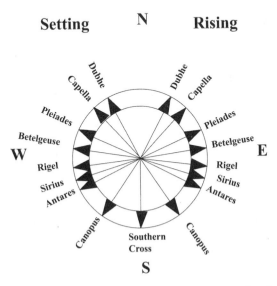

Figure 220 Star compass of Abatiku, based on the position of major rising and setting stars.

Baintabu now accompanied her father on all his voyages, both for fishing and when King Tetabo decided to make a pillaging raid. As the daughter of the navigator, her job was to organize the canoe flotillas prior to raids. On the orders of King Tetabo, she would sail to the major villages on Abemama, informing the chiefs when and where to assemble for the raids. Baintabu would ride in Tetabo's lead canoe on these raids, where her father would instruct and quiz her on all manner of things. He taught her that the navigator must always keep a mental map of the canoe in relation to the rising and setting positions of stars and nearby reference islands. Often, Maiana served as the most convenient reference island, by virtue of its central position in the atolls of the Beru warriors.

She learned to read currents by observing how the wave faces formed. She learned about the major swell, the *Nao Bangaki* or southern swell, which rarely broke and was present in all conditions at all times of the year. During the season of Rimwimata [Antares], there was a dominant swell from the east, driven by the trade winds. Also during this season the wind and currents were mainly from the east and southeast. However, during the season of Nei Auti (the Pleiades), the winds would shift, sometimes from the west, and the current would become fickle, often reversing

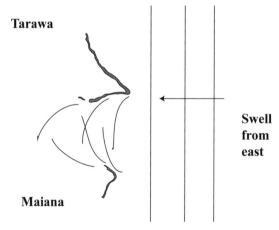

Figure 221 Swell patterns created by refraction between Tarawa and Maiana atolls.

direction and coming from the west. The current was the biggest danger, and spotting and understanding current patterns took many voyages, so many that the fishing and raiding voyages were insufficient to expose Baintabu to the full range of possibilities at sea. Baintabu's father took her out on voyages solely with the purpose of exposing her to the vagaries.

On one of these voyages, Baintabu's father took her to the passage between the atolls of Tarawa and Maiana. He said that this was a dangerous area where the unwary navigator can be fooled by *te airananti*, or "ghost currents" — eddies caused by the interaction of tides, currents, and swells. In between the two islands, the major swell was from the east (Figure 221). As they sailed past the northern tip of Maiana, they dropped sail. Her father pointed at the swell off in the distance: "The major eastern swell gets bent by Maiana toward the southwest, then straightens out as you go farther north." They sailed north toward Tarawa, and he dropped sail again. "The swell from Maiana is getting weak, but now you can see the swell from Tarawa that gets deflected north, then straightens out. In between, the deflected swells meet and form lumpy seas." Baintabu absorbed this. They then sailed north to fish for bonito on the reef north of the Western Passage into Tarawa (Figure 222).

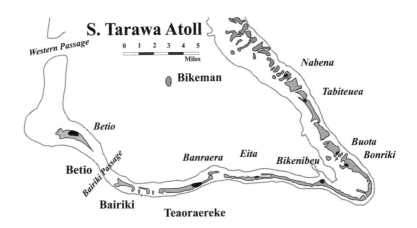

Figure 222 The southern half of the Tarawa Atoll.

As they were fishing, the wind shifted out of the southeast and into the northeast and picked up in intensity. Baintabu's father immediately stopped fishing and turned to her, asking, "Why should we now sail back to Abemama?" For Baintabu this seemed obvious. "Well, Father, if we don't take advantage of the northeast wind, we will have to sail north of Tarawa and take a series of long tacks to get back to Abemama. If we do sail with the northeast wind, we might be able to make a direct run for Abemama." They sailed into the atoll and out again through the Bairiki Passage (see Figure 222) between Betio and Bairiki. On the way through the Bairiki Passage, Baintabu saw a pair of Tarawan outrigger canoes and noticed the difference in their hull shape and the banners they flew on top. With the eye of a canoe builder, she could tell that they could not hold to windward as well as the canoes designed by her grandfather.

By the time Baintabu was twenty-six, she had already served as navigator on two of King Tetabo's raids, one to Kuria and one to Aranuka, where he returned with concubines for his two sons, Namoriki and Baiteke. Baiteke, the older of the two, was in line to inherit Tetabo's rule. Although Baintabu's father accompanied the lead ship on these raids, he said little and told King Tetabo that he was ready to step down as royal navigator to lead the quiet life of a canoe maker on Abatiku. He assured Tetabo that Baintabu was ready to assume the role of lead navigator to the Master of the Lands, which Tetabo accepted.

The Tarawan Raid

At the start of the season of Rimwimata (Antares), Baintabu and her cousin Kimaere were fishing the reef north of the Western Passage into the Abemama Atoll. There were moderate winds out of the southeast. Off in the distance Baintabu spotted a line of masts just outside the passage. She told Kimaere to pull up the gear and hoist sail. Rapidly, they sailed in the direction of the masts. There was no returning fleet that she was aware of, nor were there any groups of fishermen that large. She counted thirty masts in all. As they drew closer she could identify the hull shapes and crests characteristic of the Tarawan canoes she had seen on trips north with her father.

Baintabu immediately altered course and sailed close to the wind toward the eastern islands of Abemama and Tetabo's citadel. She could see that the Tarawan flotilla was having trouble holding to the wind and was tacking through the shallow waters of that atoll. While it would take some hours for them to reach the eastern villages, she could get there in less than an hour. Kimaere and Baintabu arrived at Tetabo's grounds in the village of Tebontebike and sounded the alarm. Tetabo dispatched them toward the southern villages to raise the warriors, while his sons prepared to sail north to alert the northern villages and do the same.

There are three large villages that constitute the "center" of Abemema: Baretoa, Tetabo's village of Tebontebike, and Kariatebike, all within a couple of miles of each other. Tetabo anticipated that the raid was destined for this strip of leeward coast and rapidly deployed his warriors for battle. Many warriors ran on footpaths, and some sailed as quickly as they could, outracing the approaching Tarawan fleet.

The battle gear of the Abemamans consisted of javelin spears, often tipped with jagged sharks' teeth, and spears also bristling with shark's teeth. The warriors took up places hidden behind huts and trees to gain the element of surprise. The Tarawans were at a distinct tactical disadvantage; they had chosen a sailing direction that required many tacks of their canoe directions, giving the Abemamans three hours of preparation. As the Tarawan warriors landed on the beaches, they began to approach the villages. All at once the Abemaman warriors leapt out from their hiding places and charged with blood-curdling screams. This caught the Tarawans by surprise. Some engaged in battle, but some ran back to their canoes to beat a quick retreat to leeward.

The Tarawans who chose to stay and fight were either killed or taken captive. Some of the Abemaman warriors took to their canoes to chase the fleeing Tarawans. In particular, they wanted to make certain that no outlying villages were attacked. The Abemaman pursuers followed the Tarawan canoes out past the Western Passage, but once it became apparent that they were fleeing toward home, the warriors broke off the pursuit.

In the aftermath of the raid, some of the captured Tarawan warriors were killed in a most gruesome manner in a public display in front of Tetabo's citadel. The bodies were left to rot in the heat of the midday Sun. A number of them who appeared to be of higher caste were taken prisoner, as surely they would be of some value. Others still were taken as slaves by relatives of the royal family.

Baintabu was invited to inspect the captured Tarawan war canoes by Tetabo and render an opinion. She saw that the keels lacked an adequate rocker (bend) and that the sharpness and depth of the hull was not so pronounced, which explained the poor windward performance of the canoe. She returned to Abatiku and concentrated on fishing for the rest of the season, but at the request of Tetabo, she commenced construction of a raiding canoe at the start of the season of Nei Auti (the Pleiades). All villages in Abemama with the capabilities of making war canoes were given similar instructions.

PLANS FOR A RAID ON TARAWA

Toward the end of the season of Nei Auti, a courier arrived on Abatiku requesting Baintabu's presence at King Tetabo's citadel. Baintabu agreed, provided her cousin Kimaere could accompany her. The envoy agreed. The next day Baintabu set out with Kimaere to see Tetabo. They pulled up their canoe and walked through his village, past the house of his concubines and his sons.

Baintabu and Kimaere entered the building and bowed before Tetabo. The Master of the Land was flanked by his soothsayer on one side and his two sons, Namoriki, the younger, and Baiteke, the elder, on the other side. Tetabo looked quite solemn.

"Baintabu, I must tell you that I have come to the decision that we are going to make a raid on Tarawa. I cannot tolerate a raid on my lands and let it go unanswered. I have asked my heir, Baiteke, to lead the raid, but I must have your assistance as navigator. Baiteke will lead, but you will be responsible for safely getting the war fleet there and back again. You can share in any plunder our warriors get."

"Oh, Master, I am willing, but please, we must discuss some details. How many canoes do you want in your fleet? Tarawa is a large land, and there are many villages and islands. Where do you want to attack?"

"Baintabu, I expect we can raise over seventy war canoes and six hundred warriors. As to the timing of departure, I leave it to you. As for the attack, the captured Tarawans tell me that there are two centers of power, one from the village of Tabiteuea on the windward side and one from the village on Betio on the southwest end of the atoll. The two chiefs combined forces for the raid on us last season."

Baintabu thought for a moment; then, "Sire, I must tell you that there are some aspects of this raid that are dangerous and must be carefully considered. If we sail after the start of the Rimwimata [Antares] season, it will be easy and fast to get there, but the return voyage will take five times as long and will require skillful navigation and strict discipline on the part of the crews. We must sail across the wind to return home. Do you understand this?"

King Tetabo nodded his head in assent and turned to Baiteke, who paused momentarily and shook his head, too. "Go on."

"As far as the target village, we will surely encounter the same kinds of winds that the Tarawans faced on their attack if we go after the village of Tabiteuea. This would leave us at a disadvantage. On the other hand, if we target the village on Betio, we can sail in through the western passage of Tarawa and in one tack sail directly to Betio, but we must land at high tide, as there is a long sloping reef that almost uncovers at low tide."

Tetabo thought for a moment and said, "So it will be. Baiteke will gather the warriors and make arrangements for the raid. You may return to Abatiku and prepare yourself and the Abatikuan canoe." Baintabu and Kimaere bowed again and left. As they left the palace they could hear the indistinct mumbles of the sooth-sayer and Baiteke talking.

Back on Abatiku Kimaere brought together the men of the village to inform them of the decision. Some days later King Tetabo's son, Baiteke, and some of his lieutenants arrived in Abatiku. Baiteke immediately hunted down Kimaere and asked him the state of his preparations. Kimaere said that they were gathering dried fish and prepared gourds for fresh drinking water for the voyage to and from Tarawa. Baiteke grew impatient: "I am not talking about food and water; I am talking about weapons. You Abatikuans have a reputation as sailors but not as warriors. While the mainlanders will lead the charge, you will surely be in the rear of the flotilla." With this, Baiteke wheeled about, jumped into his canoe, and sailed off with his entourage.

The Voyage to Tarawa

When the season of Rimwimata arrived, Baintabu sent word to King Tetabo that the canoes of the fleet should be ready to go on a day's notice at the start of the second lunar month. In the weeks leading up to the raid, she spent her nights in the sky hut and during the day walked to the south side of Abatiku. On the beaches facing out to the ocean, she looked at the way the waves broke, showing the shifts in the current. She looked at the cycles of the weather patterns, trying to determine the best time to launch the fleet.

Finally, all the signs were favorable for the voyage to Tarawa. Baintabu passed word that the fleet should be assembled in the waters near King Tetabo's citadel and be ready to sail at sunset that day. The Abatikuans loaded their war canoe with dried fish, dried pandanus fruit, gourds of water, and weapons. Kimaere, Baintabu, and the warriors of Abatiku crossed over to Tetabo's village, where the fleet was assembling. There were eighty war canoes in all. Each one carried the crest of its clan from the tops of the masts. It was the largest assemblage of war canoes ever seen in those islands.

Baintabu crossed over to the lead canoe with King Tetabo's elder son, Baiteke, in charge. The soothsayer was seated in the rear of the canoe with a headdress of feathers on. Near sunset, Baintabu passed word that they would sail when the Moon rose, some time after sunset. The order of the flotilla was dictated by Baiteke, and the lead canoes were manned by the warriors he held in the greatest esteem. As a sign of Baiteke's derision for the Abatikuans, their canoe held the last position.

In the gathering darkness the flotilla bobbed around in the water. Fires blazed on the beaches, and women chanted incantations on the water's edge. When the Moon rose in the eastern sky, a hush fell over the entourage. As soon as it cleared the highest palm tree, Baintabu indicated to Baiteke that it was time to leave.

Baiteke's canoe led out. Baintabu had it steer in the direction of the setting Regulus, which was a course toward the Western Passage. The Moon was three-quarters full and illuminated the sea and distant palm trees. In those parts low tide occurred three hours after moonrise, and the flotilla was swept through the Western Passage of Abemama on the ebb current (Figure 223).

Once clear of the western reef, Baintabu could feel the southern swell gently rolling the canoe. She ordered a change of course to the north, one hand width to the south of the setting Dubhe, which would bring them directly to Tarawa. Some time later, as they cleared the northern end of Abemama, they left its lee, and

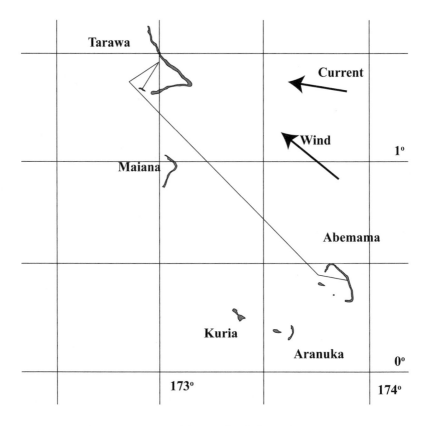

Figure 223 Outgoing voyage of the war flotilla from Abemama to Tarawa.

the swell from the east began to dominate, along with the trade wind from the southeast. This was the easy passage: the wind and the current combined to sweep the flotilla northward. The navigators in those parts often said that the world was tilted to the northwest as the winds and currents always moved in that direction, downhill. Baintabu knew the return voyage would be far more taxing.

Baintabu maintained her course to the northwest by aiming at a succession of setting stars: first, the setting Big Dipper's handle, then lesser stars. Just before sunrise Venus came up in the east. The last major star setting was Vega to the west-northwest, and Baintabu steered one-and-a-half hands' width to the north of that position. The Sun rose, and they continued on their course, using the eastern swell as a reference. Soon, Baintabu could feel the reflection of the eastern swell from Maiana, over the horizon.

By midmorning Baintabu reckoned they had made their closest approach to Maiana, which was off to the southwest. The flotilla was entering the notorious passage between Maiana and Tarawa, where the currents were strong and fickle. Baintabu gave orders to steer more to the north to take into account the current and remain on course. By early afternoon they passed a few miles to the south of Betio on the southwestern tip of Tarawa and could make out the tops of the palm trees in the distance. A landing on the ocean side of Betio was unthinkable, however, and the flotilla sailed north until it reached the Western Passage of Tarawa.

At midafternoon Baintabu gave orders to steer to east-northeast, biting hard into the wind. They crossed through the Western Passage, now pushed by the incoming tide. They sailed some distance into the waters of the atoll. When Baintabu judged that she had a good final angle into Betio, she gave orders to shunt into the wind and the fleet began its final run to the southwest toward Betio. Baintabu turned to Baiteke, saying, "On this course we will soon reach Betio; it is now your turn."

Baiteke ordered the flotilla into an attack formation with a phalanx of ten canoes abreast, and the others falling in behind them. As the flotilla approached Betio, Baintabu could first make out the tops of palm trees, then the light of fires near the water's edge. Baintabu timed the voyage so they would arrive on the shores of Betio close to high tide. The reef and beaches of Betio are quite shallow, and landing at anything but high tide would put the warriors at a disadvantage.[3]

The Landing on Betio

The Sun was getting low in the western sky, and the fires and palm trees became clearer as the flotilla bore down on the island. The canoes bristled with javelins and lances as the warriors strained their eyes toward shore. As they got closer a strange sound drifted to them: singing.

The soothsayer and Baiteke exchanged nervous glances. Baiteke asked, "Why should they be singing?" The soothsayer shrugged his shoulder: "Maybe they are preparing to attack as we land?" As the canoes landed, the warriors became silent. Women in grass skirts were dancing by the fires, and men were waving their arms in a "come hither" motion. Baiteke, the soothsayer, and Baintabu got out of the canoe. Baiteke raised his arm to tell his warriors to wait. A man in a feathered headdress flanked by tattooed but unarmed men approached them with outstretched arms. Baiteke's face had a look of disbelief on it.

The Betio king smiled and said, "Welcome, my friends. We are preparing a feast in your honor. I know you may think this is a trick, but let me assure you that we have no intention of making war. You have obviously come with superior numbers, and we only wish to honor your brave warriors." Baiteke looked crestfallen.

The king turned to the soothsayer and Baintabu. "So I see you have brought the head magician and a princess to accompany you." Baiteke, somewhat speechless, shrugged his shoulders and glanced at Baintabu, saying "*Nikara-n-roro* [concubine]." Baintabu felt stabbed in the chest by the words. The Betio king nodded his head, as if to say he understood and motioned them to come. "Tell your warriors to join the parties up and down the beach." Baiteke passed the word to the warriors of the fleet, and the Abemaman warriors were welcomed to the feasts at the many fires.

The warriors were treated to the most flavorful fish and fermented coconut sap, soon getting drunk and dancing with the Tarawan women. Baintabu, however, retreated to the beached war canoes. Sometimes, we are shown what our place in life really is in one swift moment, and the illusions are stripped away, leaving us clawing at a sense of identity. It can be dizzying, as if one is falling through space. Baintabu could only stare out to sea, watching the moonrise over the atoll as the fires burned brightly behind her and much singing and shouting drifted to the water's edge. One night before she had been the head navigator of the feared King Tetabo; now, she was dismissed as a mere concubine accompanying Baiteke. Rather than join the celebration, Baintabu returned to the simple food stored on the war canoes and fell asleep as the feast went late into the night.

The next morning at sunrise, Baintabu stretched and surveyed the human wreckage on the beach. Drunken warriors were sleeping everywhere. The Betio natives had retreated to their huts. She wandered along the shore, finding the Abatikuan canoe she helped build and erected a small shelter from the Sun out of the covers used to store the sails during storms. There was little to do but wait at this point.

Around midday Kimaere wandered back to the canoe, looking somewhat the worse for the wear. He saw Baintabu and, somewhat startled, asked, "What happened to you?"

"Baiteke called me a concubine in front of the king of Betio. One moment I was Tetabo's navigator; the next I was a whore."

Kimaere shook his head. "I was afraid something like this might happen when we got to Tarawa. The man is no good and will make a horrible king when Tetabo dies. What are you going to do?"

"What can I do? We're far from home and have a very difficult voyage back. The warriors have families who depend on them. I have a responsibility to get them home safely, regardless of Baiteke. Now I have to wait until it is time to go back."

"You are the pride of Abatiku," Kimaere said. "What can I do to help?"

Baintabu looked out across the turquoise waters of the atoll, squinting into the distance, and turned back to Kimaere. "Under the circumstances I cannot talk to Baiteke while we're on land. You must act as a go-between to convey messages. We can't set out to sail just any time. The wind must be right for a return trip;

otherwise, the currents will sweep us far out to sea to the north-west. There is no land out there. I have heard of canoes that have been storm-swept in that direction that have never returned. Find Baiteke, and tell him that I will wait until he wants the fleet to return to Abemama, and I will decide on the time for departure."

Kimaere nodded. "As you wish, Baintabu. It's the least I can do."

Many days passed, but Baintabu was not idle. She constructed a small version of the star hut out of palm fronds. The trip back to Abemama was challenging and would require many long tacks back and forth. She had to plan for the best course and condi-tions to fight the current and the prevailing winds that were allied against her, pushing everything on the ocean downhill: from the southeast to the northwest.

In her mind Baintabu ran through a number of variants of the voyage back, based on how well the war canoes could sail into the wind. Depending on the conditions, some voyages would fail miserably. If the winds were weak, or blowing from the wrong direction, the canoes couldn't fight against the current. Even if the wind was strong enough, successful navigation required good timing and a use of the atolls to guarantee a safe passage. She finally came upon a workable solution in her mind.

Baiteke and the warriors at first gorged themselves on the hospitality of the king of Betio, but they soon began to tire of the high life. One day Kimaere came to Baintabu, saying, "Baiteke has decided that he is ready to return to Abemama. He is preparing for departure."

"Kimaere, tell him that the wind is too weak now; we cannot leave until it is blowing hard from the southeast. If we leave now, we cannot fight the current." Kimaere nodded and left. When he returned, he said that Baiteke would await her decision on the departure time.

Baintabu repeated her ritual of spending nights in her star hut and wandering to the ocean side of Betio during the day to scrutinize the weather and current conditions. One day the weather gave signs of a strong, sustained southeast wind. The proper wind will sound like a monster chewing through the coconut trees, bending them to its force. Baintabu returned to the main Betio village, telling Kimaere, "Have Baiteke make ready to leave just before sunrise tomorrow; there is only a small opportunity to set sail, and we must take it."

THE VOYAGE BACK TO ABEMAMA

A map of the return voyage is shown in Figure 224.

Over the night the wind had picked up from the southeast and was blowing the coconut fronds sideways. Baintabu passed instructions to have provisions for at least a weeklong return voyage in the canoes. In the early morning twilight, the warriors dragged the canoes of the flotilla down to the water's edge and leapt in.

Baintabu made her way to Baiteke's canoe and climbed on board. Baiteke said nothing. As soon as the Sun shone over the atoll to the east, Baintabu motioned to the lieutenants to raise sail. She said, "Sail to the northwest, and have the other canoes fall in behind."

The Sun had climbed two hand widths into the sky above the horizon when they cleared the reefs of the Western Passage. Baintabu signaled to the lieutenants to change course to due north. The steering pilot began to grumble that due north was taking them away from Abemama, not toward it. What was she doing? Baintabu said, "Don't ask questions; steer north."

She could again feel the southern swell under the canoe and lay down to see if she could get any hint of current from the rocking of the canoe. In late morning she had the pilot again alter course, this time toward the northeast. "Steer just to the north of the northern point of Tarawa," was the command. The pilot rolled his eyes, and the lieutenants began to mumble among themselves. What was

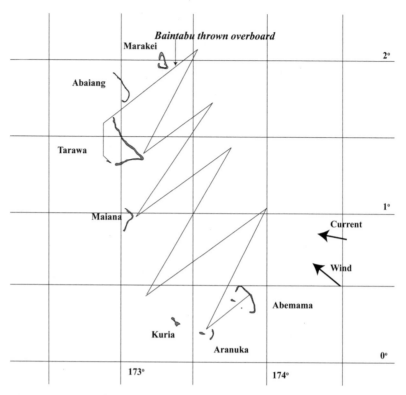

Figure 224   Map of the voyage from Tarawa back to Abemama.

she doing? She was going to get them all killed for Baiteke's slight. Baiteke seemed worried.

The canoes were now close-hauled, fighting for every inch against the wind and current. When they cleared the northern tip of Tarawa, the winds hit in full force. There was little doubt of the power of the eastern swell at this point, rolling the canoes as they made their way to windward. Somewhat past midday the southern end of Abaiang came into view on the port side. Only one man on Baiteke's canoe had been to Abaiang and knew that they were on a northeasterly course. He was alarmed, and the men grumbled more about Baintabu's motives. They asked her what she planned, and she shouted above the wind, "Do as the concubine says, and sail on." They looked askance at each other, not knowing what to make of this.

Soon Abaiang disappeared in the wake of the flotilla, and as the Sun moved lower in the sky, the atoll of Marakei came into view. Baintabu insisted that they follow a course as close to the southern tip of Marakei as they could possibly make. The men continued to grumble. This woman was a suicidal witch and was taking them away from their home, not toward it. Baintabu lay on the canoe, trying to feel the direction and strength of the current by the way it interacted with the eastern and southern swell, but to the men it seemed she was sulking.

The crew asked Baintabu what she was doing, but she just motioned, saying, "Keep to this course; sail on. I will bring you home to Abemama; don't worry."

When Marakei fell astern, the men on Baiteke's canoe were panicked. Surely, this woman would take them to their doom for Baiteke's slight. They huddled at the back of the canoe and in one swift, coordinated action moved forward; grabbed Baintabu from the hull of the rolling canoe; and heaved her overboard into the ocean.

Baintabu hardly had a sense of what was happening until she was flying through the air and landed in the waves. She hit the warm waters in a plunge. When she came to the surface, she maneuvered herself to face the fleet now passing by. The Sun was dropping in the sky. She shouted at them to come rescue her. Some didn't notice; some saw her but, out of fear of offending Baiteke, refused to give her any assistance. Some shouted words about "feeding the sharks" and laughed derisively.

At the end of the fleet, Baintabu spied the crest of the Abatikuan canoe and waved, shouting frantically to get its attention. Kimaere was on the bow and shouted orders to the crew to ease off the sheets and let the sail go slack. They dug out their paddles and moved toward Baintabu. She swam toward the canoe, and Kimaere and the others pulled her out of the water and into the canoe. She was breathless. "Baiteke's lieutenants threw me overboard; they thought I was committing suicide by sailing far out to sea and taking them down with me in revenge."

Kimaere looked around. The last of the canoes were disappearing over the horizon to the northeast. Catching up with them was out of the question. "What do we do?" Kimaere asked.

"What can we do? Sail on to the northeast; maybe some will drop sail and let us catch up."

As the Sun reached the western horizon, there was no sign of the fleet. In the deepening twilight Baintabu instructed the crew to sail one palm width to the north of the rising Deneb in the eastern sky. When Schedar in Cassiopeia rose, they steered directly toward it. When Cassiopeia rose higher, Baintabu gave instructions to shunt back and steer a course southwest toward the setting Alpha Centauri in the western sky. They continued on through the night. Kimaere wanted to know how Baintabu was determining the course, but she said that she'd tell him after sunrise. She lay on the bottom of the boat and stared up past the mast at the stars, feeling the pitch and roll of the canoe underneath.

At sunrise they continued on the southwesterly course, but soon, they could spy waves crashing in the distance and the tops of palm trees. The wind continued unabated from the southeast. Kimaere asked, "What land is that?"

"It should be the southern end of Tarawa, if I'm not mistaken," said Baintabu.

"You mean to say that we're basically where we started out yesterday morning, but the only difference is that the entire fleet is gone missing? What did you have in mind?"

Baintabu gave orders to shunt again to the northeast, again holding tight to the wind.

While the helmsman and crew were holding the course, Baintabu grabbed some pumice stones that were among the ballast for the canoe and some sticks and laid them on the hull. "Kimaere, come over here, and I'll explain."

She laid out a line of pumice stones in front of her and pointed to them. "You see this stone? It's Tarawa. These stones to the north are Abaiang and Marakei. This one to the south is Maiana, then farther south are Kuria, Aranuka, and Abemama."

Kimaere nodded, "Go on."

Baintabu swept her hand across the line of stones. "The wind and the current move in this direction, from the east. The canoe can only travel so tightly into the wind, and we cannot sail directly to Abemama. We have to shunt many times to reach Abemama, but it would have been unwise to sail directly from Tarawa. If we had started out sailing southwest from Tarawa, we would be on the wrong side of the atolls."

"I don't understand."

"You always want to sail on the windward side of a cluster of atolls. If a storm comes up, you can run downwind to the nearest island for shelter, but if you're on the leeward side and a storm hits, there's no chance for shelter, and you'll get swept far away where you can't find your way back. Also, in the tacks I want to make as many turns as I can near a known atoll. I can only steer so precisely. Small changes in the wind and currents are something I can't control, but if I know that I can aim roughly to an atoll, I can correct my position. That's why I wanted to head north from Tarawa and then northeast past Abaiang and Marakei; these were both safe havens and markers as I started my shunts on the windward side of the atolls. After I make the next turn, we should reach the windward coast of Maiana tomorrow morning."

They continued sailing on to the northeast throughout the day. In the distance towering cumulus clouds indicated the atolls of Tarawa to the west and Abaiang and Marakei to the northwest and west. Three porpoises were swimming alongside the canoe, occasionally leaping out of the water as it skidded along its tack.

An hour before sunset Baintabu ordered another shunt to the southwest, and they began the next leg to Maiana. Baintabu kept a mental picture of her location in her head, but the signs of land helped reinforce the position. After sunset the course to

Maiana was again steering toward Alpha Centauri and following the stars on that heading throughout the night. Just before sunrise, in the twilight of dawn, Baintabu could feel the swell reflections from Maiana. Shortly after sunrise, the shoreline of Maiana came into view, and they began the next shunt back to the northeast.

Unlike western sailing vessels, the Gilbertese outrigger canoes did not have to swing through the wind to change direction: They reverse direction. The crew takes the sail and shifts it from one end of the vessel to the other. Once the wind fills the sail on the opposite side, the navigator has the crew adjust the sheets and helm to get on the desired heading. The outrigger will always ride on the windward side of the canoe to provide balance against heeling from wind pressure.

After sunset they again steered a heading a hand's width north of Deneb, and when Cassiopeia appeared, they tacked again toward the setting Alpha Centauri, this being the reliable route Baintabu had worked out during the debauchery on Tarawa. Just after sunrise the next day, frigate birds could be seen flying out to sea. Baintabu guessed that these were signs of Kuria just to the southeast and ordered another tack to the northeast.

Although shunting against the wind and current is a long and somewhat painful experience, it actually improves the chances of finding an atoll, as the canoe crisscrosses a larger region of the sea than if sailing with the wind and current. After coming about again, signaled by the rising Cassiopeia, they knew there was a good chance of landfall the next morning.

During the night the slow southern swell disappeared, indicating a large atoll, presumably Abemama itself, to the south. A few hours after sunrise, they came on Aranuka. Having traveled there many times on fishing parties and on one of Tetabo's raids, Baintabu knew that one more shunt to the northeast would bring her through the southern passage and into Abemama.

"Abemama, oh, Abemama, what am I going to do with you?"

wondered Baintabu aloud. During the last two days, the sea had been her familiar friend, banishing the memory of Baiteke and the lost fleet, but the prospect of landfall brought this all back to the fore. What was she going to tell Tetabo? What about the families who lost warriors through no fault of their own? Would the fleet find its way back on its own?

Epilogue — Twenty Years Later

Timote, Kimaere's son, stared out from Maiana toward Tarawa. Baintabu waved her hand in the direction of the rolling swells. "Do you see how the swells near the atoll are distorted toward the southwest closest to the island, then straighten out?"

"Yes."

Baintabu continued, "That is from the effect of the shallow reefs from the atoll. Which way do you think the swell will bend if we were on the south shore of Tarawa?"

Timote thought for a moment and drew an imaginary set of lines in the air with his index finger; then, "I suppose they would bend toward the northwest in that case. I would also expect to see a weak swell bent by Maiana still sneaking in if I was some distance to the south of Tarawa."

"Good. You must remember that no one sign in navigation is reliable, but several signs in combination are. You must always ask yourself, 'Are there other ways I can prove to myself where I am, and where I am going?'"

Timote nodded.

Twenty years before, the fleet never returned, as Baintabu had feared. Although it may not have been deliberate on the part of the Tarawans, they could not have done a better job of decimating the warrior population of their enemies: Abemama was devastated by the loss. We deal with loss and a sense of mortality in our own ways. For Baintabu the act of teaching was a way of creating a kind of immortality: knowledge is preserved, and a sense of safety emerges. She had waited many years for an apprentice navigator to teach and let her pay off her emotional debt. Kimaere's son showed an amazing aptitude, and at Baintabu's suggestion the elders of Abatiku decided to make him heir to the tradition of navigation.

Appendix 1: Major Star Coordinates and Mapping onto Earth

This is a listing of major navigational stars and their locations in my declination/longitude scheme. I've paired the declination and celestial longitude of each star with a location on earth that should be easy to remember. In some cases I could not find a precise "memorable" location but found one within a degree.

FORMAT

Star Name (Constellation)

Declination	Celestial Longitude
Location on Earth (Lat, Long)	Location on Earth (Lat, Long)

EASTERN CLUSTER

Aldebaran (Taurus)

17° N	69° E
Montserrat (17° N, 63° W)	Dwarka, India (22° N, 69° E)

Alphard (Hydra)

9° S	142° E
Port Moresby, Papua New Guinea (9° S, 147° E)	Sapporo, Japan (141° E, 45° N)

Alpheratz (Andromeda)

29° N	2° E
New Delhi (29° N, 77° E)	Paris (49° N, 2° E)

Betelgeuse (Orion)

7° N	89° E
Georgetown, Guyana (7° N, 58° W)	Calcutta (23° N, 88° E)

Canopus (Carina)

53° S	96° E
South Georgia Island (54° S, 80° W)	Rangoon, Burma (96° E, 17° N)

Capella (Auriga)
46° N   79° E
Geneva (46° N, 6° E)   New Delhi (29° N, 77° E)

Pleiades
24° N   57° E
Key West (24° N, 82° W)   Pt. btw. Str of Hormuz, Gulf of Oman

Pollux (Gemini)
28° N   116° E
New Delhi (29° N, 77° E)   Beijing (40° N, 116° E)

Procyon (Canis Minor)
5° N   115° E
Bogota (5° N, 74° W)   Beijing (40° N, 116° E)

Regulus (Leo)
12° N   152° E
Djibouti (12° N, 43° E)   Brisbane (27° S, 153° E)

Rigel (Orion)
8° S   78° E
Port Moresby, Papua New Guinea   New Delhi 29° N, 77° E)
  (9° S, 147° E)

Sirius (Canis Major)
17° S   101° E
La Paz, Bolivia (16° S, 68°˙ W)   Bangkok (14° N, 101° W)

WESTERN CLUSTER

Altair (Aquila–Summer Triangle)
9° N   62° W
Panama City (9° N, 80° W)   New Glasgow, Nova Scotia (46° N, 63° W)

Antares (Scorpio)
26° S   113° W
Asunción, Paraguay (25° S, 58° W)   Drummond, Montana (45° N, 113° W)

Arcturus (Bootes)
19° N   146° W
Mexico City (19° N, 99° W)   Fairbanks, Alaska (65° N, 148° W)

Deneb (Cygnus–Summer Triangle)
45° N   50° W
Calais, Maine (45° N, 67° W)   Belem, Brazil (1° S, 49° W)

Fomalhaut (Southern Fish)
30° S   16° W
Perth, Australia (32° S, 115° E)   Canary Islands (28° N, 16° W)

Spica (Virgo)
11° S   159° W
Lima, Peru (12° S, 77° W)   Aniak, Alaska (62° N, 159° W)

Vega (Lyra–Summer Triangle)
39° N   81° W
Zanesville, Ohio (40° N, 82° W)   Zanesville, Ohio (40° N, 82° W)

NORTHERN CLUSTER

Dubhe (Big Dipper)
62° N   166° E
Reykjavik (64° N, 22° W)   Wake Island (20° N, 166° E)

Schedar (Cassiopeia)
54° N   10° E
Copenhagen (56° N, 12° E)   Hamburg (54° N, 10° E)

Appendix 2: Some Significant Events in Latitude and Longitude

500 BC	Spherical nature of Earth realized by Greeks
300 BC	Pytheas's voyages, concept of latitude
200 BC	Eratosthenes's measurement of Earth's radius
AD 200	Ptolemy's *Almagest*, latitude and longitude tables
850	Caliph al-Ma'mun's measurement of Earth's radius
1000	Al-Bīrūnī's work on geodesy, astronomy, mathematics
1000	Al-Jayyani's work on spherical trigonometry
1000	Norse voyages to North America
1050	Al-Zarqali's work on astronomy, Toledo Tables of latitude and longitude
1200–1300	First use of magnetic compasses in Europe, Arab Empire, portolan charts
1460	First latitude measurements on coast of Africa
1488	Bartholomeu Dias measures latitude of Cape of Good Hope
1492	Columbus's first experiments with sightings of Polaris
1504	Columbus's measurement of Saint Ann's Bay, longitude from lunar eclipse
1527	Diogo Ribeiro produces first world map with latitude and longitude lines
1569	Mercator produces a world map with a projection preserving rhumb lines
1599	Edward Wright publishes *Certaine Errors in Navigation*
1600–1700	Slow adoption of Mercator chart and use of sighting with dead reckoning
1708	Shovell fleet disaster on the Scilly Islands
1714	Longitude Act
1730	Hadley and Godfrey develop the octant
1761	First sea trial of Harrison's marine chronometer
1802	Publication of *American Practical Navigator* by Bowditch
1843	Sumner's publication on use of lines of position on Mercator charts
1875	Marcq St. Hilaire publishes his intercept method

Appendix 3: Toledo Tables

Below is a full listing of the Toledo Tables, as published by J. K. Wright in "Notes on the knowledge of latitudes and longitudes in the Middle Ages."[1] All values of latitude are north, and all longitudes are east of the prime meridian used in the table. A comparison of modern values of longitude with reliable entries in the table indicates a prime meridian for the table 23° east of the Greenwich prime meridian. This corresponds to the meridian of the Cape Verde islands. All values are positive, as they are both north of the equator and east of the prime meridian used for this table.

I note that the typical accuracy in latitude is approximately 1 degree and in longitude is approximately 6 degrees in comparison to modern values. I've included the reported values in arc minutes for completeness, not because I believe in any accuracy below 1 degree. Where a reading lists "00'" for arc minutes, the reader can safely infer that the value in the table was only entered to the nearest degree. The inclusion of 00' is merely for formatting purposes and should not be taken as an accuracy beyond 1 degree.

Location	Latitude	Longitude
Tangea	35° 15'	6° 30'
Cepta	35° 20'	8° 00'
Corduba	38° 30'	9° 20'
Toletum	40° 00'	11° 00'
Sigdmessah	22° 00'	15° 00'
Gana	10° 45'	15° 30'
Sedes regis Francorum	45° 50'	23° 45'
Insula tule	58° 10'	10° 00'
Cartago	37° 00'	27° 00'
Tuniz	38° 00'	29° 00'

Location	Latitude	Longitude
Emerita	45° 55′	8° 00′
Balgh	38° 10′	108° 45′
Albeyt	38° 00′	130° 00′
Aracah	36° 0′	73° 36′
Mecha	21° 00′	67° 00′
Gedda	20° 15′	66° 30′
Almedina	25° 00′	65° 20′
Algoz	24° 00′	63° 50′
Yspaen	34° 30′	75° 00′
Alre	37° 30′	86° 00′
Fargana	36° 00′	86° 00′
Goarizmi	42° 10′	91° 50′
Chebil	28° 00′	100° 00′
Albahra	31° 00′	74° 00′
Hamen	19° 45′	84° 30′
Adramauht	12° 30′	71° 00′
Sanaa	14° 30′	63° 40′
Armenia	41° 00′	77° 00′
Buchare	36° 50′	107° 20′
Cerendin	3° 00′	125° 00′
Almedia	36° 00′	32° 00′
Cireneti	35° 30′	31° 00′
Insula sardania	38° 00′	31° 00′
Roma	41° 50′	35° 25′
H'abiz	32° 00′	36° 00′
Insula morelani	37° 00′	37° 00′
Insula sicilie	39° 00′	36° 00′
Malta	36° 00′	38° 00′
Trabuluz	33° 00′	40° 00′
Barca	31° 00′	47° 03′

Location	Latitude	Longitude
Alexandria	31° 00'	51° 20'
Dimiath	31° 00'	54° 40'
Eraclia	46° 35'	52° 25'
Urbs a nuba	14° 30'	53° 00'
Bagdeth	33° 25'	80° 00'
Messera	30° 00'	51° 00'
Alcuzum	28° 20'	56° 30'
Assuen	22° 30'	56° 30'
Alcarme	31° 30'	55° 40'
Ashalem	32° 00'	55° 40'
Aranida	32° 15'	56° 00'
Jerusalem	32° 00'	56° 00'
Sur	33° 00'	57° 00'
Alconstantina	45° 00'	44° 00'
Damascus	33° 10'	60° 00'
Trabuluz	34° 00'	60° 35'
Alcufa	31° 50'	64° 30'

Table 3: Latitudes and Longitudes Found in the Toledo/Marseilles Tables

Appendix 4: Sailing Capabilities in Baintabu's Story

There are a number of sources of information on the capabilities of native Pacific island sailing vessels, the earliest of which were reports by Captain Cook. Arthur Grimble reports that the Gilbert canoes were capable of sailing as close as 70 degrees to the wind, and racing canoes could make as fast as twelve knots.

Ben Finney measured the capabilities of a twin-hulled catamaran, the *Nalehia*, built to imitate voyaging canoes in the Pacific Islands.[1] In Figure 225 I show a graph from Finney's characterization of the sailing capabilities of *Nalehia*. Finney concluded that the double-hulled vessel could sail as close as 75 degrees into the wind with a canoe speed that was roughly 30 percent of the wind speed.

The *Nalehia* has a rounded keel design and uses a pair of crab-claw sails. In part the design of the *Nalehia* was chosen for long-distance voyaging but not necessarily for superior windward performance. The Gilbert canoes were, in contrast, single hulled

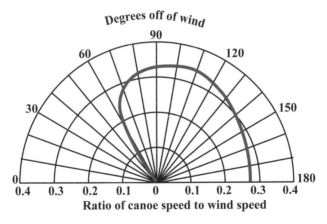

Figure 225 Sailing capabilities of a double-hulled Polynesian voyaging canoe. After Bradley Evans, "Simulating Polynesian Double-Hulled Canoe Voyaging: Combining Remotely-Sensed and Experimental Data" (master's thesis, Geography Department, University of Hawai'i at Manoa, 1999), 8; used with permission.

with an outrigger and had much sharper and asymmetric keels.[2] In addition, they used an ocean lateen sail rather than the crab-claw design. All of these factors would give the Gilbertese canoes added windward capabilities, consistent with Grimble's estimate of a beat 70 degrees to windward and with a higher canoe-to-wind-speed ratio than in Figure 225.

To estimate Baintabu's voyaging possibilities, I used the weather conditions reported for Abemama and Tarawa in mid-July 2011. I found that the options for the difficult return voyage from Tarawa to Abemama were limited using the *Nalehia*'s capabilities, with the conditions that occur on most days. The difference between 75 degrees and 70 degrees to windward and an additional 10 percent in canoe-to-wind-speed ratio was enough to allow possible return voyages. Even in this case, however, a minimum sustained wind speed of 16 knots was required to successfully fight the 1.2-knot current in that region of the Gilberts.

The tides in the Gilberts are semidiurnal, and I chose July 17, 2011, as the departure date. In looking for a return date, I had to wait until July 25 for a departure date to get a sufficient wind speed to create the return voyage sketched out in Figure 224.

GLOSSARY

Age of the tide   The difference between the time of a full or new Moon and the largest tidal range of the lunar cycle.

Altitude   The vertical angle above the horizon to an object in the sky. By convention the horizon is 0 degrees, the zenith +90 degrees, and the nadir –90 degrees.

Altocumulus   Clouds showing a heaped-up structure formed of water droplets, typically found at midlevels in the troposphere.

Amphidromic point   A fixed point around which rotary tides circulate. The amphidromic point itself shows no tidal range.

Amplitude   (1) For waves, the maximum displacement from the midpoint; (2) for the Sun, the departure of the rising azimuth from due east.

Antarctic Circle   The latitude at which the Sun becomes circumpolar during the austral summer at the solstice, approximately 66.5 degrees S.

Anticyclone   A large, contained atmospheric region of *high pressure* with rotation about its center that sheds air to its surroundings.

Aphelion   The point of farthest distance from the Sun of an object in an elliptical orbit.

Apogee   The point of farthest distance from the Earth of an object in an elliptical orbit.

Arctic Circle   The latitude at which the Sun becomes circumpolar during the boreal summer at the solstice, approximately 66.5 degrees N.

Austral   Pertaining to the Southern Hemisphere (e.g., austral winter).

Autumnal equinox   The time when the Sun crosses the celestial equator moving into the southern celestial hemisphere. Days and nights are of equal length.

Azimuth The horizontal angle between true north and a direction. By convention the direction of positive increments in angle is taken to be clockwise.

Back bearing (see also *Reciprocal bearing*) The angle in the opposite direction to a bearing; e.g., 180 degrees (bearing in degrees).

Backing   The tendency to shift counterclockwise, as in wind direction.

Ballast   Weight that is added low in a vessel to improve its stability during a rolling motion.

Bathymetry   The measurement and representation of the profile of the seafloor underlying bodies of water. The measurements are portrayed as depths, typically from some measurement datum.

Beam   The width of a vessel at its widest point.

Beam sea   Waves impinging on a vessel on its side.

Bearing   The horizontal angle from due north to an object.

Beating   The practice of sailing at an angle into the wind.

Beitass   A staff used by Norse sailors to stiffen the leading edge of a square sail, improving performance into the wind.

Boat   A small open vessel (i.e., smaller than a ship).

Boreal   Pertaining to the Northern Hemisphere (e.g., boreal winter).

Bow   The front of a vessel on water.

Bow wave   The wave generated by the bow of a vessel moving through the water.

Buoyancy A force pushing upward on an object or vessel equal to the weight of water displaced.

Caravel A small ship, typically Portuguese in origin, powered by lateen sails.

Cat's paw Gentle ripples in mostly still water caused by wind across the surface.

Celestial equator A line encircling the celestial sphere that is directly above the Earth's Equator.

Celestial sphere The projection of celestial objects onto a sphere for the purpose of assigning coordinates to them.

Center of buoyancy A point in a vessel that represents the sum of all upward forces of buoyancy on the hull.

Center of gravity A point in a vessel that is the concentration of all downward forces of gravity.

Chart datum The reference sea level used for the report of water depths on nautical charts.

Circumpolar The path of a star orbiting the North or South Pole without setting, as seen by an observer.

Cirrocumulus clouds Cirrus clouds consisting of very small, often regular, elements.

Cirrus clouds High-level clouds consisting of ice crystals.

Cognitive map A representation of a region that is contained in the mind of an individual, usually synonymous with "mental map."

Cold front The transition formed where cool dry air impinges on warm moist air.

Colure A principal line of celestial meridian. The vernal equi-noctial colure represents the origin of the system of right ascension, associated with the point where the Sun crosses the celestial equator moving toward the northern celestial hemisphere.

Conduction   Heat transfer through direct thermal contact.

Convection   The process of heat transfer through fluid motion.

Convection cell   Fluid moving in a circular motion between a hot and a cold region, transferring heat in the process.

Coriolis effect   The systematic deflection of an object to the east if it is moving away from the Equator. The object will systematically deflect to the west if moving toward the Equator. This is named after Gaspard-Gustave Coriolis, who described it mathematically.

Crescent Moon   Quarter phases of the Moon when it is less than half illuminated.

Cumulonimbus   A rain cloud associated with a thunderstorm, typically showing much vertical development.

Cumulus   A heaped-up cloud, often found in fair weather, sometimes called a puffball cloud.

Cyclone   A large, contained region of low pressure where air gets pulled in along the surface, accompanied by rotation about the center of the air mass.

Datum   A standard position or level used as a reference in measurements reported on maps and charts. In hydrographic charts, the datum is typically a mean low-tide water level (U.S. charts).

Dead reckoning   Assessing position from a knowledge of the starting point and a history of travels — time, speed, and direction.

Declination (1) Position of a celestial object in the sky, taken as the north–south angle from the celestial equator; (2) the angular difference between true north and north as indicated by a magnetic compass needle (also called magnetic declination or variation).

Dip angle (or dip) The angle from the horizontal down to the horizon created by the height of the observer and the curvature of the Earth. (Note dip is also used to describe any angle below horizontal; e.g., direction of magnetic field lines or geologic strata.)

Displacement The cargo-carrying capacity of a vessel.

Diurnal tide A tide when there is one high and one low tide in a tidal day.

Doldrums A zone of latitudes encircling the Earth near the equator, characterized by large upward convection and little wind (see also Intertropical convergence zone).

Downdraft Downward airflow, typically seen in thunderstorms.

Drag A force or forces retarding forward motion, typically from friction.

Drift Speed of an ocean current in knots, usually determined from the difference between dead reckoning and a celestial fix.

Dryline The boundary between warm, moist maritime air and cold, dry continental air in the central United States.

Ebb tide The flow of water associated with an outgoing (dropping) tide.

Ecliptic The path the Sun takes through the sky over the course of a year.

Ekman transport   The generation of surface currents from wind over water combined with the Coriolis effect.

Entorhinal cortex   A part of the limbic system in the mammalian brain where grid cells are located.

Ephemeris (*pl.* ephemerides)   A table giving the position of major celestial objects in the sky as a function of date and time.

Equal altitude method   A technique used for finding longitude by taking the midpoint of two times of equal altitude of the Sun (or other celestial body) to find the time of its meridian passage, and hence calculate longitude.

Equation of time   The difference between mean solar time and time reckoned directly from the position of the Sun in the sky.

Equirectangular projection (or map)   A map of the Earth with a representation of lines of latitude and longitude with equal spacing throughout.

Etak system   A navigational scheme employed by traditional navigators in the Caroline Islands in which their progress from one island to another is measured by the path of a reference island "moving" against a fixed background of stars as viewed by a moving canoe.

Farsakh   Arab unit of distance, approximately equal to the distance covered in one hour of walking, roughly three miles.

Ferrel cells   Global convection cells encircling the Earth in temperate climes.

Fix   A determination of position.

Flood tide   The flow of water associated with an incoming (rising) tide.

Following sea   Waves striking a vessel from the stern.

Force (1) (physics) an influence that causes an object to undergo a change, typically in its motion; (2) (meteorology) a range of wind speeds on the Beaufort scale.

Fortunate Isles The westernmost islands with respect to Europe, Asia, and Africa in the Atlantic Ocean. These were often taken as the reference for the prime meridian in tables of latitude and longitude in the Middle Ages.

Freeboard The height of a vessel's hull above its waterline.

Front A boundary between air masses, typically in the mid-latitudes.

Full Moon A fully illuminated Moon.

Furlong English unit of distance equal to 220 yards, or one-eighth of a statute mile.

Gibbous Moon Quarter phases of the Moon when it is more than half illuminated.

Gnomon The vertical post or stick used to cast a shadow for a sun compass or sundial.

Great-circle route The shortest path between two points on a globe.

Grid cells Nerve cells residing in a region of the brain called the entorhinal cortex. These will fire at a high rate when the subject is located at a periodic set of points in an environment.

Gust front Leading line of strong downdrafts associated with an advancing squall line.

Gyre A loop of current in the ocean.

Hadley cells Global convection cells encircling the Earth, extending roughly from the Equator north and south to 30 degrees latitude, named after George Hadley (1685–1768).

Heading An intended direction of travel, expressed as a horizontal angle from due north.

Heat capacity The amount of heat per unit mass that a body can absorb for a fixed temperature change.

Heel (or heeling angle) The angle of roll of a sailing vessel when the force of the wind is balanced by the righting forces of center of gravity and center of buoyancy.

Height (or wave height) The distance from the base to the crest of a wave.

Hippocampus A part of the limbic system in the mammalian brain where place cells are located.

Homing bird A bird that normally rests at night on land but flies out to sea to fish during the day.

Hull speed The speed of a vessel equal to the speed of a wave as long as its waterline.

Hyperbola A geometric curve that has two intersecting lines as asymptotes.

Insolation The amount of solar energy radiating onto an area.

Intertropical convergence zone (ITCZ) A zone of latitudes encircling the Earth near the Equator, characterized by large upward convection of air and little wind. See also *Doldrums*.

ISRID An abbreviation for "International Search and Rescue Incident Database."

Isobar Line of equal pressure.

Jet stream A zone of high-speed winds in the upper reaches of the troposphere, typically at the boundary between global convection cells. Wind speeds are typically 100 miles per hour blowing from west to east.

Keel   The bottom of the hull of a vessel, often shaped into a "V" to increase its ability to move in a straight line and resist forces that would move it sideways.

Knarr   A Viking vessel used to transport crew and cargo over long distances.

Knockdown   When a boat heels to 90 degrees and the mast is approximately parallel to the water's surface.

Knot   A speed of one nautical mile per hour.

Laminar flow   Fluid flow characterized by well-defined streamlines.

Land breeze   Airflow from land toward a body of water during the night resulting from unequal cooling of land and water.

LAT   Lowest astronomical tide. This is a chart datum used by the UK and most other countries. It is typically the lowest water level found over the course of the nineteen-year Metonic cycle.

Lateen sail   A triangular or trapezoidal sail with an angled or curved spar on its leading edge (origin from French *latine* for Latin).

Latent heat   Heat absorbed or released by a body during a phase change (e.g., from liquid to gas).

Latitude   The angle of a position on the surface of the Earth taken north or south from the Equator.

Lead (or lead line)   A weight with a rope tied to it to measure water depth.

League   English unit of distance, approximately equal to the distance covered in one hour of walking, roughly three miles.

Lee   A region that is sheltered or on the far side of the wind of an obstacle, such as an island, vessel, or sail.

Leeward   In the direction of the lee, opposite of windward.

Leeway   The sideways motion of a sailing vessel in reaction to the force of the wind on the sail and the vessel itself.

Lift   A force generated by differential pressure from fluid flow past an object, such as a sail, wing, or rudder.

Line of position (LOP)   A line along which the position of a subject is known to lie.

Local meridian   See *Meridian.*

Longitude   The angle of a position on the surface of the Earth taken east or west from the prime meridian.

Looming   (1) Sky glow associated with the presence of land over the horizon, usually associated with bright lights; (2) refractive effects near the horizon creating distortions (archaic).

Lunars or lunar method   The use of the angular distance between the Moon and another celestial object to estimate longitude.

Mean solar time   The time standard used for most purposes that puts the Sun, on average, at the meridian at noon.

Mental map   A representation of a region that is contained in the mind of an individual, usually synonymous with "cognitive map."

Mercator projection (or map)   A way of representing features on a map with latitudes and longitudes indicated, yet preserving angles of constant heading (rhumb lines). It is named after Flemish mapmaker Gerardus Mercator (1512–1594).

Meridian   (1) An imaginary arc in the sky for an observer, extending from due north, through the zenith, then through due south, sometimes called "local meridian"; (2) a line of longitude.

Meridian passage (or transit)   The moment when a celestial object crosses the local meridian of an observer.

Meter   SI (*Système International d'Unités* — International System of Units) fundamental unit of distance, originally taken to be one ten-millionth of the distance from the equator to the North Pole.

Mil   One one-thousandth of a radian (see also *Radian*).

Midheaven   The high point in the sky along the ecliptic, typically associated with one of the house systems for natal astrology.

Mixed tide   A combination of diurnal and semidiurnal tides.

MLLW   Mean lower low water. This is the chart datum used by the United States, for the average lowest tide found during a lunar cycle.

Nadir   A point directly underneath an observer.

Nautical mile   A unit of distance of one minute of latitude, equal to approximately 1.15 statute miles.

Navigation lights   Lights on an airplane or vessel that give clues to its heading. By convention red lights are on the left side of the vessel, green are on the right, and white is projecting behind.

Neap tides   A low tidal range created when the Sun and Moon are aligned at 90 degrees with respect to each other (quadrature).

New Moon   A phase of the Moon when the dark side points toward Earth.

Nimbostratus   A rain cloud typically associated with warm fronts.

Nor'easter   A cyclonic storm occurring along the northeast coast of the United States and the Maritime Provinces of Canada.

North celestial pole   A point on the celestial sphere that is at the zenith of the North Pole.

Northern temperate zone   The range of latitudes on the Earth between the Tropic of Cancer and the Arctic Circle.

Oecumene   A term used by Ptolemy and others to describe the "occupied half" of the world.

Pace   The distance for two steps of a walking person (left and right).

Pandanus   A genus of trees typically found in the tropics. The dried leaves of pandanus trees are often woven by Pacific Islanders to make sails.

Parallel   A line of constant latitude.

Perigee   The point of closest approach to the Earth of an object in an orbit.

Perihelion   The point of closest approach to the Sun of an object in an orbit.

Persistence   The forecast that weather in the near future is like the weather at the moment.

Pitch   A rotational motion of a vessel when the bow moves up and the stern moves down and vice versa.

Place cells   Nerve cells residing in a region of the brain called the hippocampus. These will fire at a high rate when the subject is located at a specific location in an environment.

Planing   A vessel skimming across the water above hull speed.

Point of departure   The starting point of a journey or voyage.

Polar cells   Global convection cells encircling the Earth in polar climes.

Polarize   To produce one direction of vibration of light.

Port   Left side of a vessel.

Portolan chart   A style of map emerging in the thirteenth and fourteenth centuries that depicted coastlines with high accuracy, displayed rhumb lines, and having a well-defined scale.

Pressure gradient   The change in pressure over a distance.

Prevailing wind   The dominant or most probable wind direction at a given location or region.

Prime meridian   The line of longitude taken to be the origin or "zero" of longitude. Note that, historically, there have been a number of prime meridians used. The current convention is to take the prime meridian as the meridian going through Greenwich, UK.

Qibla   The direction in which to face Mecca in prayer from any location.

Quadrant   A device used to measure the altitude of celestial objects, using a weight on a string to establish horizontal. The name derives from the use of one-quarter of a full circle to measure angles.

Quadrature   An alignment of two celestial objects at 90 degrees with respect to a central object or observer.

Quartering sea   Waves striking a vessel from an angle of approximately 45 degrees.

Radian   An angular measure equal to approximately 57.3 degrees.

Radiation   Heat transfer through the emission of electromagnetic radiation, typically in the infrared part of the spectrum.

Random walk   A route taken that involves random changes in direction.

Range   Distance to an object.

Reciprocal bearing   The angle in the opposite direction to a bearing; e.g., 180 degrees (bearing in degrees). (See also *Back bearing.*)

Reefing   The practice of reducing the area of a sail to adapt to conditions of strong wind.

Refraction   The bending of light at the boundary between two transparent media, such as air and water.

Retrograde motion   The apparent backward motion of planets against the fixed stars as the Earth overtakes an outer planet in its orbit, or as an inner planet overtakes Earth in its orbit.

Rhumb line   A line of heading over a long distance that keeps a constant angle with respect to geographic north.

Right ascension   The position of a celestial object in the sky, taken as the angle eastward from the vernal equinoctial colure, but expressed as time from zero to twenty-four hours.

Roll   A rotational motion of a vessel around an axis extending from the bow to the stern.

Rotary tide   A high and low tide that rotates around a central fixed point, called an amphidromic point.

Route knowledge   An awareness of an environment based on an understanding of the traversal of known paths.

SAR   Abbreviation for "search and rescue."

Sea breeze   Airflow from a body of water onto land during the daytime resulting from unequal heating of land and water.

Sea state (or *sea*)   The appearance or condition of the surface of a body of water, typically the wave heights, wavelengths, and periods.

Secular variation   The change in magnetic variation over time.

Semidiurnal tide   A tide that has two high and two low tides in a tidal day.

Set   The direction of an ocean current with respect to true north.

Sheet   A line or rope holding tension on a sail.

Shoal   A region of water with relatively shallow depths.

Shoaling (or wave shoaling)   Waves that acquire steepened faces or break when crossing shoals.

Ship   (1) An oceangoing vessel capable of making long, unassisted voyages and capable of carrying another vessel; (2) in the age of sail, a square-rigged vessel with three or more masts.

Sine wave   (1) A smooth, repetitive oscillation; (2) a traveling wave shaped like a sine function (see *Sinusoid*).

Sinusoid   A curve representing a smooth repetitive oscillation; the figure itself is the projection of the path of point on a circle onto a line as the point moves around the circumference.

Skerry   A small rocky island, typically too small for habitation.

Sounding   A measurement of the depth of water from the surface to the seafloor.

South celestial pole   A point on the celestial sphere that is at the zenith of the South Pole.

Southern temperate zone   The range of latitudes on the Earth between the Tropic of Capricorn and the Antarctic Circle.

Spring tide   A large tide range created when the Sun and Moon are aligned (see *Syzygy*).

Squall line   A stretched-out band of thunderstorms forming a single unit.

Stade (*pl.* stadia)   Ancient Egyptian and Greek unit of distance, approximately equal to a furlong (220 yards).

Stagnation point   The dividing point where laminar fluid flow bifurcates around an object.

Stall   The condition created on a wing when the attack angle increases to the point at which drag forces begin to exceed lift.

Star compass   A conceptual system that divides the horizontal angle into a set of regions associated with the rising and setting azimuths of prominent stars.

Star path   A "path" in the sky formed from stars all rising or setting on the same general azimuth.

Starboard   Right side of a boat or a ship.

Statute mile   5,280 feet, defined by an act of the British Parliament.

Stern   Rear of a vessel.

Stern wave   A wave generated from the stern of a vessel moving through the water.

Stick charts   A latticework of palm fronds and pandanus roots assembled to form a chart indicating wave patterns in the Marshall Islands.

Stratus   A flat or layered cloud.

Summer solstice   The time when the Sun reaches its maximum declination; days are the longest in the Northern Hemisphere.

Surfing   Planing down a wave face.

Survey knowledge   An awareness of an environment based on an internal two-dimensional representation that allows for shortcutting.

Swell   A smooth wave that propagates well away from the wind or storm that created it.

Syzygy   An alignment or conjunction between two celestial bodies. In particular the alignment of the Moon and Sun on the same side or on opposite sides of the Earth constitutes a syzygy.

Tacking   (1) Sailing against the wind using a zigzag path, in which each leg of the path is at some angle with respect to the wind; (2) a maneuver in which a sailing vessel's bow crosses the wind direction in the process of moving from one direction to another.

Tangent   (1) The ratio of the opposite to adjacent sides of a right triangle, (2) a line that is parallel to a curve at one point.

Thermal conductivity   A material's ability to conduct heat.

Tidal day   The length of time it takes for the Moon to make successive transits (twenty-four hours, forty-eight minutes).

Tide (or tidal) lag   The difference between the time of lunar transit and the high tide.

Torque   A force times a distance at which it acts relative to some center of rotation, creating rotational motion.

Trade winds   Prevailing winds in the Tropics that blow from east to west, typically out of the northeast north of the doldrums, and out of the southeast south of the doldrums.

Transit   The passage of a celestial object over the local meridian (near transit) or 180° opposite to the local meridian (far transit).

Triangulation   The process of finding a position from the intersection of two or more lines of position.

Trim   (1) The distribution of the load of a vessel along the line connecting the bow and stern; (2) the configuration of a sail, in terms of the tension held by ropes, also called sail trim.

Trochoid   A curve with a periodic shape that is the same as one created by the point on the rim of a wheel as it rolls along the ground.

Tropic of Cancer   The latitude where the Sun passes over the zenith during the summer solstice, approximately 23 degrees N.

Tropic of Capricorn   The latitude where the Sun passes over the zenith during the winter solstice, approximately 23 degrees S.

Tropical cyclone   A cyclone found in the tropics. Typhoons and hurricanes are tropical cyclones.

Tropics   A region of Earth bounded by the Tropic of Cancer and the Tropic of Capricorn.

Troposphere   The lowest layer of Earth's atmosphere, where most weather systems develop.

Turbulent flow   Semichaotic fluid flow characterized by constantly shifting vortices.

Variation   The angular difference between true North and North as indicated by a compass needle. Also called magnetic variation; see also *Declination*.

Veering   Tendency to shift clockwise, as in wind direction.

Vernal equinoctial colure   The celestial line of longitude running from the north celestial pole through the point where the Sun crosses the celestial equator in the spring and extending to the south celestial pole.

Vernal equinox   The time when the Sun crosses the celestial equator in the spring. The length of days and nights are the same.

Viscosity   The thickness or stickiness of a fluid.

Visible range   The distance at which an object of some height is first visible over the horizon (also called geographic range).

Waning Moon   The period during which the amount of illumination of the Moon is decreasing with time.

Warm front   A transition formed when warm, moist air impinges on cool, dry air.

Water wave   A wave at the boundary between water and air.

Wave height   The vertical distance from crest to trough in a wave, typically a water wave.

Wavelength   The longitudinal distance from one crest of a wave to the next.

Waxing Moon   The period during which the amount of illumination of the Moon is increasing with time.

Westerlies   Prevailing winds in the temperate climes that blow from west to east.

Wind compass   An orientation scheme based on the direction of recognizable winds.

Windward   In the direction from which the wind is coming; opposite of leeward.

Winter solstice   The time when the Sun reaches its lowest declination. The length of days is the shortest in the Northern Hemisphere.

Wiwijet   A Marshallese term for anxiety at being lost at sea.

Woods shock   A state of anxiety induced by being lost in a wilderness setting.

Xebec   A sailing vessel with a protruding bow powered by lateen sails and sometimes oars.

Yard   A spar or beam attached to the mast of a sailing vessel that holds a sail.

Yaw   A rotational motion of a vessel when the bow moves left and the stern moves right and vice versa.

Zenith   The point directly overhead in the sky as seen by an observer.

Notes

Chapter 1

1. See chapters 15 and 17 for more discussion.

2. Birgitta Ferguson, "L'Anse aux Meadows and Vinland, Approaches to Vinland," proc. of *A Conference on the Written and Archaeological Sources for the Norse Settlements in the North-Atlantic Region and Exploration of America, Reykjavik, 9–11 August 1999,* ed. A. Wawn and T. Sigurdardottir (Reykjavik, Iceland: Sigurdur Nordal Institute, 2001), 134–146.

3. Geoffrey Irwin, *The Prehistoric Exploration and Colonisation of the Pacific* (Cambridge: Cambridge University Press, 1992), 7–9.

Chapter 2

1. Robin Lord, "Hope wanes for missing kayakers," *Cape Cod Times* (October 14, 2003): 1.

2. Knud Rasmussen, "The Netsilik Eskimos: Social Life and Spiritual Culture," in *Report of the Fifth Thule Expedition 1921–24*, 8:1–2 (Copenhagen: Gyldendalske Boghandel, Nordisk Forlag, 1930), 91.

3. Colin Irwin, "Inuit Navigation, Empirical Reasoning and Survival," *Journal of Navigation* 38 (1985).

4. David Pelly, "Lost? Never! Understanding Traditional Inuit Navigation," *Above and Beyond* (2001), 39–43. http://www3.sympatico.ca/dpelly/Lost%20-%20Never.pdf.

5. Irwin, "Inuit Navigation," 184.

6. Ibid., 183.

7. Pelly, "Lost? Never," 41.

8. David Lewis, *We the Navigators*, 2nd ed. (Honolulu: University of Hawaii Press, 1994), 171–191.

9. Thomas Gladwin, *East Is a Big Bird* (Cambridge, MA: Harvard University Press, 1960), 181–195.

10. Over the course of thousands of years, Polaris wanders from its position at true north. For example, in 1500 it was approximately 4 degrees off true north.

11. Peter Sawyer, *The Oxford Illustrated History of the Vikings* (Oxford: Oxford University Press, 1997), 198.

12 Vidar Hreinsson and Liefur Eiriksson, eds, *The Complete Sagas of Icelanders* (Reykjavik: Liefur Eiriksson Publishing, 1997), 20.

13. Edward Tolman, "Cognitive Maps in Rats and Men," *The Psychological Review,* 55:4 (1948): 189–208.

14. Roger Peters, "Communication, cognitive mapping, and strategy in wolves and hominids," in *Wolf and Man: Evolution in Parallel*, ed. Roberta Hall and Henry Sharp (New York: Academic Press, 1978), 95–112.

15. J. O'Keefe and J. Dostrovsky, "The hippocampus as a spatial map: preliminary evidence from unit activity in the freely moving rat," *Brain Research* 34 (1971): 171–175.

16. A. Saint-Exupéry, *Wind, Sand and Stars*, trans. L. Galantiere (New York: Harcourt, Inc., 1967), 45.

17. Terrazas, A. et al., "Self-motion and the hippocampal spatial metric," *J. Neurosci.* 25 (2005): 8085–8096.

18. T. Hafting, M. Fyhn, S. Molden, M.-B. Moser, E. I. Moser, "Microstructure of a spatial map in the entorhinal cortex," *Nature* 436 (2005): 801–806.

19. C. Doeller, C. Barry, and N. Burgess, "Evidence for grid cells in a human memory network," *Nature* 463 (2010): 657–661.

Chapter 3

1. K. A. Hill, *Lost Person Behavior* (Ottawa, Canada: The National SAR Secretariat, 1998), 2.

2. J. Genz, *Marshallese Navigation and Voyaging: Re-learning and Reviving Indigenous Knowledge of the Ocean,* PhD dissertation (University of Hawaii, May 2009), 136.

3. J. Genz, personal communication (2012).

4. www.dbs-sar.com.

5. William Syrotuck, *Analysis of Lost Person Behavior: An Aid to Search Planning* (Mechanicsburg, PA: Barkleigh Productions, Inc., 2000), 22.

6. Cary J. Griffith, *Lost in the Wild* (St. Paul, MN: Borealis Books, 2006), 90–91.

7. J. Souman et al., "Walking Straight into Circles," *Current Biology* 19:18 (2009): 1538–1542.

8. F. Lund, "Physical asymmetries and disorientation," *American Journal of Psychology* 42 (1930): 51–62.

9. Robert Koester, private communication (2011).

10. Syrotuck, *Analysis of Lost Person Behavior*; Hill, *Lost Person Behavior*; A. Koestler et al., *Lost Person Behavior* (Charlottesville, VA: dbS Productions, 2008).

11. Syrotuck, *Analysis of Lost Person Behavior*, 29

12. Koestler, et al., *Lost Person Behavior,* 155, 164, 184, 194.

13. Koester, et al, *Lost Person Behavior*, 53–56.

14. R. Descartes, trans. J. Veitch, *Discourse on the Method* (New York: Cosimo, 2008), 25.

15. A. P. Low, "Report on Exploration in the Labrador Peninsula along the East Main, Koksoak, Hamilton, Manicuagan and Portions of Other Rivers in 1892–93–94–95," *Annual Report of the Geological Survey of Canada* 8L (Ottawa: Geological Survey of Canada, 1895).

16. Dillon Wallace, *The Lure of the Labrador Wild* (White River Junction, VT: Chelsea Green Publishing, 1990), 1–2.

17. Wallace, *The Lure of the Labrador Wild,* 53.

18 Ibid., 53, 54.

19 Ibid., 188.

Chapter 4

1. M. Wittlinger, R. Wehner, and H. Wolf, "The Ant Odometer: Stepping on Stilts and Stumps," *Science* 312 (2006): 1965–1967.

2. David Lewis, *We the Navigators*, 2nd ed. (Honolulu: University of Hawai'i Press, 1994), 277–278.

3. James Pinsep, "Note on the Nautical Instruments of the Arabs," *Journal of the Asiatic Society of Bengal* 5 (1836): 784.

4. Peter Ifland, *Taking the Stars: Celestial Navigation from Argonauts to Astronauts* (Malabar, FL: Kreiger Publishing, 1998), 38.

5. James Pinsep, "Note on the Nautical Instruments of the Arabs", 784.

6. Harold Gatty, *Finding Your Way Without Map or Compass* (New York: Dover Books, 1983), 150.

7. Pierre Berton, *The Arctic Grail* (London: Penguin Books, 1988), 42.

8. John Vigor, *The Practical Mariner's Book of Knowledge* (Camden, ME: International Marine, 1994), 67.

9. See, for example, L. Boroditsky, "How Language Shapes Thought," *Scientific American* , February 2011 (2011): 64.

10. Cecil Brown, "Where Do Cardinal Directions Come From?,", *Anthropological Linguistics*, 25:2 (1983): 121–-161.

11. Ibid., 138.

12. John Lawson, "An Account of the Indians of North Carolina," in *A New Collection of Voyages and Travels*, John Stevens, ed. (London: Karpian, Bell, Midwinter, Will-Taylor, Collins and Baker, 1711), 204.

13. C. Frake, "A Reinterpretation of the Micronesian 'star compass,'" *Journal of the Polynesian Society* 104:4 (1995): 147–148.

14. G. Irwin, *The Prehistoric Exploration and Colonisation of the Pacific* (Cambridge: Cambridge University Press, 1992), 51.

15. D. Lewis, *We the Navigators,* 246.

16. W. Kyselka, *An Ocean in Mind* (Honolulu: University of Hawai'i Press, 1987), 22–234.

17. Roland Huntsford, *The Last Place on Earth* (New York: Atheneum, 1985), 484.

Chapter 5

1. John Stevens, ed., "An Account of the Indians of North Carolina," in *A New Collection of Voyages and Travels* (London: Karpian, Bell, Midwinter, Will-Taylor, Collins and Baker, 1711), 204.

2. Robert Walsh, "Vegetable Physiology," *American Quarterly Review* 21 (1837): 141.

3. A. Willich, *The Domestic Encyclopedia,* (London: Murray and Highley, 1802), 232.

4. George Thompson, *The Negro's Flight from American Slavery to British Freedom* (London: John Snow, 1849), 10–11.

5. W. M. Mitchell, *The Under-ground Railroad* (London: William Tweedie, 1850), 36.

6. Francis Galton, *The Art of Travel: Shifts and Contrivances Available in Wild Countries* (London: John Murray, 1855), 90–91.

7. Horace Kephart, "How to How to Tell Direction in Forest and on Prairie," *Outing* 42 (1903): 79.

8. Caspar Whitney and Albert Britt, "Does Moss Point North? Discussion of a Time Honored Belief," *Outing* 74, (1919): 34.

9. Ideas column, *Boys' Life* 8 (1968): 12.

10. Marilyn Lichman, "A Closer Look: The Moss on the North Side of the Tree," *Garden* 3:6 (1979): 4.

11. Earl Conrad, *Harriet Tubman* (Washington, DC: Associated Publishers, 1943), 62.

12. Judith Nies, *Nine Women: Portraits from the American Radical Tradition* (Berkeley: University of California Press, 1977), 41.

13. Aubrey Burl, *Prehistoric Astronomy and Ritual* (Oxford: Shire Publications, Bucks, 1983).

14. Harold Williams, "Monuments and the past in early Anglo-Saxon England, The Past in the Past: The Reuse of Ancient Monuments," *World Archaeology* 30:1 (1998): 90–108.

15. Tertullian, *Apologeticus* c. xvi, quoted in Francis Bond, *An Introduction to English Church Architecture from the Eleventh to the Sixteenth Century* (London: Oxford University Press, 1913), 959.

16. This study was carried out by my students, using Google Earth and listings of medieval churches in England and Normandy: Catherine Ntube, Cerianne Robertson, Laura Hinton, Abigail Brown, Veselin Kulev, and Pilar Mayora (2010).

17. Cory Doctorow, "Urban navigation technique," *Boing boing*, http://www.boingboing.net/2002/02/27/urban-navigation-tec.html (2002).

18. For example, the website www.lyngsat.com gives the TV channels available on various satellites. The website www.dishpointer.com allows you to find the orientation for the particular satellite of interest.

Chapter 6

1. P. Utrilla, C. Mazo, M. C. Sopena, M. Martinez-Bea, R. Domingo, "A palaeolithic map from 13,660 calBP: engraved stone blocks from the Late Magdalenian in Abruntz Cave (Navarra, Spain)," *Journal of Human Evolution* 57:2 (2009): 99–111.

2. A. R. Millard, "Cartography in the Ancient Near East," from *The History of Cartography*, Vol. 1, eds. J. B. Harley and D. Woodward (Chicago: Univ. of Chicago Press, 1987), 113–114.

3. G. Aujac, "The Foundations of Theoretical Cartography in Archaic and Classical Greece," Ibid., 134–135.

4. This is highly controversial, and I don't have anything to add to the debate. I only point out that a substantial body of opinion points to this as a forgery.

5. For example, consult geomag.usgs.gov.

6. A. R. T. Jonkers, *Earth's Magnetism in the Age of Sail* (Baltimore, MD: Johns Hopkins University Press, 2003), 42.

Chapter 7

1 See, for example, Jeffrey Gettleman, "With Aid of Forgotten Bolt, Frenchman Escapes Somalis," *New York Times* (August 26, 2009): 1. This was covered extensively in the news media on this date. Mr. Aubriere was part of a news conference in Nairobi, Kenya, on August 26, after his escape. Some details of his account have been disputed.

2. http://www.rfi.fr/actufr/articles/116/article_84044.asp (trans. by the author).

3. Hesiod, *Works and Days*, trans. Hugh Evelyn-White (Cambridge, MA: Harvard University Press, 1914), 49.

4. Bruce Cartwright, "The Legend of Hawaii-loa," *Journal of the Polynesian Society,* 38:150 (1927): 108.

5. Homer, *Odyssey*, trans. Robert Fagles (New York: Penguin Classics, 1997), 161–162.

6. Geoffrey Irwin, *The Prehistoric Exploration and Colonisation of the Pacific* (Cambridge: Cambridge University Press, 1992), 48.

7. I should note that astronomers use "right ascension" (RA) for the equivalent of celestial longitude, and navigators often use "Greenwich Hour Angle" (GHA). My convention is to allow for a straightforward

mapping of celestial onto terrestrial longitude. There is also the use of "longitude" to denote the position of the Sun or a planet in the elliptic, but this usage is largely archaic or limited to use in the astrological community.

8 The astronomically inclined reader will immediately jump out of his or her seat, screaming, "No, this is called right ascension or Sidereal Hour Angle." The astronomically inclined reader is correct. However, the conventions for this coordinate perpendicular for declination is confusing for the casual reader, so I am going to use "celestial longitude."

9. Sir Arthur Grimble, *Migrations, Myth and Magic from the Gilbert Islands* (London: Routledge and Kegan Paul, 1972), 218.

10. Teuira Henry, originally recited by Ruanui in 1818, "Tahitian astronomy," *Journal of the Polynesian Society*, 62 (1907): 101.

11. David Lewis, *We the Navigators*, (Honolulu: University of Hawaii Press, 1994), 280–281.

12. Irwin, *Prehistoric Exploration and Colonisation of the Pacific, 8-9.*

Chapter 8

1. The value of 23 degrees is rounded off to the nearest degree.

2. Matthew Schneps and Philip Sadler, *A Private Universe* (Cambridge, MA: Harvard–Smithsonian Center for Astrophysics, 1987), DVD.

3. Alexander Stephen, *The Hopi Journal of Alexander M. Stephen*, Contributions to Anthropology 23, ed. E. C. Parsons (New York: Columbia University Press, 1936), 155–58, 287–89.

4. Department of the U.S. Air Force, *Search and Rescue Survival Training* AFR 64-4 (Washington, DC: U.S. Air Force, 1985), 345.

5. Ebenezer Henderson, *Iceland, or the Journal of a Residence in that Island, During the Years 1814 and 1815*, Vol. 1 (Edinburgh, Scotland: Oliphant, Waugh and Innes, 1818), 186.

6. Thorsteinn Vilhjalmsson, "Time and Travel in Old Norse Society," *Disputatio* II (1997): 89–114.

7. *The Vinland Sagas*, trans. Keneva Kunz (New York: Penguin Classics, 2008), 7.

8. Søren Thirslund, *Viking Navigation* (Rothskilde, Denmark: Viking Ship Museum, 2007).

9. T. Ramskou, "Solstenen," *Skalk* 2 (1967): 16.

10. Leif Karlsen, *Secrets of the Viking Navigators* (Seattle, WA: One Earth Press, 2003).

11. Vilhjalmsson, "Time and Travel in Old Norse Society," 94.

Chapter 9

1. Robert Greenler, *Rainbows, Halos and Glories* (Cambridge: Cambridge University Press, 1980), 165.

2. William Scoresby Jr., *An Account of the Arctic Regions* (Edinburgh, Scotland: Archibald Constable and Co., 1820), 386.

3. Robert E. Peary, *Nearest the Pole* (New York: Doubleday, 1907), 202.

4. Peary, *Nearest the Pole*, 207.

5. Rollin Arthur Harris, *Arctic Tides* (Washington, DC, U.S. Government Printing Office, 1911), 91.

6. Donald MacMillian and Walter Ekblaw, *Four Years in the White North* (New York: Harper and Brothers, 1918), 80.

7 David Lewis, *We, the Navigators* (Honolulu: University of Hawaii Press, 1972), 222.

8. Ibid., 222.

9. S. Rizvi, "A Newly Discovered Book of Al-Biruni 'Ghurrat-ua-Zijat' and Al-Biruni's Measurements of Earth's Dimensions," in *Al-Biruni, Commemorative Volume, Proceedings of the International Congress held in Pakistan on the Occasion of the Millenary of Abu Raihan Muhammad ibn Ahmad al-Biruni, Nov. 28–Dec. 12, 1973,* ed. Hakim Mohammed Said (Karachi, Pakistan: Hamdard Academy, 1979), 605–690.

10. Jim al-Khalili, *Science and Islam, episode 2, The Power of Reason,* BBC4, first aired Jan. 12, 2009.

11. J. O'Connor and E. Robertson, in MacTutor History of Mathematics Archives, http://www-history.mcs.st-andrews.ac.uk/Biographies/Al-Biruni.html.

Chapter 10

1. John K. Wright, "Notes on the knowledge of latitudes and longitudes in the Middle Ages," *Isis* 5:1 (1923): 75–98.

2. Yaqut al-Hamawai, *The Introductory Chapters of Yaqut's Mu-Jam Al-Buldan,* trans. Wadie Jwaideh (Washington, DC: the George C. Keiser Foundation, 1955), 60.

3. Other combinations are needed for other climes. For a full treatment the interested reader should consult a book on celestial navigation.

4. George Nunn, *The Geographical Conceptions of Columbus* (New York: American Geographical Society, 1922), 7.

5. Ibid., 8.

6. Samuel Eliot Morison, *Admiral of the Ocean Sea* (Boston: Little, Brown and Co., 1942), 258.

7. Ibid., 186–187.

8. Sir John Narborough, Captain Jasmen Tasman, Captain John Wood, Frederick Marten Well, *An Account of Several Late Voyages and Discoveries to the South and North* (London: S. Smith and B. Walford, 1694), 160.

9. See, for example, Morison's account in *Admiral of the Ocean Sea,* Vol. 2.

Chapter 11

1. Joshua Slocum, *Sailing Alone Around the World* (New York: Barnes and Noble Classics, 2005, original 1900), 59.

2. Peter Kemp, *The Oxford Companion to Ships and the Sea* (London: Oxford University Press, 1976), 233, 399.

3. I thank my mother for this one, Carol Monnik Huth, who would often repeat this saying to me. It has many folkloric variants.

4. David Lewis, *We the Navigators* (Honolulu: University of Hawaii Press, 1994), 221.

5. Nick Ward, *Left for Dead* (New York: Bloomsbury USA, 2007), 40.

6. Frank Gallagher, "Green Thunderstorms" (Ph.D. diss., University of Oklahoma, 1997).

7. Gene Ammarell, *Bugis Navigation* (New Haven, CT: Yale University Southeast Asia Studies, Ernest Beaglehole and Pearl Beaglehole, 1999), 98.

8. David Lewis, *We the Navigators*, 111–115.

9. Richard Feinberg, *Polynesan Seafaring and Navigation* (Kent, OH: Kent State University Press, 1988), 92–100.

10. Ibid., 98.

Chapter 12

1. Joshua Slocum, *Sailing Alone Around the World* (New York, Barnes & Noble Classics, 2005), 169.

2. www.sailing.org.

3. Many details of Abby's attempt and subsequent controversy can be found in much of the news media during this period; e.g., Paul Harris, "Parents of rescued teenage sailor Abby Sunderland accused of risking her life," *The Guardian*, July 13, 2010, 1, http://www.guardian.co.uk/world/2010/jun/13/abby-sunderland-lone-sailor-rescued.

4. Marianne George, "Polynesian Navigation and *Te Lapa* — 'The Flashing,'" *Time and Mind* 5 (2012): 159–166.

5. David Lewis, *We the Navigators* (Honolulu: University of Hawai'i Press, 1994), 224–251.

6. Joseph Genz et al., "Wave Navigation in the Marshall Islands," *Oceanography* 22 (2009): 234–245.

7. David Lewis, *We the Navigators*, 225.

8. Ibid., 225.

9. James Cook, *A Voyage Towards the South Pole and Round the World* (London: Strahan and Cadell, 1777), 316.

10. Genz et al., "Wave Navigation in the Marshall Islands," 238.

11. Genz et al., "Wave Navigation in the Marshall Islands," 241.

12. Joseph Genz, "Marshallese Navigation and Voyaging: Re-learning and Reviving Indigenous Knowledge of the Ocean" (PhD diss., University of Hawai'i, 2008), 189.

13. Genz et al., "Wave Navigation in the Marshall Islands," 238.

14. Genz, PhD diss., 162–163.

15. Genz, PhD diss., 155–158.

16. Genz, PhD diss., 158.

17. M. George, "Polynesian Navigation and *Te Lapa*," 163–165.

18. Genz, et al., "Wave Navigation in the Marshall Islands," 242.

19. Winkler, Capt., "On sea charts formerly used in the Marshall Islands, with notices on the navigation of these islanders in general," *Smithsonian Institute Report for 1899*, 54 (1901).

20. Genz, PhD diss., 166–176.

21. Genz et al., "Wave Navigation in the Marshall Islands," 240.

Chapter 13

1. Ole Crumlin-Pedersen, "Skibe, sejlads og ruter hos Ottar og Wulfstan," in *Ottar og Wulfstan to rejsebeskrivelser fra vikingetiden* (Roskilde, Denmark: The Viking Ship Museum, 1983) 32–44.

2. See, for example, George Indruszewski and C. M. Barton, "Simulating Sea Surfaces for Modeling Viking Age Seafaring in the Baltic Sea," in *Digital Discovery: Exploring New Frontiers in Human Heritage, Proceedings of 34th Conference on Computer Applications and Quantitative Methods in Archaeology, April 2006*, ed. J. Clark and E. Hagenmeister (Budapest, Hungary: Archaeolingua, 2007), 617–618.

3. Pliny, *Natural History, Book II*, 99. See, for example, translation in Perseus Digital Library: www.perseus.tufts.edu/hopper.

4. Hermann Diels, *Doxographi Graeci* (Berlin, Germany: G. Reimer, 1879), 383.

5. John Kirtland Wright, *Geographic Lore of the Time of the Crusades*, V2 (New York: Dover Books, 1965) 82–100.

6. L. Arnaudon et al., "Effects of terrestrial tides on the LEP beam energy," *Nuclear Instruments and Methods* A 357 (1995): 249–252.

7. Brian Arbic, Pierre St-Laurent, Graig Sutherland, and Chris Garrett, "On the resonance and influence of the tides in Ungava

Bay and Hudson Strait," *Geophysical Research Letters* 34 (2007), doi:10.1029/2007GL030845.

8. Gabriel Godin, "The resonant period of the Bay of Fundy," *Continental Shelf Research* 8 (1988): 1005–1010.

Chapter 14

1. H. A. Marmer, "The Gulf Stream and its Problems," *Geographical Review* 19:2 (1929): 457.

2. Silas Bent in a letter to Judge Daly, published in the *Journal of the American Geographical and Statistical Society* in 1870, quoted by J. K. Wright, "The Open Polar Sea," *Geographical Review* 43:3 (1953): 356.

3. F. Nansen, "Some Results of the Norwegian Arctic Expedition, 1893-96," *The Geographical Journal* 9:5 (May 1897): 495.

4. See, for example, Roemmich et al., "Decadal Spinup of the South Pacific Subtropical Gyre," *Journal of Physical Oceanography* 137 (2007): 162–173.

5. Ernest Sabatier, *Astride the Equator,* trans. U. Nixon (Melbourne, Australia: Oxford University Press, 1977), 13.

6. David Lewis, *We the Navigators* (Honolulu: University of Hawai'i Press, 1984), 154.

7. http://jboatnews.blogspot.ch/2012/06/bermuda-race-sailing-preview.html.

8. Steve Thomas, *The Last Navigator* (New York: International Marine/Ragged Mountain Press, 1987), 29.

9. R. Firth, "Anuta and Tikopia: Symbiotic elements in social organization," *Journal of the Polynesian Society* 63 (1954): 91.

10. Steve Thomas, *The Last Navigator,* 32.

11. Joshua Slocum, *Sailing Alone Around the World,* 2nd ed. (New York: The Century Co., 1919), 58.

12. Lin and Larry Pardey, *Storm Tactics Handbook* (Arcata, CA: Paradise Cay Publications, 1996), 5.

Chapter 15

1. http://www.physorg.com/news/2011-01-cretan-tools-year-old-sea.html (from Associated Press).

2. Geoffrey Irwin, *The Prehistoric Exploration and Colonisation of the Pacific* (Cambridge: Cambridge University Press 1992), 39.

3. Herodotus, trans. Aubrey De Sélincourt, John Marincola, *Herodotus: The Histories* (London: Penguin Books, 2008), 253.

4. http://vikingeskibsmuseet.dk/en/exhibitions/the-skuldelev-ships/.

5. Ole Crumlin-Pedersen, Olaf Olse, ed., *The Skuldelev Ships I* (Roskilde, Denmark: Viking Ship Museum, 2002).

6. D. S. Noble, "The Coastal Dhow Trade of Kenya," *The Geographical Journal* 129:4 (1963): 500.

7. Birgitta Wallace Ferguson, "L'Anse aux Meadows and Vínland," in *Approaches to Vínland*, proceedings of the 1999 Conference on the written and archaeological sources for the Norse settlements in the North-Atlantic region and exploration of America, The Nordic House, Reykjavik, 9–11 Aug. 1999, ed. A. Wawn and Þórunn Sigurðardóttir (Reykjavik, Iceland: Sigurður Nordal Institute, 2001), 140–141.

8. W. Barron and G Burgess (ed.), *The Voyage of Saint Brendan, Representative Versions of the Legend in English Translation* (Exeter, UK: University of Exeter Press, 2002), 29.

9. Henry Colman Folkard, *The Sailing Boat: A Description of English and Foreign Boats* (London: Longman, Green and Roberts, 1870), 243.

Chapter 16

1. Geoffrey Irwin, *The Prehistoric Exploration and Colonisation of the Pacific* (Cambridge: Cambridge University Press 1992), 39.

2. David Lewis, *We the Navigators* (Honolulu: University of Hawai'i Press, 1996), 157–158.

3. Ben Finney, *Voyage of Rediscovery* (Berkeley: University of California Press, 1994), 38.

4. I. C. Campbell, "The Lateen Sail in World History," *Journal of World History* 6:1 (1995): 13.

5. Zaraza Friedman, Levent Zoroglu, "Kelenderis Ship — Square or Lateen Sail?," *The International Journal of Nautical Archaeology* 35:1 (2006): 108–116.

6. Patrice Pomey, "The Kelenderis Ship: A Lateen Sail," *The International Journal of Nautical Archaeology* 35:2 (2006): 326–329.

7. *Saga of the Greenlanders*, from *The Complete Sagas of Icelanders*, ed. Vidar Hreinsson (Reykjavic, Iceland: Leifur Eiriksson Publishing, 1997), 23–24.

8. See, for example, http://lexicon.ff.cuni.cz/html/oi_cleasbyvigfusson/b0056.html.

9. Leo Block, *To Harness the Wind* (Annapolis, MD: Naval Institute Press, 2003), 132.

10. Ibid., 132.

11. C. A. Marchaj, "Planform Effect of a Number of Rigs on Sail Power," in *Proceedings of Regional Conference on Sail-Motor Propulsion,* Manila, Philippines, 1985, ed. C. Mudie, *Journal of Navigation* 41 (1988): 71.

Chapter 17

1. "Pilot fatigue cited in Air Canada mid-flight dive," CBC News release, April 16, 2012, www.cbc.ca/news/canada/toronto/story/2012/04/16/air-canada-zurich-flight-incident-report.html.

2. James Hornell, "The Role of Birds in Early Navigation," *Antiquity* 20:79 (1946): 142–149.

3. J. Frank Stimson, *Songs and Tales of the Sea Kings*, ed. Stanley Marshall (Salem, MA: Peabody Museum of Salem, 1957).

4. David Lewis, *We the Navigators* (Honolulu: University of Hawai'i Press, 1996), 324.

5. Bruce Cartwright, "The Legend of Hawaii-Loa," *Journal of the Polynesian Society* 38 (1927): 150.

6. Hornell, "The Role of Birds in Early Navigation," 144.

7. Samuel Eliot Morison, *Admiral of the Ocean: A Life of Christopher Columbus* (Chicago: University of Chicago Press, 1942), 214.

8. www.ancienttexts.org/library/mesopotamian/gilgamesh/tab11.htm.

9. Le-Qing Wu, J. David Dickman, "Neural Correlates of a Magnetic Sense," *Science* 336:6084 (2012): 1054–1057.

10. From the website of the New Northvegr Center, text transcribed by Aaron Myer, http://www.northvegr.org/sagas%20annd%20epics/miscellaneous/landnamabok/003.html#.

11. Translation from John Watson McCrindle, *Ancient India as Described in Classical Literature* (Westminster, UK: Elbron Classics, 2005), 103.

12. Frank Reed, personal communication (2010).

13 Hornell, "The Role of Birds in Early Navigation" (1946).

14. Harold Gatty, *Finding your Way Without Map or Compass* (New York: Dover Books, 1983), 43.

15. David Lewis, *We the Navigators,* 205.

16. J. Frank Stimson, *Songs and Tales of the Sea Kings*, 76.

17. William Longyard, *A Speck on the Sea* (Camden, ME: International Marine, 2003), 78.

18. Lin and Larry Pardey, *Storm Tactics Handbook* (Arcata, CA: Paradise Cay Publications, 1996), 5.

19. Steven Callahan, *Adrift: Seventy-Six Days Lost at Sea* (New York: Houghton Mifflin, 1986).

20. William Butler, *66 Days Adrift* (Camden, ME: International Marine, 2005).

21. Lauren Hillenbrand, *Unbroken* (New York: Random House, 2010), 159.

22. Lewis, *We, the Navigators,* 253.

23. Lewis, *We, the Navigators,* 255.

24. Jeffrey Kluger, James Lovell, *Apollo 13* (New York: Houghton Mifflin, 1994), 69.

25. George Shelvocke, *A Voyage Round the World by Way of the Great Southern Sea* (London: J. Senex, 1727), 418.

26. Marianne George, "Polynesian Navigation and *Te Lapa* — 'The Flashing,'" *Time and Mind: The Journal of Archaeology, Consciousness and Culture* 5:2 (2012): 141.

27. Ibid., 153.

28. Richard Feinberg, "In Search of *Te Lapa:* A Navigational Enigma in Vaeakau-Taumako, Southeastern Solomon Islands," *The Journal of the Polynesian Society* 120:1 (2011): 57–70.

29. Ibid., 68.

30. M. George, "Polynesian Navigation and *Te Lapa*," 168.

31. http://www.youtube.com/watch?v=0b_S9C1UZd4.

Chapter 18

1. Captain Winkler, "On Sea Charts Formerly Used in the Marshall Islands, with Notices on the Navigation of These Islanders in General," *Annual Report of the Smithsonian Institution, 1899* (Washington, DC: Smithsonian Institution, 1901): 507.

2. R. G. Roberts, "The Dynasty of Abemama," *The Journal of the Polynesian Society* 62:3 (1953): 267–278.

3. During World War II marine landing craft attacking Betio ran aground on shallow reefs because of unusual tidal conditions. This made them vulnerable to shell fire from Japanese positions on the island.

Appendix 3

1. John K. Wright, "Notes on the knowledge of latitudes and longitudes in the Middle Ages," *Isis* 5:1 (1923): 75–98

Apendix 4

1. B. Finney, "Voyaging Canoes and the Settlement of Polynesia," *Science* 196:4296 (1977): 1277–1285.

2. Arthur Grimble, "Canoes in the Gilbert Islands," *The Journal of the Royal Anthropological Institute of Great Britain and Ireland* 54 (1924): 108–139.

Acknowledgments

Many people had a role in the development of this book. The initial impetus came from the tragic deaths of Sarah Aronoff and Mary Jagoda in 2003, as described in chapter 2. This episode had a profound effect on me and drove me to learn navigation techniques without instruments. I developed and taught these with the hope that instruction could help others in situations like those that the two young women encountered.

Two people stand out for their influence: David Burch and Frank Reed. Although David's book *Emergency Navigation* was intended for an audience of mariners who might lose instruments while at sea, it also is an excellent path to a deeper understanding of navigation. I worked through many of the examples in his book on my own and used this as a text for some of my seminars. David has been very helpful during consultations and shares a keen interest with me in many topics related to navigation. Frank Reed is the founder and curator of an e-mail list called NavList, which serves as a forum for people interested in navigation in general, but with an emphasis on celestial navigation. Frank himself has been extremely helpful with a large number of details that aided my understanding. Some members of NavList who were particularly helpful include George Brandenburg, Byron Franklin, the late George Huxtable, Gary LaPook, Greg Rudzinski, and Marcel Tschudin. I apologize to any helpful members of NavList whom I may have inadvertently omitted.

I am grateful to Harvard University for allowing me to teach my freshman seminar, Primitive Navigation, for two years and the General Education course SPU 26 for three years. These spurred me to develop the material in this book. Students in the course and seminar participated in many exercises, validating a lot of the material I use. A number of projects the students performed were informative, and I mention some of their conclusions in this book. I am grateful for the help provided by those in the General

Education office: Kristina Barrett, Anne Marie Calareso, Lucy Gedrites, Kendra Gray, Stephanie Kenen, and Sandy Cantave Vil. I am also grateful to Sandra Naddaff for encouraging and approving my initial seminar offering.

For help with Old Norse literature, I thank Kaaren Grimstad and Astrid Ogilvie for their support and patience with my efforts in translation. I am grateful to Aravi Samuel from the Harvard Physics Department and Peyton Greenside for looking over my discussion of grid and place cells. In the area of lost-person behavior, I benefited from discussions with Robert Koester of the International Search and Rescue Incident Database (ISRID). Maine guide Mark Schoon accompanied me on a four-day trip off the coast of Down East Maine to try some of the dead reckoning techniques I had developed.

In my research on Pacific Islands navigational strategies, David Lewis's book *We the Navigators* stands out as a major starting point. I am grateful to Rick Feinberg for inviting me to the Association for Social Anthropology in Oceania (ASAO) meeting in Portland, February 2012. My contacts there were extremely helpful. I mention in particular Joe Genz's discussions with me on Marshall Island wave piloting. Rick Feinberg himself has been extremely helpful in vetting a large number of details of Pacific Islands navigational schema, language, and culture.

Cathleen Pyrek's thesis of navigational tool kits in Vaeakau-Taumako shaped my thinking on the concept of a "navigator's tool kit." Mimi George has been helpful to me in discussions of wave piloting in the Santa Cruz Islands and observations of *te lapa*, or underwater lightning. I am grateful to Woody Hastings for discussions about dinoflagellate bioluminescence and his allowing me access to his samples of *Pyrocystis lunula*.

Sara Schechner opened up the vault of the Harvard University Collection of Historical Scientific Instruments and spent hours considering my questions and allowing me to inspect a large number of navigational instruments. Owen Gingerich of the

Harvard University Astronomy Department had numerous discussions with me about the state of ephemerides during the start of the era of Western European exploration, among other topics. Astrologer Joseph Crane graciously spent a lunch meeting with me describing the many variations on house systems and the general history of astrology.

Daniel Jacob of the Harvard University School of Engineering answered many of my questions pertaining to climatology and meteorology. Wolf Rueckner of the Harvard University Science Center aided me in many ways, but in particular with setting up my experiment in dinoflagellate bioluminescence. Doug Goodale, also with the Science Center, helped me with lecture demonstrations and allowed me to borrow equipment for my homemade experiments. Fred Abernathy of the Harvard School of Engineering answered my questions about the role of the boundary layers and vortex shedding in producing lift in wings and sails.

Christina Mairs and Adam Scherlis deserve credit for many of the illustrations; their names appear in the appropriate captions. I also thank Darin Ragozzine for his help in the trigonometric problem of mapping observer-based coordinates onto celestial coordinates and displaying this. Peyton Greenside did research on the literature on grid and place cells.

Closer to home, I convey my deep appreciation to my parents, Carol and Edward Huth, who carefully read my drafts and gave me detailed feedback on its accessibility to a general audience; they tried some of the ideas I describe. Most importantly, I am grateful for their careful proofreading of later drafts of the chapters, which gave me an education in editing. I thank my son, James, for pointing out the somewhat obvious connection of "north" to *septentriones* during a bluegrass festival and reading over my description of moss, algae, and fungi. My daughters Phoebe and Charlotte continually put my skills to the test, often asking me to predict the weather or find north from the orientation of a random church. Their questions kept me sharp and humble.

I express my particular gratitude to my dear wife, Karen Agnew, who not only encouraged me in this book but cheerfully put up with all that it entailed: my growing dinoflagellates in our basement, my making homemade quadrants during soccer tournaments, and, most importantly, enduring the endless weekends when I was perched in front of my computer writing, drafting figures, and editing.

Index